Sustainable Civil Infrastructures

Sustainable Civil Infrastructures (SUCI) is a series of peer-reviewed books and proceedings based on the best studies on emerging research from all fields related to sustainable infrastructures and aiming at improving our well-being and day-to-day lives. The infrastructures we are building today will shape our lives tomorrow. The complex and diverse nature of the impacts due to weather extremes on transportation and civil infrastructures can be seen in our roadways, bridges, and buildings. Extreme summer temperatures, droughts, flash floods, and rising numbers of freeze-thaw cycles pose challenges for civil infrastructure and can endanger public safety. We constantly hear how civil infrastructures need constant attention, preservation, and upgrading. Such improvements and developments would obviously benefit from our desired book series that provide sustainable engineering materials and designs. The economic impact is huge and much research has been conducted worldwide. The future holds many opportunities, not only for researchers in a given country, but also for the worldwide field engineers who apply and implement these technologies. We believe that no approach can succeed if it does not unite the efforts of various engineering disciplines from all over the world under one umbrella to offer a beacon of modern solutions to the global infrastructure. Experts from the various engineering disciplines around the globe will participate in this series, including: Geotechnical, Geological, Geoscience, Petroleum, Structural, Transportation, Bridge, Infrastructure, Energy, Architectural, Chemical and Materials, and other related Engineering disciplines.

SUCI series is now indexed in SCOPUS

and EI Compendex.

Chakib El Mokhi · Hanaa Hachimi ·
Anand Nayyar
Editors

The Future of Urban Living: Smart Cities and Sustainable Infrastructure Technologies

Proceedings of the 1st International Conference on Advanced Sustainability Engineering and Technology (ICASET'25)

Springer

Editors
Chakib El Mokhi
Ibn Tofail University
Kenitra, Morocco

Hanaa Hachimi
Ibn Tofail University
Kenitra, Morocco

Anand Nayyar
Duy Tan University
Da Nang, Vietnam

ISSN 2366-3405 ISSN 2366-3413 (electronic)
Sustainable Civil Infrastructures
ISBN 978-3-031-98333-7 ISBN 978-3-031-98334-4 (eBook)
https://doi.org/10.1007/978-3-031-98334-4

This Springer imprint is published by the registered company Springer Nature Switzerland AG
The registered company address is: Gewerbestrasse 11, 6330 Cham, Switzerland

Preface

This book presents the peer-reviewed proceedings of **the *International Conference on Advanced Sustainability Engineering and Technology (ICASET 2025)***, held on April 17–18, 2025, at Ibn Tofail University in Kenitra, Morocco. The conference was organized to foster academic and professional exchange on the latest advances in sustainable civil infrastructure, smart cities, green construction, and resilient urban systems.

ICASET 2025 brought together researchers, engineers, architects, urban planners, and decision-makers to share novel ideas, practical solutions, and theoretical advances related to sustainable development and infrastructure. The accepted papers in this volume reflect high-quality contributions across a wide range of topics including but not limited to:

- Sustainable materials and construction techniques
- Smart and resilient infrastructure systems
- Renewable energy integration in urban environments
- Intelligent transportation systems
- Water resources and environmental management
- Digital technologies for civil infrastructure (AI, IoT, BIM, GIS)

All submitted papers were subjected to a rigorous peer-review process to ensure technical soundness and relevance to the conference themes. We are grateful to the members of the scientific and program committees, as well as the external reviewers, for their time and commitment in maintaining the academic quality of this volume.

We also wish to acknowledge the contributions of all authors who chose ICASET 2025 as a platform to share their research. Their work adds significant value to the ongoing global dialogue on sustainable infrastructure. Our appreciation extends to the keynote speakers, organizing committe, and session chairs, whose insights and engagement enriched the scientific program.

Special thanks are due to Springer and the editors of the **Sustainable Civil Infrastructure (SUCI)** book series for their continued support and for making this publication widely accessible through their digital platform.

We hope that this volume will serve as a valuable reference for academics, practitioners, and policy-makers seeking innovative solutions for building a more sustainable and resilient future.

April 2025

Chakib El Mokhi
Hanaa Hachimi
Anand Nayyar

Contents

Advanced Technologies for Smart Cities Management

Intelligent Control Systems for Power and Energy Management

Multi-objective Metaheuristic Optimization for Reliable Hybrid Renewable Energy Systems

Driss Assioui(✉) and Rachid Elgouri

National School of Applied Sciences, University of Ibn Tofail, BP 241 Kenitra, Morocco
{driss.assioui,rachid.elgouri}@uit.ac.ma

Abstract. The Hybrid Renewable Energy Systems (HRES) are essential to provide reliable energy in areas without grid access. Designing these kind of systems faces challenges, like intermittent renewable inputs, mitigating operational uncertainties, and ensuring long-term system stability. To solve these obstacles, many researchers has been integrating advanced solutions, in controlling, predicting and optimizing algorithms, these efforts not only reduce costs but also increase sustainability of the system and the most important the reliability, to avoid the oversizing of the system and the power demand issues. Such strategies significantly advance global renewable energy adoption, by ensuring the demand with less costs. This informative paper focuses on optimizing the reliability of the HRES with multi-objective optimization using the metaheuristics algorithms. The idea is the evaluation of the system reliability with other objectives like minimizing costs, CO_2 emission by using different metaheuristic algorithm (GA, PSO). Future work will validate the framework using real-world data. Our paper focused on the effect of others objective function and algorithms on the HRES reliability.

Keywords: Energy management system · Metaheuristics · Multi-objective optimization

1 Introduction

As the world increasingly transitions toward renewable energy, there is a growing need for hybrid renewable energy systems (HRES) to meet the demand for reliable and sustainable power [1]. These systems are made by combination of many technologies like the energy sources, solar panels, wind turbines, and the energy storage units. It offers a significant potential for off-grid applications. However, optimizing their performance is complex, particularly when load demand uncertainty is a factor [2]. Variations in consumption patterns and the absence of accurate data pose significant challenges to designing effective systems.

Traditional methods often struggle with the nonlinear and multi-objective nature of HRES optimization. In contrast, metaheuristic algorithms known for their versatility and capability to discover the search space and return the best solutions [3]. By integrating techniques like genetic algorithms and particle swarm optimization, these methods enable the exploration of feasible system configurations, even in scenarios

C. El Mokhi et al. (Eds.): ICASET 2025, SUCI, pp. 3–8, 2025.
https://doi.org/10.1007/978-3-031-98334-4_1

with incomplete or uncertain load information [4]. This work outlines a framework for optimizing HRES, while addressing the complexities of multi-objective functions, and how the variation of the objective function effects the reliability of the system using different optimization algorithms. Although the core structure and computational model are not defined here. The proposed methodology prioritizes renewable energy utilization, minimizes operational costs, and ensures system reliability.

2 Methodology

The HRES contains, PV panel, Wind turbine, Storage system, Inverter, Diesel generator. Each component is modelled using the mathematical modelling, which contains the main physical, economical information (Fig. 1).

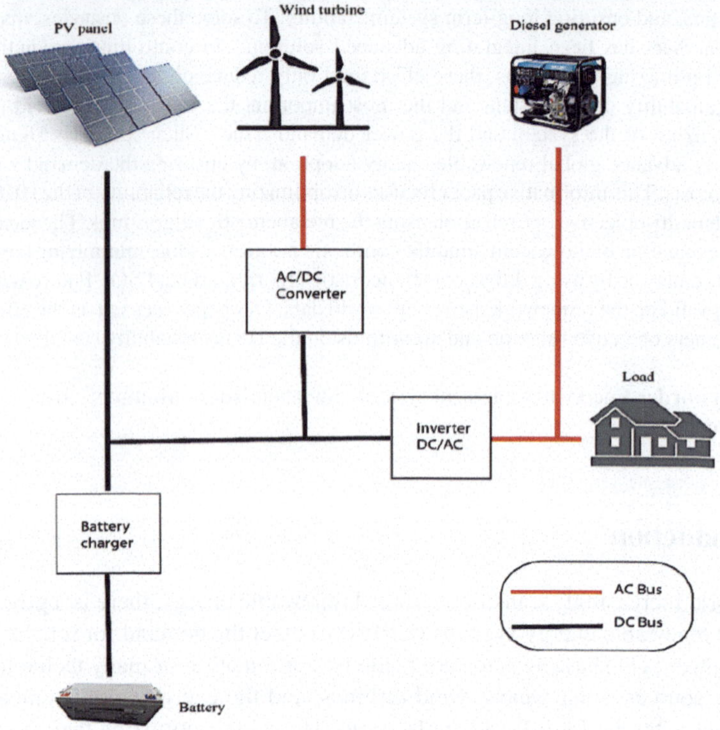

Fig. 1. HRES Components

Photovoltaic Panel. The modelling of the PV panel is made by using the equivalent circuits of the PV cell [5]. The weather data of the received irradiance into the PV surface, and the ambient temperature are the main factors affecting the PV panel performances [6], because the performance of the PV panel depend on the cell temperature, The PV efficiency η_{pv} is defined as [6]:

$$\eta_{pv} = \eta_r \cdot \eta_{pc} \cdot \left[1 - \beta\left(T_c - T_{ref}\right)\right] \quad (1)$$

where η_r is the manufactory information about the reference efficiency of the solar panel, η_{pc} is equal to 1 when MPPT is used, and β (°C) is temperature coefficient between (0.004–0.006) [6]. T_{ref} is reference cell temperature in °C which equal usually 25 °C, and T_c is the cell temperature in °C it can be calculated based on ambient temperature Ta and the global solar irradiance G (W/m^2) with the formula [7]:

$$T_c = T_a + G \cdot \frac{(NOCT - 20)}{800} \tag{2}$$

where the NOCT is nominal operating cell temperature in °C, and it is given by the manufactory. The panel declination is assumed to be equal to the latitude of the location. Finally, the power output is given by [7]:

$$P_{pv} = A_{pv} \cdot \eta_{pv} \cdot G \tag{3}$$

where η_{pv} is solar panel efficiency, A_{pv} is the panel area (m^2).

Wind Turbine. The wind turbine produces power by the extraction of the kinetic energy from the wind. The power generated depends directly on the wind velocity, it presented with the following equations depending on the wind velocity [8].

$$\begin{cases} P_r \cdot \left(\frac{V^2 - V_c^2}{V_r^2 - V_c^2}\right) & if\ V_r < V < V_c \\ P_r & if\ \ V_c < V < V_f \\ 0 & if\ \ \ otherwise \end{cases} \tag{4}$$

where V_r is the rated wind velocity, V_f and V_c is the cut off and cut in of the wind speed, respectively. Because of the variation of wind speed with the height, the following equation presents the equation which describes the variation of the wind velocity depending on the height of the measurements [8].

$$\frac{V}{V_{ref}} = \left(\frac{H}{H_{ref}}\right)^a \tag{5}$$

where H_{ref} is the reference height, V_{ref} is the wind velocity at the same height, and α is a coefficient assumed to be equal to 1/7 [8].

Diesel Generator. The diesel generator becomes essential in the moments where the load demand overtake the energy produced by renewable sources, and the storge units state is not enough. The selection of the DG is based on the maximum power that could be needed by the system.

The modelling of the diesel generated made based on the fuel consumed per unit of power need to over comes the unmet demand load.

The equation used for modelling the diesel generator consumption is [9]:

$$F = \alpha_1 \cdot P_{gen}(t) + \alpha_2 \cdot P_r \tag{6}$$

where P_r, P_{gen} represent the rated power, and the generated power at a given time t in Watt respectively. The coefficients α_1, and α_2 are equal to 0.246 and 0.0842, respectively [9].

Batteries. The batteries selected is the lead-acid batteries. It serves the storing the surplus of the generated energy and ensure the excess of demand. But for the durability reasons, the batteries stop the discharging mode in a specific level, called the maximum depth of discharge (DOD). The total capacity of the storage system is depending on factor like the load demand, autonomy duration, DOD, inverter and the batteries efficiency [8]. The following equation is the total capacity of the batteries expressed in Ah.

$$C_{batt} = \frac{D_a \cdot E_{load}}{DOD \cdot \eta_{inv} \cdot \eta_{batt} \cdot V_b} \tag{7}$$

where D_a is the autonomy of the batteries given in days, E_{load} is the daily load demand, V_b is the rated voltage of the batteries, η_{inv} and η_{batt} is the efficiency of the inverter and the batteries.

The batteries states are modelled with two equations (charging/discharging) [10]:

Charging process during the period where the system has an excess of energy from the renewable sources. The state of charge is calculated by assuming the previous state and the excess of energy.

$$SOC(t) = SOC(t-1) \cdot (1-\sigma) + \left(E_{gen} - \frac{E_l}{\eta_{inv}}\right) \cdot \eta_{batt} \tag{8}$$

where E_{gen} is the generated energy, E_l demand load, σ is the self-discharge of the batteries. During the discharging process, the batteries provide the unmet load demand. The new SOC of the batteries during the discharge is calculated using this expression:

$$SOC(t) = SOC(t-1) \cdot (1-\sigma) + \left(\frac{E_l}{\eta_{inv}} - E_{gen}\right) \tag{9}$$

Inverter. The presence and the performance of the invertor is a critical because this component ensures the performance and the power migration between DC/AC current, the selection of the inverter depends on the demand load power. The efficiency of the inverter is given by [6].

$$\eta_{inv} = \frac{E_{out}}{E_{in}} \tag{10}$$

The equipment of the HRES works based on a system management strategy (Fig. 2), starting by comparing the amount of generated energy and the demand load, and then it makes decision based on others factor such the storage system. (Fig. 2) shows this algorithm.

By using a real-world data of weather and load demand, the program code starts by calculating the energy generated and stored for each moment, then compared to the demand load any surplus of power will be stored in the storage units, and any unmet load demand will meet by using the diesel generator. The variation of the objective functions has goal to evaluate the impact of changing one of the multi-objectives functions on the other one, in our case we will focus on the reliability of the system. The configuration selected by the MO should satisfy all the constraints and ensure a good values of optimizations goals.

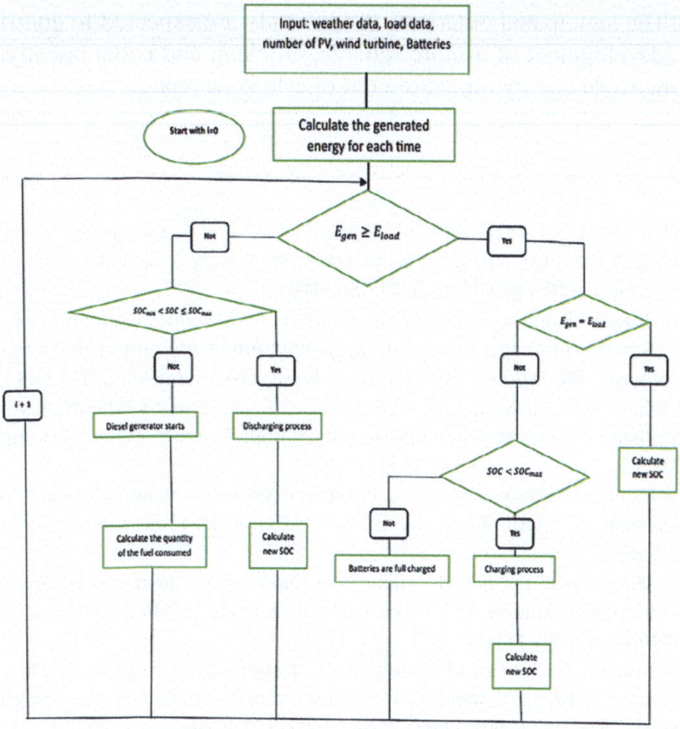

Fig. 2. Energy management strategy

3 Conclusion

The cost remains the primary optimization goal in energy management systems, which are important for advancing modern energy solutions. The significant portion of the overall expense in a particularly off-grid installation is attributed to investments in storage systems. One commonly adopted strategy for cost minimization in these cases optimizing resource utilization and operational efficiency and thereby reducing waste significantly. Our focus is to analyze the influence of additional objective functions on overall system reliability by maintaining the same reliability metric while varying other optimization objectives. This approach explores the relation between cost minimization, environmental impact reduction, and system reliability, using a uniform set of metaheuristic algorithms for multi-objective optimization. Addressing relations could increase both cost-effectiveness and operational performance and ensure long-term stability throughout the system's operational lifespan. To overcome these obstacles, a comprehensive optimization framework that integrates system constraints with the details of the computational techniques of multi-objective metaheuristic optimization is required. Future research will focus on refining this methodology, constraints, and selecting appropriate software tools for effective implementation. These enhancements will further optimize system performance. The anticipated outcomes of this study are expected to contribute significantly to the development of a more reliable, efficient, and economically, supporting sustainable renewable energy utilization out of grid dependency.

References

1. Mostafa, H.A., Shatshat, R.E., Salama, M.M.A.: A review on energy management systems. In: 2014 IEEE PES T&D Conference and Exposition, Chicago, IL, USA, pp. 1–5. IEEE April 2014. https://doi.org/10.1109/TDC.2014.6863413
2. Tajjour, S., Singh Chandel, S.: A comprehensive review on sustainable energy management systems for optimal operation of future-generation of solar microgrids. Sustain. Energy Technol. Assessm. **58**, 103377 (2023). https://doi.org/10.1016/j.seta.2023.103377
3. Rezk, H., Olabi, A.G., Sayed, E.T., Wilberforce, T.: Role of metaheuristics in optimizing Microgrids operating and management issues: a comprehensive review. Sustainability **15**(6), 4982 (2023). https://doi.org/10.3390/su15064982
4. Ezugwu, A.E., et al.: Metaheuristics: a comprehensive overview and classification along with bibliometric analysis. Artif. Intell. Rev. **54**(6), 4237–4316 (2021). https://doi.org/10.1007/s10462-020-09952-0
5. Gow, J.A., Manning, C.D.: Development of a photovoltaic array model for use in power-electronics simulation studies. IEE Proc., Electr. Power Appl. **146**(2), 193 (1999). https://doi.org/10.1049/ip-epa:19990116
6. Diaf, S., Notton, G., Belhamel, M., Haddadi, M., Louche, A.: Design and techno-economical optimization for hybrid PV/wind system under various meteorological conditions. Appl. Energy **85**(10), 968–987 (2008). https://doi.org/10.1016/j.apenergy.2008.02.012
7. Bhandari, B., Poudel, S.R., Lee, K.-T., Ahn, S.-H.: Mathematical modeling of hybrid renewable energy system: a review on small hydro-solar-wind power generation. Int. J. Precis. Eng. Manuf.-Green Tech. **1**(2), 157–173 (2014). https://doi.org/10.1007/s40684-014-0021-4
8. Kaabeche, A., Belhamel, M., Ibtiouen, R.: Sizing optimization of grid-independent hybrid photovoltaic/wind power generation system. Energy **36**(2), 1214–1222 (2011). https://doi.org/10.1016/j.energy.2010.11.024
9. Rathod, A.A., B. S.: Modified Harris Hawks optimization for the 3E feasibility assessment of a hybrid renewable energy system. Sci. Rep. **14**(1), 20127 (2024). https://doi.org/10.1038/s41598-024-70663-5
10. Javed, M.S., Song, A., Ma, T.: Techno-economic assessment of a stand-alone hybrid solar-wind-battery system for a remote island using genetic algorithm. Energy **176**, 704–717 (2019). https://doi.org/10.1016/j.energy.2019.03.131

An Integrated Framework for Remaining Range Forecasting

Zakaria Charouh[1](✉), Amal Ezzouhri[2], Hamza Mediouni[3], Mounir Ghogho[4], and Zouhair Guennoun[5]

[1] Giantlink, Fez, Morocco
z.charouh@gmail.com
[2] L3IA Laboratory, Sidi Mohamed Ben Abdellah University, Fez, Morocco
[3] EODIC Team, ENSAM, Mohammed V University, Rabat, Morocco
[4] TICLab, International University of Rabat, Rabat, Morocco
[5] ERSC Team, EMI, Mohammed V University, Rabat, Morocco

Abstract. Accurate energy consumption prediction is crucial for efficient travel planning and reliable driving range estimation, particularly for low-resource electric vehicles. Energy consumption is influenced by several dynamic factors, such as driving behavior, road topology, vehicle load, and environmental conditions, making precise forecasting a challenging task. In this paper, we propose an integrated framework that systematically incorporates these key influencing factors to improve the reliability of forecasting energy consumption for upcoming trips. Our approach leverages machine learning techniques to model complex relationships between these variables, enabling more reliable remaining range estimation. To assess the performance of the proposed framework, extensive experiments were conducted using a public real-world dataset. The results demonstrate the robustness of our approach, achieving an R-squared value of up to 95.2%, highlighting its potential for improving energy management strategies and increasing user confidence in electric vehicle adoption. Furthermore, our framework can be adapted to various driving conditions, offering a scalable and practical solution for future intelligent transportation systems.

Keywords: driving range · battery electric vehicle · electric vehicle · range anxiety

1 Introduction

The integration of intelligent transportation systems with data analytics presents a significant opportunity to address key challenges in electric vehicle (EV) adoption. One of the primary concerns for EV users is range anxiety, which arises from uncertainties in energy consumption and remaining range estimation. Research shows that most drivers recharge their EVs before the State of Charge (SoC) falls below 25% due to this anxiety [1]. In fact, recent surveys have shown that

C. El Mokhi et al. (Eds.): ICASET 2025, SUCI, pp. 9–17, 2025.
https://doi.org/10.1007/978-3-031-98334-4_2

range anxiety remains a major barrier, with one-third of potential European users hesitant to transition to EVs due to concerns about driving range [2,3].

To facilitate EV adoption, accurate Remaining Range Forecasting (RRF) is essential for enabling effective trip planning and energy management. However, RRF is a complex, multi-factorial problem influenced by numerous variables, including vehicle characteristics, road topology, driving behavior, weather conditions, and traffic patterns. Among these, the driving profile plays a crucial role, as factors such as acceleration intensity and speed fluctuations significantly impact energy consumption.

In this paper, we introduce an integrated framework for RRF that systematically incorporates these influencing factors to improve the precision of energy consumption prediction. Our approach leverages machine learning techniques to model complex dependencies and improve forecasting reliability. As part of this research, we also present a speed profile prediction module, which refines range estimation by anticipating vehicle dynamics, and a real-world dataset (EVECS) collected from Moroccan roads, offering valuable insights into energy consumption and speed variations under diverse driving conditions.

The primary contributions of this work include:

- A machine learning-based speed profile prediction module.
- An integrated framework for remaining range forecasting.
- A real-world dataset (EVECS) for EV energy consumption and speed profiles collected on Moroccan roads.

The remainder of this paper is structured as follows: Sect. 2 presents a review of related work on RRF. Section 3 details the proposed RRF framework. In Sect. 4, we analyze the datasets and report the experimental findings. Lastly, Sect. 5 summarizes the key conclusions of the paper.

2 Related Work

To reduce range anxiety, research has focused on improving EV energy consumption prediction using simulation-based and data-driven methods. This section presents the key studies in both approaches.

Simulation-Based Approaches. Authors in [4] developed a simulation using the curve fitting of the motor efficiency with a resistance model to forecast the consumed energy. Authors in [5], developed a simulator that helps estimating the remaining range of EVs. In [6] authors have proposed a synthetic dataset to build a simulation-based approach for remaining range forecasting. In [7], the authors proposed a vehicle simulation model and tested it using experimental data. The authors aimed to demonstrate how different environmental and control parameters influence energy consumption, while in [8] the authors explored the effect of vehicle configuration parameters on the EVs energy consumption in urban areas using the ADvanced VehIcle SimulatOR (ADVISOR) software and real-world driving cycle (RWDC) data.

Data-Driven Approaches. Under the assumption of constant driving parameters, data-driven approaches often rely on regression models trained on historical records to estimate the remaining range. Using data collected from 30 EVs, and applied a multiple linear regression model to estimate the energy usage of EVs. A recurrent neural network (RNN) was proposed in [9] to forecast the available driving range and determine the minimum required charge for safely completing a journey. By analyzing the outputs of the proposed RNN, the probability of a vehicle successfully reaching its destination without recharging is calculated, and if charging is required, the remaining range and the minimum amount to charge to finish the trip are calculated. A study on modeling EV energy consumption is described in [10]. MATLAB/Simulink software is used for EV modeling in accordance with a real electric vehicle. EV models incorporate both longitudinal vehicle dynamics and powertrain systems. Authors in [11] found a Weibull distribution for the correlation between speed and driving range per unit of State of Charge (DDPS). Accordingly, they predicted the DDPS by using the Weibull distribution. The remaining driving distance was then calculated based on SoC and DDPS. In [5,12,13], and [14] the authors derived the remaining range using classical models including gradient-boost decision trees and ANN.

3 Remaining Range Forecasting

The proposed RRF framework adopts a modular architecture that integrates road topology, speed profile, weather conditions, and vehicle characteristics. To enhance accuracy, the trip is segmented, and an Energy Consumption Prediction (ECP) is computed for each segment. Given the strong influence of speed profiles on energy consumption, we developed a machine learning-based algorithm to predict the mean speed and acceleration per segment. This section presents the speed profile prediction module and the energy consumption prediction module.

3.1 Speed Profile Prediction

Our study assumed a strong correlation between speed profiles and vehicle profiles, as vehicles with robust engines typically deliver higher power and respond more quickly to the driver's desired speed. To support this analysis, we utilized the public EV database (EVDB) provided in [15]. The EVDB serves as a comprehensive cheat sheet containing data on 230 EV models, including efficiency (energy consumption in kWh per kilometer), rated acceleration (time required to reach 0–100 km/h), battery capacity, top speed, towing capacity, and range. To ensure consistency, range values were adjusted across different versions of the same model.

Road topology significantly influences speed profiles, as highway speeds differ from those on local roads, and steep slopes impose additional strain on engines, particularly for EVs with limited power. Given the variability in driving behavior, we introduced a feature to characterize driving attitude. To this end, we identified four driver categories: experienced aggressive drivers, newbie aggressive drivers, newbie cautious drivers, and experienced cautious drivers.

Congestion patterns are inherently reflected in speed profiles, as typical drivers reduce speed in congested areas and adhere to freeway speed limits, while aggressive drivers exhibit high accelerations even in heavy traffic. The proposed Speed Profile Prediction (SPP) module is built using a machine learning model trained on a subset of the energy consumption dataset. The model predicts the mean speed and acceleration for each road segment based on congestion level, vehicle-rated acceleration, driving attitude, road type, and segment slope angle. (Implementation details are provided in Sect. 4).

3.2 Energy Consumption Prediction

This work aims to develop an efficient trip-planning platform. Each trip is defined by a starting point, destination, time, vehicle, and driving conditions. The Overpass API [16] is used to generate intermediate points, and road segments are constructed based on predefined length thresholds. The trip is then divided into segments, with energy consumption predicted for each.

Energy consumption on a given segment depends on factors such as speed profile, slope angle, and vehicle efficiency. Climbing inclines increases power demand, while lightweight or optimized EVs consume less energy under identical conditions. Additionally, weather significantly impacts efficiency, with EV range dropping by approximately 30% in cold conditions [15]. The proposed energy consumption prediction (ECP) method integrates these factors using a machine learning model trained on real-world data. Implementation details are provided in Sect. 4.3.

4 Experimental Results

In this section we describe the VED and the proposed dataset, and we present and discuss the obtained results on the proposed modules. We conducted experiments on SPP module and ECP frameworks using VED benchmark and the proposed energy consumption dataset.

4.1 Energy Consumption Datasets

In this section, we present the VED (a public dataset for energy consumption), and the EV energy consumption and speed profiles (EVECS) dataset we collected on Moroccan roads.

VED Dataset. In this study, we used the public Vehicle Energy Dataset (VED) proposed in [17]. The VED is a large-scale database of real-world fuel and energy use of a wide variety of vehicles. This data was collected through onboard OBD-II from November 2017 to November 2018 and provides GPS trajectories of vehicles as well as fuel, energy, trip identifier, ambient temperature, speed, auxiliary power usage data, High Voltage (HV) Battery Current, HV Battery SOC, and HV Battery Voltage. It consists of 383 personal cars (264 gasoline, 92 hybrids, 24 plug-in hybrids, and 3 electric vehicles) in Ann Arbor, Michigan, USA. Overall, VED accumulates approximately 374,000 miles. Driving scenarios range from highways to congestion-filled downtown areas during multiple driving conditions and seasons.

EVECS Dataset. This study involved two electric vehicles: a Zoe Q210 and a Kangoo ZE [18]. Both were fitted with data loggers linked to the CAN bus, enabling the recording of high-voltage battery parameters such as current, voltage, state of charge (SoC), and real-time speed. Additionally, a mobile application with GPS logging functionality was utilized to track routes and elevation changes. Data collection spanned approximately 200 km and involved six drivers—four men and two women—with diverse driving backgrounds, ranging from novices with under two months of experience to seasoned drivers with over a decade on the road. The data was gathered under diverse road conditions and driving scenarios, as shown in Fig. 1.

As depicted in Figs. 1b and 1a, HV battery voltage decreases when a high current is drawn (e.g., at time = 300 s) and recovers when the throttle is released (e.g., at time = 250 s), which correlates closely with speed, as illustrated in Fig. 1d).

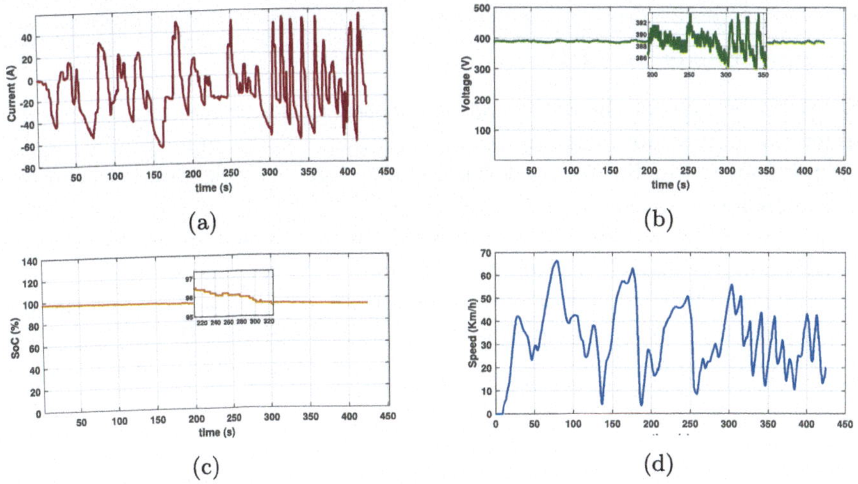

Fig. 1. Illustration of the HV battery current, voltage, state of charge, and speed from a trip. Negative values in 1a refer to pressing on the throttle pedal (power demand), while positive current values refer to the regenerative braking.

In Fig. 2, we show data from an experiment conducted on a steep slope with an altitude difference of approximately 160 m (from 158 m to 1 m), as seen in

Fig. 2b. When traveling downhill, HV battery current values became positive, and the state of charge increased slightly due to the regenerative braking feature.

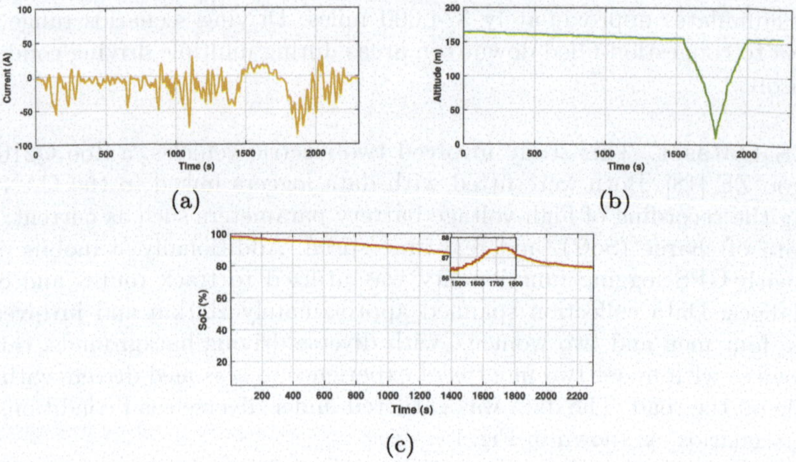

Fig. 2. Illustration of the HV battery current, state of charge, and altitude from a trip. A change of SoC in 2c is observed from 1550 s to 1700 s, with a change of altitude from 158 m to 1 m.

4.2 Experiments on SPP

In this part, we present the SPP module construction and evaluation. The module relies on a machine learning-based regression model. The model takes as input five features including road type, slope angle, vehicle-rated acceleration, driving attitude, and congestion level. The EVECS dataset contains a vehicle identifier and GPS records. The vehicle identifier is used in accordance with [15] to extract the vehicle's rated acceleration. GPS records are used to create road segments. The road type is determined using TomTom's API, and the slope angle S_a is determined using elevations from NASA's Shuttle Radar Topography Mission SRTM [19] such as:

$$S_a = sin^{-1}(\frac{|E_B - E_A|}{D_{A,B}}) \tag{1}$$

were $D_{A,B}$ denotes the distance between points A and B, and E_A and E_B denote A and B elevations. In order to determine which model would best suit this task, we considered many machine learning models, including Decision Tree (DT), Logistic Regression (LR), Naive Bayesian (NB), K-Nearest Neighbors (KNN), Random Forest (RF), and Artificial Neural Network (ANN).

We divided the data set into training (70%), validation (10%), and testing (20%). ANNs proved to be the most reliable. The results for speed prediction are: r-squared = 89.7; RMSE = 7.3; and the results for acceleration prediction are r-squared = 91.12; RMSE = 2.8, which we consider sufficient considering the type of roads and events experienced during data collection, as illustrated in Table 1.

Table 1. Result comparison on SPP module. $RMSE_{speed}$, $RMSE_{acc}$ represent the RMSE obtained for speed and accuracy, respectively.

ML	$RMSE_{speed}$	$R^2{}_{speed}$	$RMSE_{acc}$	$R^2{}_{acc}$
LR	16.33	72.56	7.11	74.52
DT	14.27	78.32	5.97	78.45
SVR	11.51	83.12	4.45	84.31
RF	13.68	79.41	5.21	80.22
ANN	7.3	89.7	2.8	91.12

4.3 Experiments on EPC

In this section, we describe the experiments conducted on the VED and EVECS datasets. As outlined in Sect. 4.1, the VED dataset contains real-world fuel and energy usage data for various vehicles. We focused on the EV subset, applying preprocessing to select accurate records and retain relevant columns such as vehicle identifier, trip identifier, timestamp, GPS data, vehicle speed, ambient temperature, air conditioning power, heater power, HV battery current, HV battery state of charge (SoC), and HV battery voltage. The next step involved creating road segments based on predefined segment length criteria. For the EVECS dataset, data collected through two mobile applications were synchronized using timestamps, allowing us to compute road segment descriptors. Both the VED and EVECS datasets were formatted to include segment-level information, with records describing the amount of energy consumed per segment. The energy consumption was computed as follows:

$$Q = \Sigma(C_i \times V_i - Aux_i) \times (t_i - t_{i-1}) \tag{2}$$

where t_i, C_i, V_i denote respectively the timestamp, the current, and the voltage at instant i, Aux = Air Conditioning power + Heater power (Table 2).

We introduced also an average instantaneous energy consumption per 100 Km that represents a standardized metric we used in eliminating aberrant values (usual EVs consume under 60 KWh/100 km), we set a threshold of 150 KWh/100 km as a maximum valid value. We also computed segment length, slope angle, average speed, mean acceleration, and average temperature, and we extracted road type and EV efficiency.

Table 2. A model investigation for the task of ECP on VED and EVECS datasets.

ML	$RMSE_{VED}$	$R^2{}_{VED}$	$RMSE_{EVECS}$	$R^2{}_{EVECS}$
LR	14.22	56.31	13.79	58.37
DT	11.71	69.87	12.02	65.1
SVR	9.51	71.87	9.33	72.45
RF	11.89	65.28	12.11	63.08
ANN	3.1	95.2	3.88	94.61

5 Conclusion

In this work, we have proposed an integrated framework for remaining range forecasting (RRF) that effectively incorporates key influencing factors in energy consumption through abstract feature representation. By leveraging a machine learning-based approach, our method enhances the accuracy and reliability of EV range estimation, addressing real-world variability in driving conditions. As part of this research, we introduced a novel dataset (EVECS) specifically collected from Moroccan roads, providing a valuable benchmark for EV energy consumption studies in diverse driving environments. Additionally, we developed a speed profile prediction module, which plays a crucial role in refining energy estimation by anticipating vehicle dynamics. Extensive experiments conducted on the proposed RRF framework demonstrated its robustness and high predictive performance, achieving an R-squared value exceeding 95%. These promising results underscore the potential of our approach in improving range estimation accuracy, thereby contributing to more efficient energy management and enhanced user confidence in electric vehicle adoption.

References

1. Togun, H., et al.: Review on recent advances on improving fuel economy and performance of a fuel cell hybrid electric vehicle. Int. J. Hydrogen Energy **89**, 22–47 (2024)
2. Pei, M., Huang, Z., Zhang, Z., Wang, K., Ye, X.: Range anxiety and willingness to pay: psychological insights for electric vehicle. J. Renew. Sustain. Energy, **17**(1) (2025)
3. Tao, S., Zhu, J., Li, Y., Chang, W., Dai, H., Wei, X.: A data-driven framework for battery capacity estimation in real-world electric vehicles using virtual impedance and incremental capacity analysis (No. 2025-01-8561). SAE Technical Paper (2025)
4. Sangeetha, E.P., Subashini, N., Santhosh, T.K., Uma, D.: Validation of EKF based SoC estimation using vehicle dynamic modelling for range prediction. Electr. Power Syst. Res. **226**, 109905 (2024)
5. Wang, J., Tu, R., Wang, A., He, D.: Trust in range estimation system in battery electric vehicles–a mixed approach. IEEE Trans. Hum. Mach. Syst. (2024)
6. Mediouni, H., Ezzouhri, A., Charouh, Z., El Harouri, K., El Hani, S., Ghogho, M.: Energy consumption prediction and analysis for electric vehicles: a hybrid approach. Energies **15**(17), 6490 (2022)

7. Xie, Y., et al.: Microsimulation of electric vehicle energy consumption and driving range. Appl. Energy **267**, 115081 (2020)
8. Wu, H.: Modeling and simulation of electric city bus power system based on multi-objective genetic algorithm. In 2024 IEEE 13th International Conference on Communication Systems and Network Technologies (CSNT), pp. 1–7. IEEE (2024)
9. Eagon, M.J., Kindem, D.K., Panneer Selvam, H., Northrop, W.F.: Neural network-based electric vehicle range prediction for smart charging optimization. J. Dyn. Syst. Meas. Control, **144**(1) (2022)
10. Miri, I., Fotouhi, A., Ewin, N.: Electric vehicle energy consumption modelling and estimation—a case study. Int. J. Energy Res. **45**(1), 501–520 (2021)
11. Li, F., Zuo, W., Zhou, K., Li, Q., Huang, Y.: State of charge estimation of lithium-ion batteries based on PSO-TCN-attention neural network. J. Energy Storage **84**, 110806 (2024)
12. Miao, L., Han, W.: Simulation analysis of electrical vehicle's remaining discharge energy based on driving profile prediction. In: 2024 IEEE 10th International Power Electronics and Motion Control Conference (IPEMC2024-ECCE Asia), pp. 4883–4888. IEEE (2024)
13. Peng, S., Zhang, D., Dai, G., Wang, L., Jiang, Y., Zhou, F.: State of charge estimation for LiFePO4 batteries joint by PID observer and improved EKF in various OCV ranges. Appl. Energy **377**, 124435 (2025)
14. Sun, S., Zhang, J., Bi, J., Wang, Y.: A machine learning method for predicting driving range of battery electric vehicles. J. Adv. Transp. (2019)
15. Electric Vehicle Database (EVDB). https://ev-database.org/. .Accessed 26 Jan 2025
16. Overpass API User's Manual. https://dev.overpass-api.de/overpass-doc/en/. Accessed 29 Jan 2025
17. Oh, G., Leblanc, D.J., Peng, H.: Vehicle energy dataset (ved), a large-scale dataset for vehicle energy consumption research. IEEE Trans. Intell. Transp. Syst. (2020)
18. Ezzouhri, A., Charouh, Z., Mediouni, H., Ghogho, M., Guennoun, Z.: EV Energy consumption and speed profiles dataset (EVECS). IEEE Dataport. (2022). https://doi.org/10.21227/t4e1-w222
19. Shuttle Radar Topography Mission. www.earthdata.nasa.gov/data/instruments/srtm. Accessed 29 Jan 2025

Impact Study of Harmonic Analysis of a Grid-Connected Photovoltaic System Feeding Nonlinear Load

Soukaina Essaghir[1](✉), Tajeddine Khalili[2], and Mustapha Oulcaid[1]

[1] Sultan Moulay Slimane University, National School of Applied Sciences, ENSA – Beni-Mellal, LGS Laboratory, Beni-Mellal, Morocco
s.essaghir@usms.ma

[2] Sultan Moulay Slimane University, National School of Applied Sciences, ENSA – Beni-Mellal, LICPM Laboratory, Beni-Mellal, Morocco

Abstract. This research performs harmonic examination of a PV system when attached to grid power that distributes non-linear loads through various control instruments. Three control techniques are used which include conventional PI regulator together with proportional resonant controller (PR) and Fuzzy Logic regulator. The analyzed system refers to a photovoltaic model integrated with grounded in a DC-DC converter followed by an MPPT device based on Perturb and Observe algorithm connected to an electrical grid supplying a DC-AC inverter which controls a nonlinear load through Pulse Width Modulation (PWM). The study concludes with a comparative analysis of Total Harmonic Distortion (THD) levels across the three control techniques.

Simulation outcomes obtained with Matlab/Simulink toolset validate the robustness of this regulation loop based on the Fuzzy Logic Controller which guarantees more effective energy transmission to the grid with fewer harmonic disturbances especially when using a non-linear load.

Keywords: PV System · Grid · PR Controller · PI Controller · Fuzzy Logic Controller · THD · Non-Linear Load

1 Introduction

Today, the rapid transformation of the power grid by photovoltaic (PV) systems [1] which link to the electrical network presents major challenges for system controllers in keeping the grid reliable and stable. Moreover, photovoltaics has emerged as one of the most promising renewable energies worldwide and is expected to grow significantly in the future [1]. Currently, more and more research focus on photovoltaic systems and their interconnection with the grid [2, 3], however this connection presents critical challenges including the shift in phase between the current and voltage delivered to the grid and decreased power quality which results from rising power electronic usage. In addition, the use of non-linear loads [4], causes numerous disturbances on the utility side, such as sags, ripples and harmonics. Despite the existence of many researches

C. El Mokhi et al. (Eds.): ICASET 2025, SUCI, pp. 18–29, 2025.
https://doi.org/10.1007/978-3-031-98334-4_3

on grid-connected PV Systems, there remains a gap in developing a control strategy that ensures simultaneously a unit power factor and mitigates current harmonics under dynamic load variation.

To address these challenges, this study provides a comparative analysis of various control strategies, including - the PI controller, PR regulator and Fuzzy PI current controller - within the context of a photovoltaic system subjected to grid experiencing load fluctuations. This study makes two keys contributions:

- Enhancing current control performance by integrating advanced control techniques in order to improve grid stability.
- Reducing harmonic distortions caused by power electronic devices and especially non-linear loads to improve power quality and reliability.

Through this study, we aim to boost the development of this grid connected photovoltaic system through two essential factors starting with better current control using diverse control approaches. Power electronic devices together with non-linear load sources generate harmonics that require reduction as the second vital factor.

The research structure allocated in this document begins with a comprehensive system configuration explanation in Sect. 2. Section 3 defines the strategy for each current controller including PI, PR and Fuzzy PI controllers. Section 4 analyzes the measurement data along with operational inputs followed by the closing statements in the final part.

2 PV Grid-Connected System Model

2.1 Structure

The Fig. 1 illustrates a three-phase grid-connected photovoltaic (PV) system which supplies power to a nonlinear load. A photovoltaic array together with DC/DC voltage regulator and DC-AC power inverter allows connection to the electrical grid.

The photovoltaic array in this study consists of 66 strings, each containing five series-connected modules configured in a parallel arrangement. The system uses a DC voltage booster as an MPPT controller [5–7], Meanwhile, a three-stage voltage inverter is incorporated to increase system adaptability and enhance the quality of the produced waveform [8].

Through its three-level configuration the VSI system performs DC voltage control and performs power factor correction to unity when dealing with shifting loads. The Regulation framework contains two operation loops with an outer loop that controls DC link regulation and an inner loop that regulates active and reactive current components (I_d and I_q). The PWM modulator permits the inverter and converter to achieve output voltage regulation as part of their operational functions [9, 10].

The model development used Matlab/Simulink as its platform and Table 1 provides complete specifications.

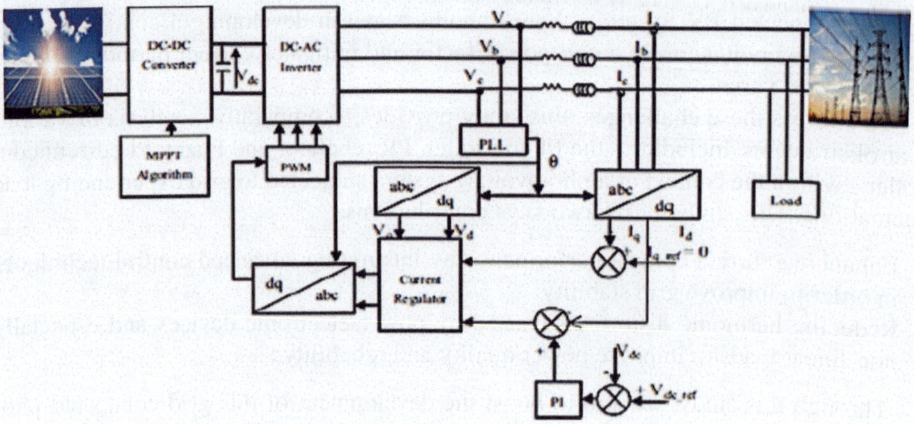

Fig. 1. Topology of the modelled system

2.2 Current Control Strategies

The central part of inverter construction depends on its control system. The controller system contains a Phase-Locked Loop (PLL) that performs output alignment of the inverter with grid voltage phase and frequency [11]. The direct-quadrature (d-q) reference frame together with its inverse version allow the conversion of three-phase data consisting of a, b, c currents and 3 voltages into two-variable d-q terms. Our study primarily focuses on comparing three types of controllers: PI, PR, and FLC. We began by introducing the PI controller, which combines proportional and integral control actions to minimize the gap between the system response and the reference signal. To enhance the system's efficiency and responsiveness, we propose the use of PR and FLC current controllers and compare their respective impacts on system performance.

a. PI Regulator

The Proportional-Integral (PI) regulator is a standard current corrector used to ensure Optimal power factor. Figure 2 shows the Flowchart diagram of this controller, where KP and Ki represent the proportional and integral gains. These factors are calculated based on the system's parameters [12].

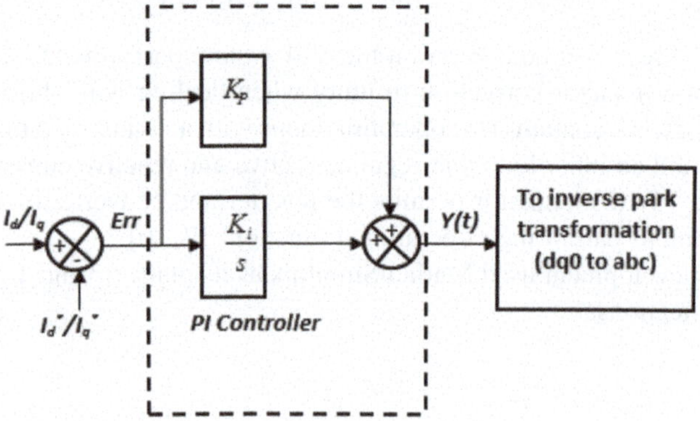

Fig. 2. Flowchart diagram of PI controller

b. PR Controller

The proposed paper conducts harmonic analysis of a PV system connected to the grid starting from a basic PI regulator. While this controller regulates the system's overall performance, it has some limitations especially in targeting specific harmonic frequencies generated by nonlinear loads. In contrast, the PR controller is intended to focus on eliminating particular harmonics in power systems.

In our research, we opted for an approximate non-ideal proportional-resonant (PR) controller to address several challenges. An ideal PR controller, for example, can result in infinite gain, which complicates its use for effectively managing harmonics caused by fluctuations in grid voltage. Additionally, such a level of control is impractical to achieve in both analog and digital systems [13, 14].

The PR current controller is mathematically described as follows:

$$G_{PR} = K_P + K_I \frac{2\omega \cdot s}{s^2 + 2\omega_c \cdot s + \omega_0^2} \tag{1}$$

The resonant peak, depicted in Fig. 3, achieves a gain of 40 dB-high enough to ensure effective voltage tracking [15].

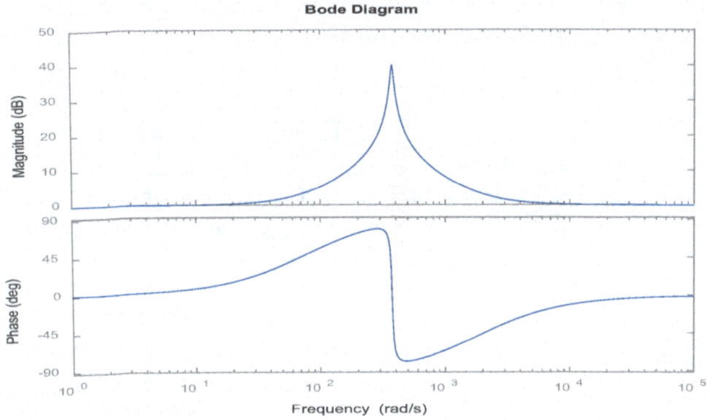

Fig. 3. Frequency behavior of a non-ideal PR corrector with $K_P = 1$, $K_I = 100$, $\omega_c = 10$(rad/s)

c. Fuzzy Logic Corrector

The Fuzzy Logic PI Corrector functions as a non-linear system which applies datasets to single numerical outputs.

In many nonlinear applications, conventional controllers may not perform effectively. Consequently, fuzzy logic controllers have proven to be a suitable solution for various nonlinear problems [16, 17]. The FLC operates through three main stages: Fuzzification, Rule-Based Table Lookup, and Defuzzification.

A FLC follows three fundamental operational steps which are Fuzzification followed by Rule-Based Table Lookup and finishing with Defuzzification. Fuzzy logic controllers require an implementation of membership functions which represent both inputs and outputs. Linguistic descriptors emerge from numerical input variables through the fuzzification stage by using these membership functions as shown in Fig. 4. Multiple fuzzy categories provide detailed specifications for this context including NB (Negative Big), NS (Negative Small), ZE (Zero), PS (Positive Small), and PB (Positive Big) [18].

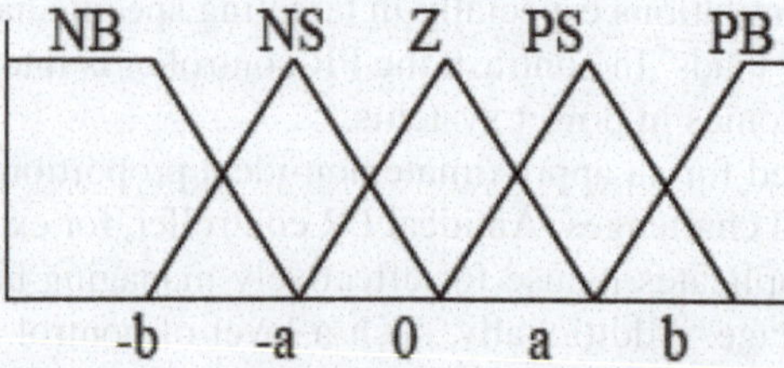

Fig. 4. The input and output membership functions operate in the fuzzy controller

This research utilizes the PLL to synchronize the Park transformation with grid voltage fluctuations. It also uses the FLC for direct and quadrature current control. Under stability conditions the Park transformation V_d amplitude equals measured voltage amplitude and V_q retains a value of zero. The active power component equals Id while the reactive power component corresponds to Iq according to Eqs. (2) and (4) from [19, 20]. The figure below displays the bloc diagram of the fuzzy logic regulator which was implemented during this research work (Fig. 5).

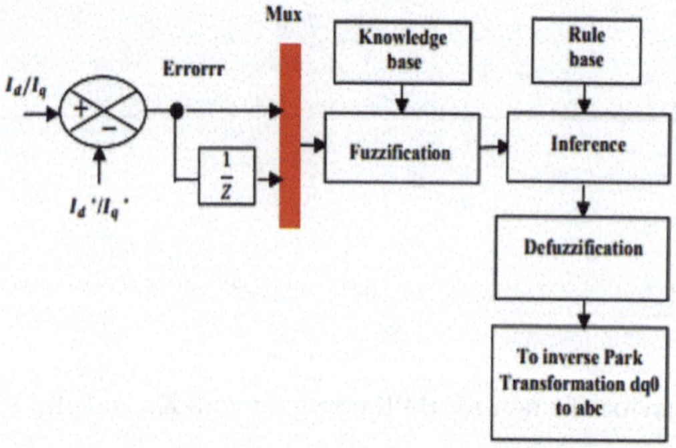

Fig. 5. Schematic diagram of fuzzy logic corrector

3 Simulation Results and Discussion

The results obtained from Matlab/Simulink simulations are outlined in this section. Figure 6 illustrates the comprehensive system, that includes a photovoltaic (PV) module that supplies a maximum output of 100 kW at an irradiance of 1000 W/m^2.

The system model includes the following modules:

- A PV system providing an output of 100 kW at a solar irradiance of 1000 W/m^2.
- A boost converter that increases the voltage from the PV array's output of 273.5 V_{DC} to 500 V_{DC}.

- An MPPT tracker, implementing the P&O approach, fine-tunes the boost converter's switching duty cycle.
- A three-level, three-phase inverter with a 1980 Hz frequency that transform the 500 V_{DC} into 260 V_{AC} while maintaining a unity power factor.
- A current corrector, integrating PI, PR, and FLC controllers, ensures a unit power factor.
- A three-phase coupling transformer rated at 100 kVA, with 260V/25kV voltage levels.
- Grid parameters, 25 kV distribution feeders alongside 120 kV equivalent transmission networks.

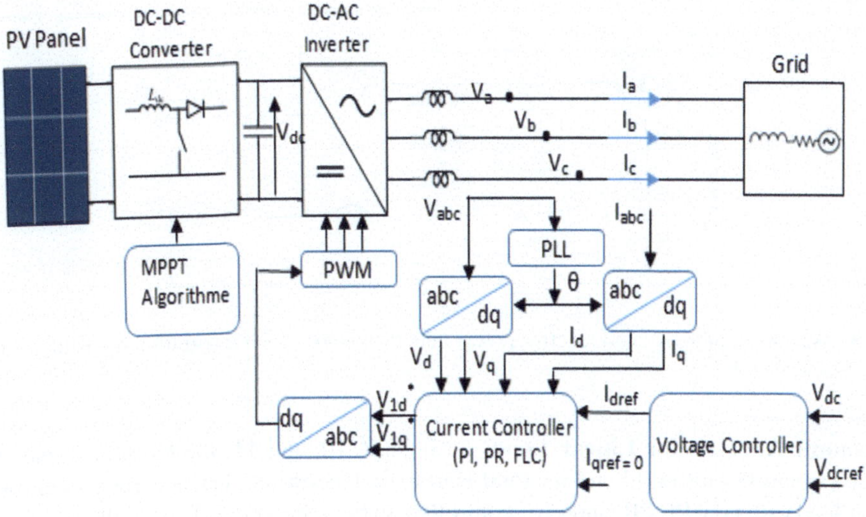

Fig. 6. Detailed system modeling using MATLAB/Simulink

3.1 Modeling Simulation Under Inductive and Capacitive Load Integration

Active and Reactive Power Under Linear Load Variation. In the figure below and at (t = 0.3 s), a reactive load is introduced. Based on our previous publication and simulation [20], we conclude that the PR controller achieves a unit power factor more effectively than both the PI and FLC controllers when reactive loads are added, without impacting the reactive power rate (Figs. 7 and 8).

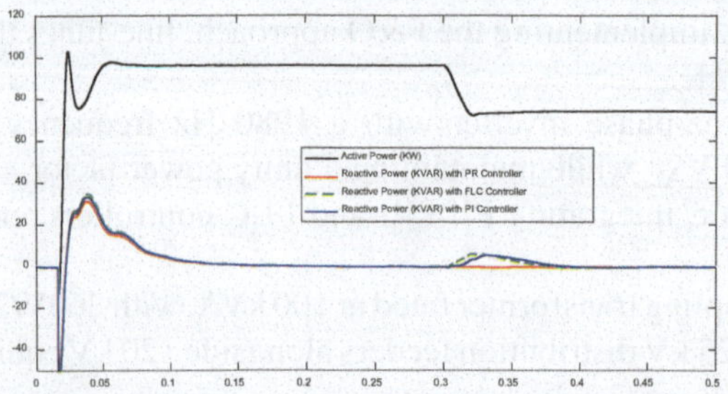

Fig. 7. Response of Active and reactive power under inductive load variation by using different regulators

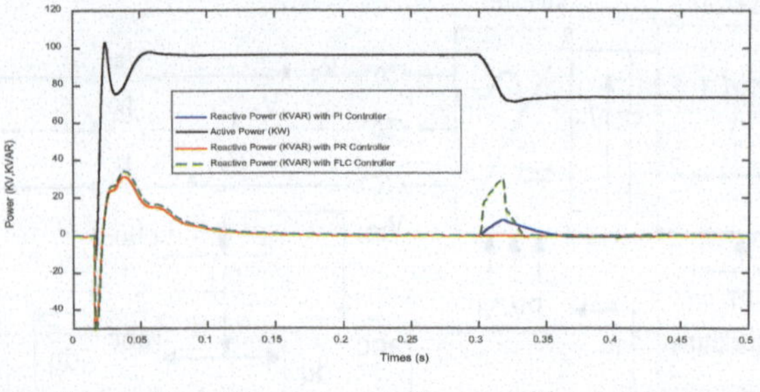

Fig. 8. Waveform of active and reactive power under capacitive load integration by using PI, PR and FLC corrector

Harmonic Analysis for Linear Load. In Figs. 9, 10, and 11, the harmonic analysis of the current fed into grid and its Total Harmonic Distortion (THD) is presented for a linear load using the PI, PR, and FLC controllers in the same order. The results outcomes demonstrate that the Fuzzy Logic PI controller delivers a notably lower THD compared to both the PR and PI controllers. The THD for the FLC corrector is 1.10%, in contrast to the higher THD values for the PR and PI controllers.

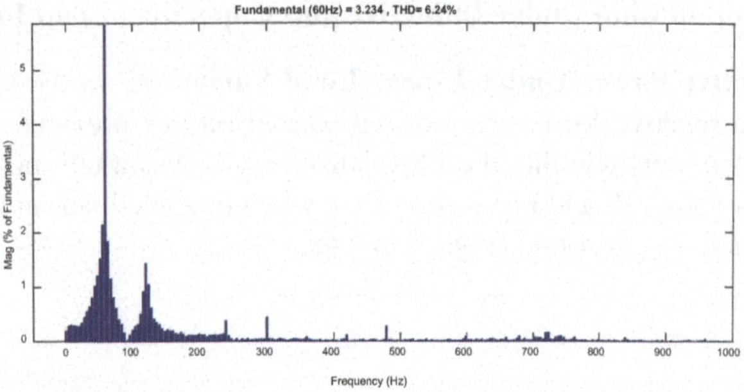

Fig. 9. Harmonic behavior and THD measurement of grid current with the PI corrector

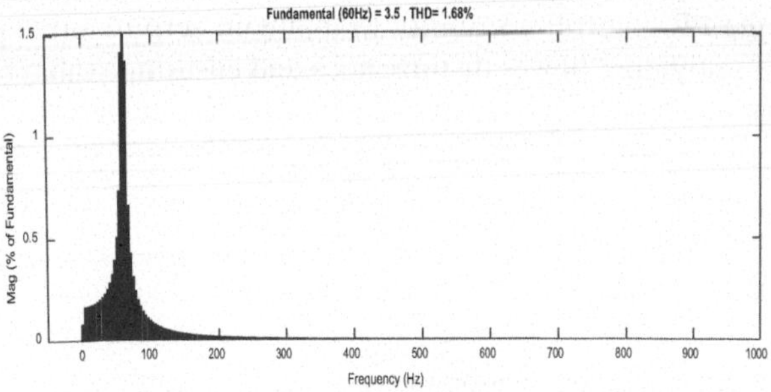

Fig. 10. Harmonic study and THD value while using PR regulator for the grid

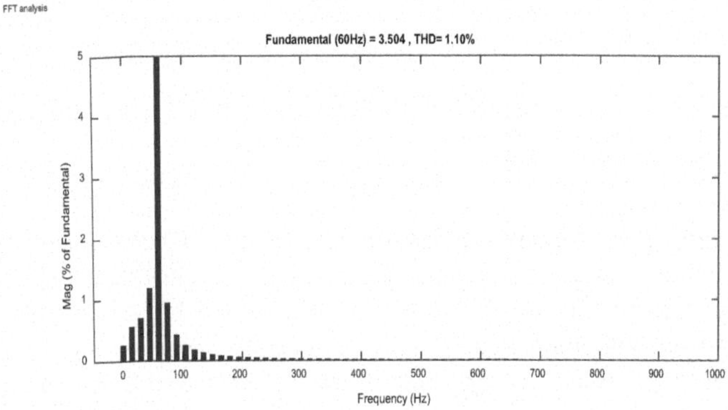

Fig. 11. Harmonic characteristics and THD evaluation of grid current with the FLC corrector

3.2 Simulation Under Addition of Nonlinear Load

Active and Reactive Power Under Nonlinear Load Addition. Figure 12 illustrates both active and reactive power controlled by PI, FLC and PR controller under addition of non-linear load.

At (t = 0.3 s), a nonlinear load is introduced into the system, causing noticeable differences in the controllers 'responses. The figure clearly shows that the PI controller struggles to accommodate the added load and the rate of reactive power has many fluctuations. This instability indicates that the PI controller lacks the robustness required to maintain unit power factor under dynamic load addition.

On the other hand, the FLC controller exhibits more adaptive behavior. Initially, the reactive power briefly shifts into a positive range for a few milliseconds before returning to its original value. This response suggests that the FLC controller can react dynamically to disturbances, but it still delivering minor oscillations before reaching a steady-state condition.

Contrary, the PR controller exhibits superior performance, it ensures that only active power is supplied into the grid which is the main outcome for maintaining both efficiency and power quality.

To conclude, the PR controller's ability to maintain zero reactive power highlights its effectiveness in mitigating power disturbances and ensuring stable operation in grid-connected PV systems.

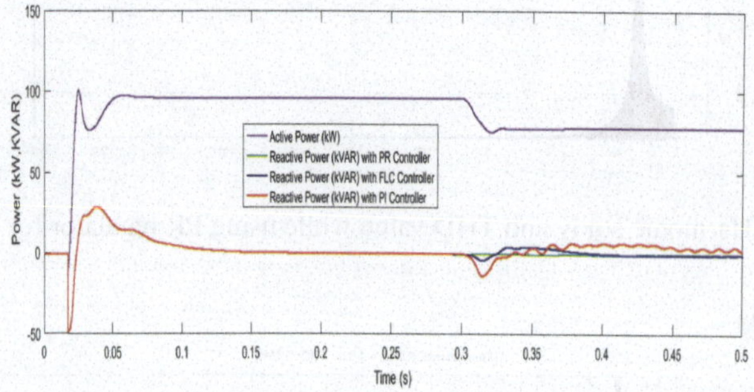

Fig. 12. Active and reactive power controlled using different controller under addition of non-linear load

Harmonic Analysis for Non-linear Load. The frequency behavior of the current fed to the grid together with the THD value when driving a non-linear load controlled the different types of correctors are shown in Figs. 13, 14 and 15 respectively.

The simulation outcomes clearly indicate that the Total Harmonic Distortion for the Fuzzy Logic PI corrector is significantly lower compared to the one of PR and PI controllers.

Specifically, the THD value obtained with the FLC controller is 7,04% which is considerably lower than the 12,39% recorded with the PR controller and the 36.07% observed with the PI controller.

This reduction in THD value demonstrates the FLC controller's superior ability to maintain power quality factor and to filter out harmonics.

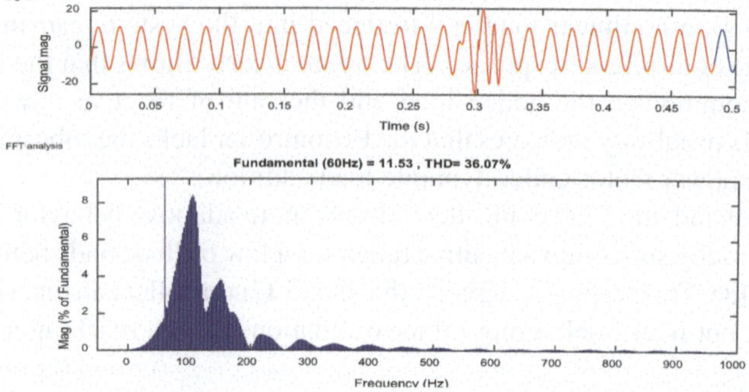

Fig. 13. Harmonic study and THD value of grid current by using a PI regulator

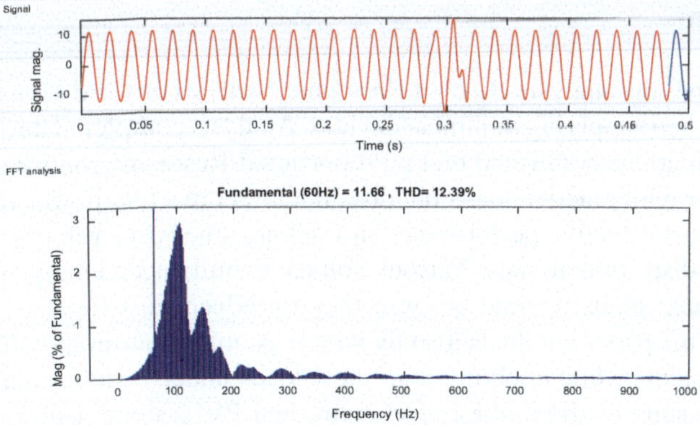

Fig. 14. Harmonic examination and total harmonic distortion of grid current with the PR corrector

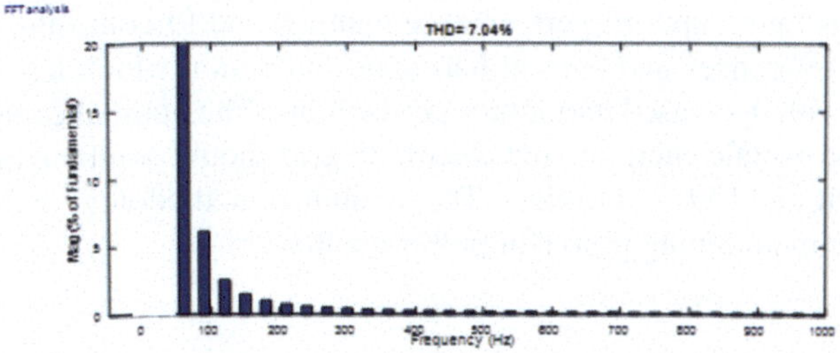

Fig. 15. Harmonic study and grid current's THD with the FLC controller

We can therefore conclude that the FLC controller not only accurately tracks the sinusoidal reference but also reduces harmonics more effectively than both the PI and PR correctors. By contrast, the PR regulator moderate harmonic mitigation, while the PI controller provides the highest THD value indicating poor harmonic elimination.

The table below summarizes the different THD values obtained by the harmonic analysis:

Table 1. Summary table comparing THD results

Load	THD Value		
	PI Controller	PR Controller	FLC Controller
Linear load	6,24%	1,68%	1,10%
Non-linear load	36,07%	12,39%	7,04%

As a result, the simulation yielded the expected outcomes, demonstrating that both the PR and FLC controllers are essential for ensuring a stable grid connection with improved energy transfer and reduced harmonics.

4 Conclusion

This paper examines the analysis effect of Harmonic Analysis on Grid-Connected Photovoltaic Systems when supplying nonlinear devices. Analysis completed through previous research and simulations confirmed that a Proportional-Resonant controller guarantees unity power factor with reactive loads but does not affect reactive power rate. The FLC controller maintains effective performance in tracking sine wave references as it minimizes harmonic distortion signals. Various voltage disturbances emerge on the utility side when nonlinear loads operate because they introduce sags together with ripples and harmonics. This paper conducts a study which examines harmonic effects on grid-connected PV systems while applying non-linear loads under various control methods. The simulation results of three-phase grid-connected PV systems using PR and FLC regulator show that this method successfully matches the analyzed PV framework's dynamic performance. Through linear and non-linear load conditions PR regulators maintain a unit power factor yet show no impact on reactive power fluctuations. A FLC corrector demonstrates superior performance than PR and PI controllers when dealing with sinusoidal references and current harmonics reduction which leads to better grid energy transfer with decreased harmonics productions. The optimal control strategy for transferring photovoltaic energy to the electricity grid should combine the benefits provided by both PR and FLC controllers. The combined methodology works to decrease harmonics while maintaining powerful power quality.

Appendix

(Table 2)

Table 2. Photovoltaic system variables

Temperature	T	25	°C
Open circuit voltage	V_{oc}	64.2	V
Short circuit current	I_{sc}	5.96	A
Voltage maximum power	V_{max}	54.7	V
Current maximum power	I_{max}	5.58	A
Maximum power	P_{max}	305	W

References

1. Khanna, V., Das, B.K., Bisht, D.: Matlab/simelectronics models based study of solar cells. Int. J. Renew. Energy Res. **3**(1), 30–34 (2013)
2. Ouatman, H.: Modeling and control of a grid-connected PV energy conversion system. **10**, 484–492 (2015)

3. Wail Rezgui, M.D.M., Leila, H.M.: Modeling of a photovoltaic field in malfunctioning. In: International Conference on Control, Decision and Information Technologies (CoDIT), p. 2013
4. Tsai-Fu Wu, H.-S.N., Chih-Lung, S.: A single-phase inverter system for PV power injection and active power filtering with nonlinear inductor consideration. In: IEEE Transactions on Industry Applications (2005)
5. Mollah, P.K.S.A.H., Panda, G.K.: Three phase grid connected photovoltaic system with maximum power point tracking. Int. J. Adv. Res. Electr. Electron. Instrum. Eng. **4**(5), 2278 –8875. ISSN (Online)
6. Sayal, S.M.A., Sekhar, P.C.: A review on maximum power point tracking techniques for photovoltaic power generating systems. In: National Electrical Engineering Conference (NEEC-2011) on Power and Energy Systems for Tomorrow (2011)
7. Laagoubi, T., Bouzi, M., Benchagra, M.: Analysis and comparison of MPPT nonlinear controllers for PV system. In: Proceedings of 2015 IEEE International Renewable and Sustainable Energy Conference, IRSEC 2015, November 2016. https://doi.org/10.1109/IRSEC.2015.7455012
8. Ozdemir, S.: Z-source T-type inverter for renewable energy systems with proportional resonant controller. Int. J. Hydrogen Energy **41**(29), 12591–12602 (2016). https://doi.org/10.1016/J.IJHYDENE.2016.01.140
9. Gebreel, A.A.G., et al.: Simulation and implementation of two- level and three-level inverters By Matlab and Rt-Lab, MSc Thesis, p. 155 (2011)
10. Zelici, C.: A three three - phase pwm inverter with with (2012)
11. Alhamrouni, I., Hairullah, M.A., Omar, N.S., Salem, M., Jusoh, A., Sutikno, T.: Modelling and design of PID controller for voltage control of AC hybrid micro-grid. Int. J. Power Electron. Drive Syst. **10**(1), 151 (2019). https://doi.org/10.11591/ijpeds.v10.i1.pp151-159
12. Kadri, G.C.R., Gaubert, J.: An Improved maximum power point tracking for photovoltaic grid-connected inverter based on voltage-oriented control. IEEE Trans. Ind. Electron. **58**(1), 66–75 (2011). https://doi.org/10.1109/TIE.2010.2044733
13. Mari Devi, D.K.P.S.: Resonant current controller based THD reduction in ac micro. Eng. Electron. Eng. Electron. **22**(2), 425–429 (2016)
14. Hanju, C., Trung-kien, V., Jae-eon, K.: Design and control of proportional-resonant controller based photovoltaic power conditioning system. In: IEEE, pp. 2198–2205 (2009)
15. Cha, H., Vu, T.K., Kim, J.E.: Design and control of proportional-resonant controller based Photovoltaic power conditioning system. In: 2009 IEEE Energy Conversion Congress and Exposition, ECCE 2009, pp. 2198–2205 (2009). https://doi.org/10.1109/ECCE.2009.5316374
16. Letting, L.K., Munda, J.L., Hamam, Y.: Optimization of a fuzzy logic controller for PV grid inverter control using S-function based PSO. Sol. Energy **86**(6), 1689–1700 (2012). https://doi.org/10.1016/j.solener.2012.03.018
17. Chebre, M., Zerikat, M., Bendaha, Y.: Adaptation des Paramètres d ' un Contrôleur PI par un FLC Appliqué à un Moteur Asynchrone. In: 4th International Conference on Computer Integrated Manufacturing CIP'2007, pp. 2–7, November 2007
18. Omar, M.S., Amgad El-Deib, A.L.E.S., El.rhman, M.E.A.M.A.: Comparative study between PI and fuzzy-logic controllers for three-phase grid-connected photovoltaic systems photovoltaic systems. Proc. IEEE **81**(3), 365–377 (1993). https://doi.org/10.1109/5.241491
19. Essaghir, S., Benchagra, M., El Barbri, N.: Power factor control of a photovoltaic system connected to grid under load variation. In: International Conference on Electrical and Information Technologies (ICEIT) (2017)
20. Essaghir, S., Benchagra, M., El Barbri, N.: Comparison between PI and PR current controllers of a grid-connected photovoltaic system under load variation. Int. J. Power Electron. Drive Syst. **9**(3), 1311–1320 (2018). https://doi.org/10.11591/ijpeds.v9n3.pp1311-1320

Remote Area Electrification Using PV/Fuel Cell/Diesel Hybrid System in Oman – Techno Economic and Environment Viability Analysis

Nadir Al Shibli[1(✉)], Arun S. Gopinath[1], Maryam Al Balushi[1], Khalifa Al Quri[1], and Joseph Sekhar Santhappan[2]

[1] Engineering Department, College of Engineering and Technology, University of Technology and Applied Sciences, Shinas, Oman
nader.alshibli@utas.edu.om

[2] University of Technology and Applied Sciences, Musandam, Oman

Abstract. Diesel generators are the indispensable generation mode for most of the off-grid operations in countries that have humid and dry climate conditions. Greenhouse gas emission is higher for the diesel plants compared to natural gas in their main grid scenario. Oman, having high solar irradiance, is trying to improve the penetration of solar electricity to replace natural gas from the grid or diesel generators, especially. This study models various scenarios involving PV, diesel generators, fuel cells, electrolyzers, and hydrogen tanks using HOMER Pro. The hydrogen produced by utilizing the excess energy involved in the renewable energy system is collected for the generation of green hydrogen. We compared the results across different scenarios. In the current market scenario, we found that the PV/fuel cell/DG is expensive in terms of the payback period. However, they effectively neutralize CO_2 emissions. Among the cases considered, we found the PV/DG combination to be more effective, with a simple payback period of less than 7 years. Advanced cost-effective techniques adopted in the fuel cell technologies will improve the viability of the PV/FC/DG combinations in the near future.

Keyword: Green Hydrogen · Fuel cell · Simple payback · Diesel Generators

1 Introduction

Electrification of remote areas using electricity from microgrids involving renewable energy technologies is found to be an ideal option for replacing conventional diesel generators in most countries [1]. The selection of renewable energy sources is contingent upon the local meteorological data. Presently, diesel generators are observed as sources of emitters of harmful greenhouse gases in the atmosphere, including CO_2, CO, NO_x, and other particulate matters. Several research projects in Oman, including one by AER Oman, have looked into possible renewable energy sources. The results show that solar energy has a high ambient potential, with an average solar irradiance ranging from 4.5 to 6.1 kWh/m^2/day. Irradiance data indicates the strong possibility of green hydrogen generation and the possibility of a hybrid power generation facility keeping the fuel cell as one power source [2].

C. El Mokhi et al. (Eds.): ICASET 2025, SUCI, pp. 30–40, 2025.
https://doi.org/10.1007/978-3-031-98334-4_4

A hybrid energy system combines different energy generating and/or storage methods, or it uses a generator that is powered by two or more different forms of fuel. A hybrid energy system is an effective strategy for moving away from economies that rely on fossil fuels. The employment of traditional thermoelectric production alongside renewable generating can help increase the usage of renewable energy sources, especially in the near term while new technologies to better integrate renewable energy sources are still being developed. The most viable and accessible form of climate change mitigation is renewable energy [3]. Also, in the Gulf region, renewable energy is the most competitive source of electricity production. In addition, the GCC region has significant potential for renewable energy that goes beyond solar [4].

This paper envisages the possibility of a hybrid system partially replacing the usage of diesel generators. In this study [5] a hybrid system made up of photovoltaics (PV), fuel cells (FC), and diesel generators (DG) is tested for remote residential loads in different parts of the Sultanate. The locations considered are Duqm, Dhofar, Sohar, and Muscat. We collected meteorological data to determine the location and develop this model. HOMER Pro is the software used in this attempt. The monthly average of solar irradiance, wind speed, and temperature is available for various locations through NASA's database and is utilized through HOMER Pro software in this analysis. In the irradiance profile, it is found that the Duqm region has the highest irradiance level among the locations considered, and due to the grid's shortcomings, a large zone of this area is away from the country's main grid, the Main Interconnected System (MIS). This calls for a study on the feasibility of implementing a hybrid system that incorporates renewable energy technologies in this area. We conduct the analysis by taking the existing DG system as the base case and proposing PV with DG as the alternative scenario. In addition to PV and DG, the possibility of including the fuel cell as another source utilizing the green hydrogen generation from PV is also included as a proposed case in this study.

2 Methodology

This study looked at meteorological data from certain places, like Duqm, Muscat, Sohar, and Salalah. We then compared the monthly averages of important factors like solar radiation and the clearness index in those places. Table 1 shows the clearness index variation, and Table 2 lists the monthly average solar radiation of all selected locations. Firstly, in Duqm, the highest solar radiation shows in May, which is 7.4 kWh/m2/day; however, it starts to decrease gradually to reach the lowest point in December, which is 4.5 kWh/m2/day. In temperature characteristics, the highest degree shows in June, which is 30 °C; however, the lowest is in December, which is 24 °C. Secondly, in Dhofar, the highest solar radiation appears in May, which is 7.4 kWh/m2/day; however, it starts to decrease gradually to reach a minimum in December, which is 4 kWh/m2/day. June records the highest temperature of 30 °C, while December records the lowest temperature of 20 °C.

In Sohar, the highest solar radiation grift shows in two months, May and June, which is 7 (kWh/m2/day); however, it started to decrease gradually to reach the lowest point in December, which is 3.8 (kWh/m2/day). Temperature characteristics show that the highest recorded temperature is in two months, June and July, which is 35 °C;

however, the lowest is in December, which is 20 °C. On the other hand, in Muscat, the highest solar radiation happens in two months, May and June, which is 7 kWh/m2/day; however, it starts to decrease gradually to reach the lowest point in December, which is 4 kWh/m2/day. Wind speed usually reaches its highest point in February, which is 3.4 (m/s), but the lowest point of wind speed was shown in November, which is 3 (m/s). In temperature grift, the highest degree shows in two months, June and July, which is 34 °C; however, the lowest is in January, which is 20 °C. From the meteorological data, it is very clear that the Salalah region is having the maximum solar radiation, but the clearness index seems to be less compared to the location Duqm. The annual average monthly solar radiation in Duqm is found to be 6.12 kWh/m2/day with a clearness index of 0.648 compared to Salalah's 6.11 kWh/m2/day with an average clearness index of 0.641. For further analysis of the system, the Duqm with a high clearness index and high irradiance is considered as a lot of local dwellings are coming up as a result of the high expectations of huge industrial expansions happening in that zone.

The HOMER Pro (Hybrid Optimization of Multiple Energy Resources) is the software tool utilized for the feasibility analysis of various hybrid combination cases. This analysis utilizes the modules of energy modelling, economic parameter analysis, and emission calculation. The optimal hybrid system solution is obtained from the software using a cycle charging algorithm by keeping in mind the objective function of minimizing the net present cost and the levelized cost of energy [6]. The hydrogen circuit modelling function in the HOMER pro software including hydrogen component library of the PEM electrolyzer, hydrogen tank and PEM fuel cell is opted for this viability analysis. The excess energy from the off-grid combination was utilized for the generation of the hydrogen using the electrolyzer and stored in the hydrogen tank. This stored hydrogen was utilized for the power generation using the PEM fuel cell. The software tool shown great scalability and repeatability in the analysis due to the strong built-up platform and are experienced in various scales of hybrid combination systems [7].

Table 1. Clearness Index of the Locations selected

Month	Duqm	Muscat	Salalah	Sohar
Jan	0.639	0.607	0.654	0.58
Feb	0.661	0.639	0.653	0.606
Mar	0.669	0.6	0.65	0.569
Apr	0.673	0.625	0.662	0.601
May	0.679	0.641	0.688	0.629
Jun	0.625	0.621	0.623	0.62
Jul	0.582	0.57	0.579	0.565
Aug	0.61	0.589	0.584	0.587
Sep	0.663	0.617	0.641	0.611
Oct	0.693	0.659	0.669	0.643
Nov	0.656	0.632	0.655	0.62
Dec	0.626	0.609	0.635	0.581

Table 2. Monthly average solar radiation of the selected locations

Month	Duqm	Muscat	Salalah	Sohar
Jan	4.82	4.21	5.18	3.96
Feb	5.63	5.13	5.76	4.8
Mar	6.47	5.61	6.4	5.29
Apr	7.09	6.53	6.99	6.27
May	7.4	7.06	7.43	6.94
Jun	6.85	6.94	6.73	6.95
Jul	6.34	6.3	6.23	6.26
Aug	6.47	6.25	6.18	6.23
Sep	6.58	5.98	6.44	5.89
Oct	6.12	5.53	6.08	5.34
Nov	5.08	4.53	5.31	4.37
Dec	4.51	4.01	4.83	3.75

2.1 Load Details

The monthly load in Fig. 1 illustrates the pattern of load variation for residential loads. It reached the peak load requirement in July, which is 2.4 kW at 7 pm, and it decreased to reach the starting point, which is approximately 0.1 kW. However, the lowest monthly average load is in December.

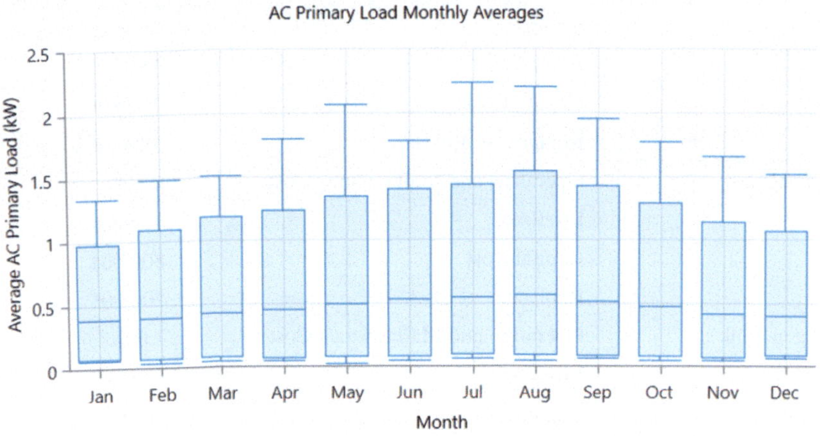

Fig. 1. Monthly load curve for the selected system

2.2 Inputs to the System Model

The photovoltaic panel, diesel generator, electrolyzer, PEM fuel cell, inverter, and hydrogen tank that make up this system were all modelled using the HOMER Pro program. Solar PV generates power during the day to supply the load and the electrolyser's energy requirements. The hydrogen generated by the electrolyzer is kept in a hydrogen tank. The technical details for various components used in Homer analysis are presented in Table 3.

Table 3. Technical data of DG, PV, Hydrogen Tank, Inverter, FC and Electrolyzer [8]

Device	Specifications	Data
Photovoltaic Panel	Capital Cost	1000.00$
	Replacement Cost	1000.00$
	Operation and Maintenance Cost	10.00$
Diesel Generator	Lifetime	25 years
	Capital Cost	250.00$
	Replacement Cost	250.00$
	Lifetime	15,000.00 h
	Capital Cost	1500.00$
	Replacement Cost	1000.00$
Electrolyzer	Operation and Maintenance Cost	30.00$
	Lifetime	25 years
	Efficiency	65%
	Capital Cost	2000.00$
	Replacement Cost	1800.00$
PEM Fuel Cell	Operation and Maintenance Cost	0.10$
	Lifetime	15 years
	Capital Cost	800.00$
	Replacement Cost	700.00$
Hydrogen Tank	Operation and Maintenance Cost	3.00$
	Lifetime	25 years
	Capital Cost	500.00$
	Replacement Cost	500.00$
Inverter	Lifetime	15 years
	Efficiency	95%

2.3 Cases Analyzed

In the base case, we only have a diesel generator and an electric load, as shown in Fig. 2. The proposed case 1 comprises a diesel generator, PV, inverter, and electric load, as connected in Fig. 3. In case 1, the power directly takes it from the diesel generator to the load. This case represents the existing system with DG as the available option. This configuration will consume conventional diesel fuel to produce electricity. Figure 4 shows the monthly electric production for the base case, and as you can see in the summer season, particularly in July and August, the highest power production requirement is observed. We take the diesel generator equivalent to the peak load of 2.4 kW as the base case.

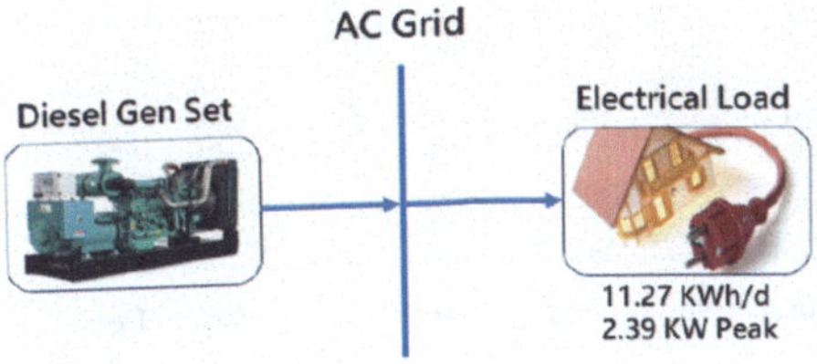

Fig. 2. Schematic diagram base case

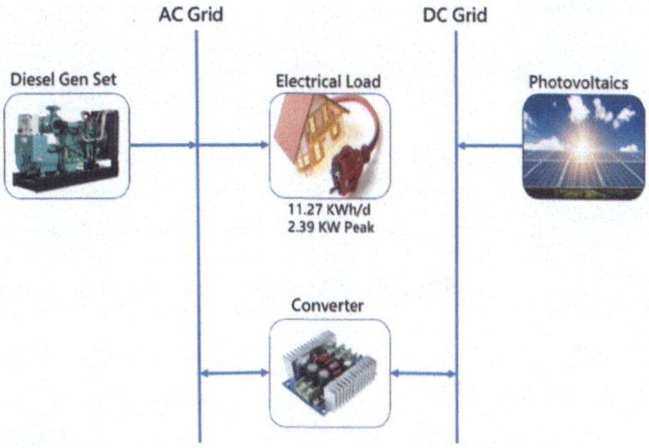

Fig. 3. Schematic diagram of proposed case 1 (DG-PV)

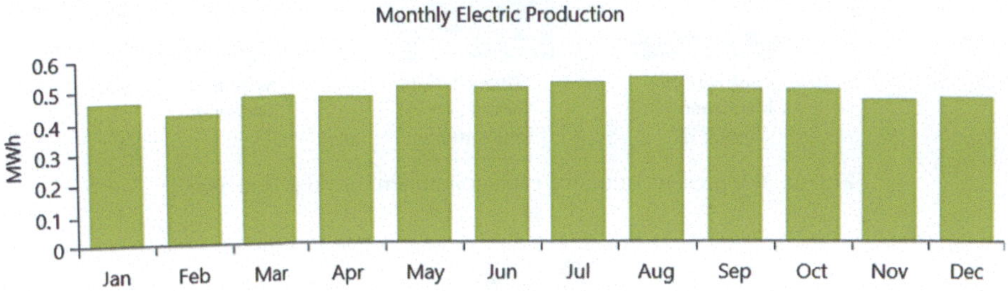

Fig. 4. Monthly electric production for base case

The PV panel with an integrated bidirectional converter or the diesel generator will directly supply the power. Figure 5 shows the monthly electric production from proposed case 1, and as seen, the power production comes from both DG and PV. Figure 6 displays the net present value of each part of proposed case 1. The data indicate that the DG part has a very high net present cost when compared to the PV and the converter.

The proposed case 2 (Fig. 7) consists of a hydrogen tank, a PEM electrolyzer, and a PEM fuel cell circuit along with a PV panel, a converter, and a diesel generator to meet the load. Multiple energy sources such as PV, FC, and diesel generators can directly meet the power requirement of the load in this case. Moreover, the converter transforms the DC power from the DC busbar into the AC power necessary for the load. DC power fed by PV to the DC busbar will be taken to the PEM electrolyzer for splitting the water into oxygen and hydrogen and then use that hydrogen in the fuel cell to produce electricity. Additionally, a hydrogen tank can store this hydrogen under high pressure. Figure 8 shows the monthly electric production for the proposed case 2; the power comes from 3 sources: DG-PV-FC. Figure 9 shows the net present cost for each component of proposed case 2, and the diesel generator is the highest.

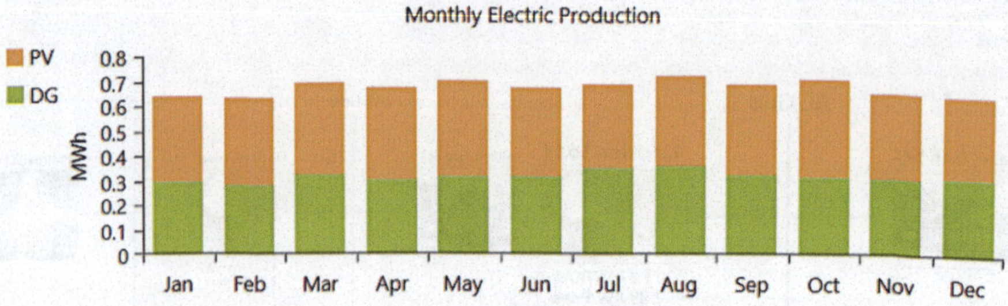

Fig. 5. Monthly electric production for Proposed case 1 (DG-PV)

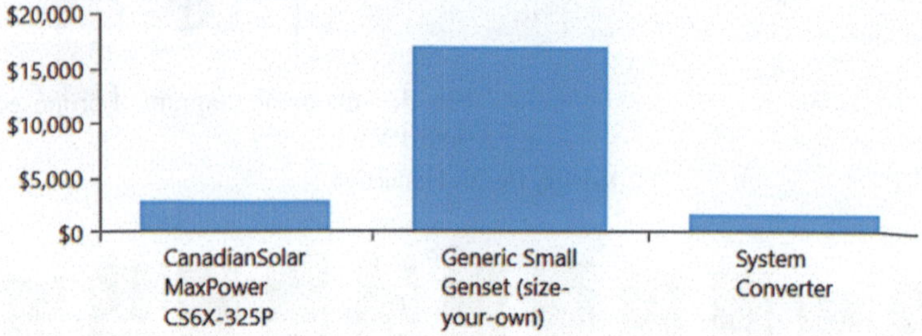

Fig. 6. Net present value for each component of proposed case 1

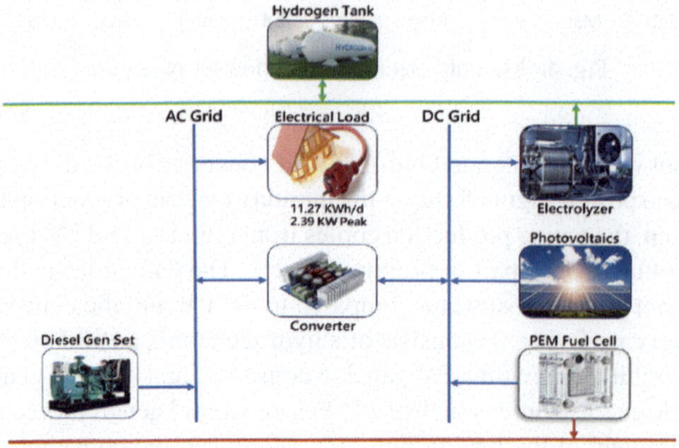

Fig. 7. Schematic diagram of proposed case 2 (DG-PV-FC)

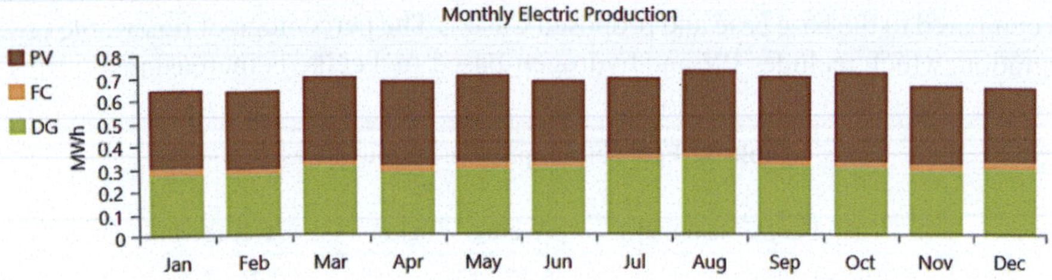

Fig. 8. Monthly Electric production for Proposed case 2 (DG-PV-FC)

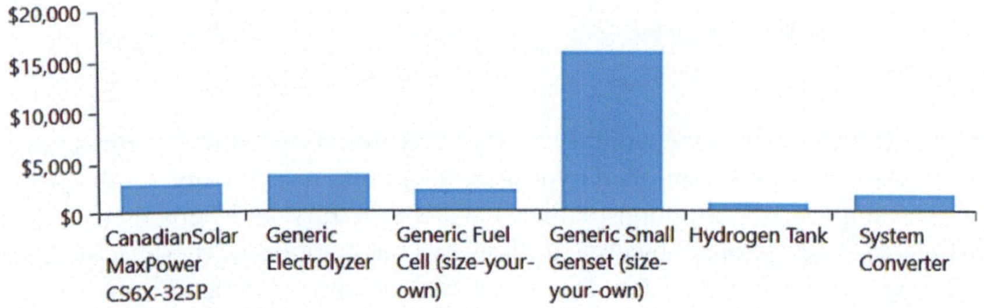

Fig. 9. Net present value for each component of proposed case 2

3 Results and Discussion

The diesel generator's equivalent peak load of 2.4 kW is taken as the base case and compared with the two proposed cases. Optimized ratings for the proposed case 1 are a PV of 2.6 kW, a converter of 2.5 kW, and a DG of 2.4 kW. The optimized rating of the proposed case 2 includes a PV of 2.6 kW, a converter of 2.5 kW, and a DG of 2.4 kW, in addition to the hydrogen components of a 2 kW PEM electrolyzer, a 1 kg hydrogen tank, and a 1 kW PEM fuel cell. Table 4 shows the comparison of energy generation between the base case and proposed cases. The total power that comes from the base case is less compared to the total power that comes from the proposed cases, and the proposed case 2, which consists of DG, PV, and FC, has more total power than proposed case 1.

Table 4. Electricity consumption in various cases

Cases	Power (PV) (kWh/yr)	Power (DG) (kWh/yr)	Power (FC) (kWh/yr)	AC Primary Load (kWh/yr)	Excess Electricity (kWh/yr)	Renewable Fraction (%)
Base Case	0	5841	0	4113	1728	0
Case 1	4336	3817	0	4113	3956	7.2
Case 2	4336	3546	285	4113	192	13.8

Table 5 represents the fuel consumed by the base case and proposed cases. It is clearly shown that proposed Case 2 (DG, PV, and FC) consumed a smaller amount of fuel compared to the base case and proposed Case 1. The percentage of renewable power integration, which includes PV and hydrogen-based fuel cells, is increasing.

Table 5. Fuel consumption in various cases

Cases	Total Fuel Consumed	Average Fuel in	Average Fuel Diesel
Base Case	Diesel 2288 L	6.27 L/day	0.261 L/hr
Case 1	Diesel 1488 L	4.08 L/day	0.170 L/hr
Case 2	Diesel 1387 L	–	–
	Hydrogen 59.9 kg	0.164 kg/day	0.00684 kg/hr

Figures 10 and 11 show the monthly average consumption of diesel in liters/hour and the monthly average consumption of hydrogen in kg/hour, respectively, for the proposed case 2. The highest consumption of diesel will be in July and September, which is approximately 0.68 L/hour; however, the lowest consumption of diesel is in January, which is approximately 0.31 L/hr. The diesel fuel consumption pattern of the various cases considered is shown in Fig. 10. The consumption of hydrogen is represented in Fig. 11; the highest consumption will be in three months (January, June, and December); it reached 0.15 kg/hr. However, in March, August, and October, it reached the lowest point, which is approximately 0.11 kg/hr.

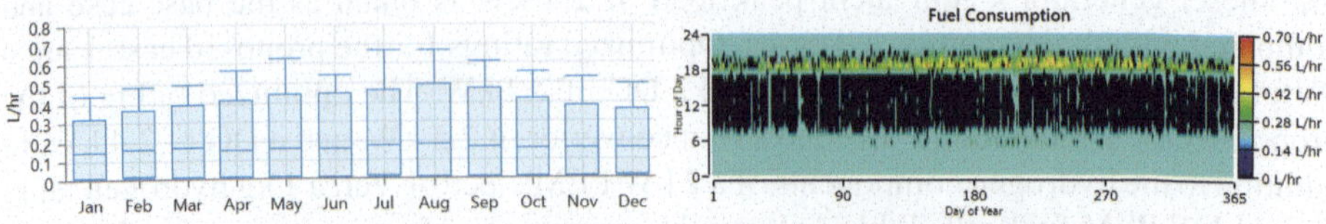

Fig. 10. Monthly average consumption of diesel by Liter/hour

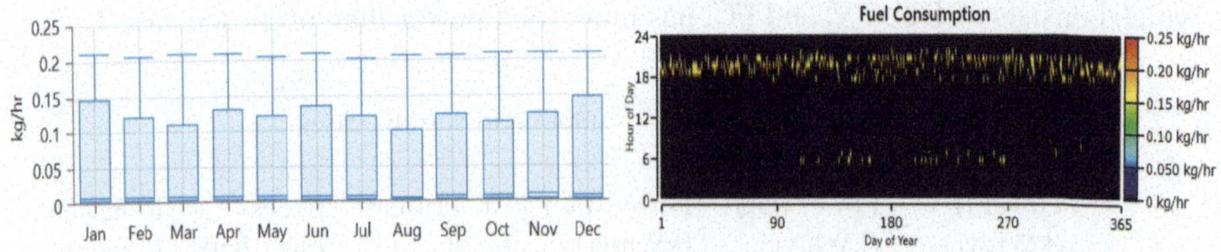

Fig. 11. Monthly average consumption of hydrogen by kilogram/hour

Table 7 clearly mentions the economic comparison between the base case and proposed cases. As it's showing, no payback comes from the base case; however, the payback of proposed case 1 is 6.17 years, and the payback of proposed case 2 will be after 17.86 years.

Base case emissions are higher than those from proposed cases 1 and 2. Proposed case 2 emissions are lower than those from the base case, as shown in Table 7's comparison

Table 6. Comparison of Economical parameters between base case and proposed cases

	Present worth ($)	Roi (%)	Irr (%)	Simple payback (years)
Base Case	0	0	0	0
Case 1	4345	11.6	15.4	6.17
Case 2	1450	2.4	3.8	17.86

of case emissions. Table 8 compares the base case, proposed case 1, and proposed case 2. The capital cost of the base case is less compared to the proposed case 1 and proposed case 2; however, the replacement and maintenance cost for the base case is very high compared to proposed case 1 and proposed case 2.

Table 7. Emission Comparison between base case and proposed cases

Cases	CO_2 kg/yr	CO kg/yr	UBH kg/yr	PM kg/yr	SO_2 kg/yr	NO_X kg/yr
Base Case	5991	37.4	1.65	0.224	14.7	35.1
Case 1	3895	24.3	1.07	0.146	9.54	22.9
Case 2	3632	22.7	0.999	0.136	8.89	21.3

Table 8. Comparison between base case and proposed cases

Cases	Fuel consumption	Renewable fraction (%)	Cost of energy	Simple payback (year)
Base Case	2288 L	0	0.4643 $/kWh	–
Case 1	Diesel 1488 L	7.2	0.3863 $/kWh	6.17
Case 2	Diesel 1387 L Hydrogen 59.9 kg	13.8	0.4903 $/kWh	17.86

4 Conclusion

This study models various scenarios involving PV, diesel generators, fuel cells, electrolyzers, and hydrogen tanks for rural electrification in Oman. The analysis of various scenarios shows that the base case is consuming 2288 L of diesel, and the fuel consumption of the proposed case 1 is reduced to 1488 L. The fuel consumption of proposed case 2 is less than other cases and makes hydrogen fuel in addition, which can be utilized by using a fuel cell. Table 6 mentioned that the proposed case 2 has the highest renewable fraction, which equals 13.8%. In addition, the base case's energy cost was $0.463, while the proposed case 2's price is $0.4903. However, the proposed case 1 has the lowest energy cost, at $0.3863. The payback period for proposed case 1 will be after 6.17 years, and the simple payback for proposed case 2 will be after 17.86 years compared to the base case. The proposed case 2, the PV-FC-DG hybrid system, has many benefits, even though it takes a long time to pay for itself. These benefits include using more renewable energy, less diesel, and fewer greenhouse gas emissions. The Levelized cost of energy will reduce once the cost of the proton exchange membrane electrolyzer and proton exchange membrane costs of fuel cells decrease. To reduce the complexity of the study, we assumed constant values for the parameters like fuel costs, PV efficiency, and diesel prices. Future studies can analyze these sensitive parameters. Such sensitive analysis can enhance the accuracy in feasibility assessments.

References

1. Akbarzadeh Niaki, A., Jamil, M.: An efficient hydrogen-based water-power strategy to alleviate the number of transmission switching within smart grid. Int. J. Hydrogen Energy **70**, 347–356 (2024). https://doi.org/10.1016/j.ijhydene.2024.05.024
2. Dawood, F., Shafiullah, G.M., Anda, M.: Stand-alone microgrid with 100% renewable energy: a case study with hybrid solar PV-battery-hydrogen. Sustainability **12**(5), (2020). https://doi.org/10.3390/su12052047
3. Rao, C.K., Sahoo, S.K., Yanine, F.F.: A literature review on an IoT-based intelligent smart energy management systems for PV power generation. Hybrid Adv. **5**, 100136 (2024). https://doi.org/10.1016/j.hybadv.2023.100136
4. Kiriakidis, P., Christoudias, T., Kushta, J., Lelieveld, J.: Projected wind and solar energy potential in the eastern Mediterranean and Middle East in 2050. Sci. Total Environ. **927**, 172120 (2024). https://doi.org/10.1016/j.scitotenv.2024.172120
5. Bacha, B., Ghodbane, H., Dahmani, H., Betka, A., Toumi, A., Chouder, A.: Optimal sizing of a hybrid microgrid system using solar, wind, diesel, and battery energy storage to alleviate energy poverty in a rural area of Biskra, Algeria. J. Energy Storage **84**(PA), 110651 (2024). https://doi.org/10.1016/j.est.2024.110651
6. Kavadias, K.A., Triantafyllou, P.: Hybrid renewable energy systems' optimisation. A review and extended comparison of the most-used software tools. Energies **14**(24), (2021). https://doi.org/10.3390/en14248268
7. Wang, Y., He, X., Liu, Q., Razmjooy, S.: Economic and technical analysis of an HRES (Hybrid Renewable Energy System) comprising wind, PV, and fuel cells using an improved subtraction-average-based optimizer. Heliyon **10**(12), 1–22 (2024). https://doi.org/10.1016/j.heliyon.2024.e32712
8. Al-Badi, A., Al Wahaibi, A., Ahshan, R., Malik, A: Techno-economic feasibility of a solar-wind-fuel cell energy system in Duqm, Oman. Energies **15**(15), (2022). https://doi.org/10.3390/en15155379

Energy Analysis of a Trigeneration and Hydrogen Plant Driven by Concentrated Photovoltaic Thermal System for Residential Application Purposes

Salim Khammal(✉) and Ahmed Khouya

Energy and Materials Team - Innovative Technologies Laboratory - ENSA, Abdelmalek Essaadi University, Tangier, Morocco
salimhome45@gmail.com

Abstract. This study focuses on analyzing a Trigeneration system designed to supply cooling and heating for residential purposes and electricity to produce green hydrogen using concentrated photovoltaic thermal system (CPVT). The CPVT provide the thermal energy for cooling and for the production of electricity. The residual heat generated from the CPVT system are used in buildings and also in the electrolysis plant. The technology used to generate hydrogen is the alkaline water electrolysis (AWE). The refrigerant proposed in the system is the R124. The results obtained indicate that the trigeneration plant achieves a 12% of maximum exergy efficiency and 42% of thermal efficiency. The outlet water from generator exits with the temperature of 65 °C and it is utilized to heat the alkaline water electrolysis (AWE) in the electrolysis plant. Overall, the proposed system can produce cooling and heating with the objective to be utilized in residential applications and electricity for green hydrogen.

Keywords: Trigeneration system · concentrated photovoltaic thermal system · Efficiency · Green hydrogen

1 Introduction

The impacts of global warming are dramatically reshaping the whole ecosystem, primarily due to rising carbon dioxide levels in the atmosphere. To mitigate climate change, experts have developed innovative ways to produce efficient energy and with few environmental impact [1]. One such method is cogeneration later advanced into trigeneration systems. It can produce heat and electricity. Trigeneration, or combined cooling, heat, and power (CCHP), extends cogeneration by adding cooling capabilities, making it more efficient. Trigeneration systems can utilize many sources of energy, such as fossil fuels, biomass or renewable energy. For example, trigeneration systems have been developed using alternative energy like geothermal and biomass [2]. Bellos et al. studied focusing in energy and exergy assessment of a Trigeneration plant using solar energy [3]. Green hydrogen plays a crucial role in decreasing the dependence on fossil fuels, which enhance the climate change. For hydrogen production, the best environmental method is Alkaline Water Electrolysis (AWE) when powered by renewable energy [4].

C. El Mokhi et al. (Eds.): ICASET 2025, SUCI, pp. 41–50, 2025.
https://doi.org/10.1007/978-3-031-98334-4_5

This study proposes a trigeneration system using concentrated photovoltaic thermal system designed to produce heat and cooling in residential buildings and electricity for green hydrogen production. The analysis will evaluate the performance of the system from energetic, exergetic and financial perspectives. Simulations will be conducted to examine various outlet temperature levels from the CPVT, aiming to identify the best conditions to increase electricity production. Furthermore, waste heat will be repurposed for residential use. The study will conclude by summarizing the key findings and results of the research.

2 Description of the Proposed System

The schematic diagram in Fig. 1 shows a novel Trigeneration system (CCHP) system. This system incorporates multiple components and processes. The water flows into the concentrated photovoltaic thermal (CPVT) system. Heated water from the CPVT system is transferred to the Generator to provide thermal energy to R124 and LiBr (Absorber). Once the water outs the Generator, is driven into the electrolysis plant to heat the solution and after it will be used for residential applications. Inside the Generator, the heated LiBr solution then passes to the Turbine, where electricity is generated and sent to the inverter for use in electrolysis plant. The refrigerant fluid R124 is used to provide heat and cool for building applications.

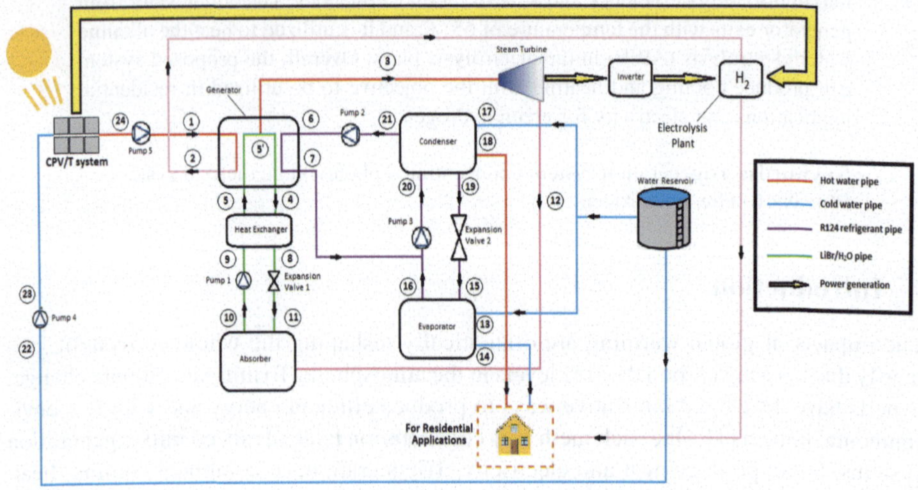

Fig. 1. Schema of the Trigeneration system with electrolysis plant

3 Theoretical Study of the Trigeneration Plant

The system will be analyzed theoretically with the equations for each component of the CCHP system.

3.1 CPVT Model

The CPVT system uses the triple junction InGaP/GaAs/Ge to produce electrical and thermal performances. The net electricity is calculated using the next equation [5]:

$$P_{el,net} = Cr\, I\, A_{PV}\eta_{op}\eta_{refl}\eta_{PV}\eta_{inv}\eta_{mod} - Q_{par}\eta_{inv} \tag{1}$$

To calculate η_{PV} and Cr we need to use the next equations [6, 7]:

$$\eta_{PV} = 0.298 + 0.0142\ln(Cr) + (0.000069\ln(Cr) - 0.00075)x(T_{PV} - 298) \tag{2}$$

$$Cr = \frac{A_{ap}}{A_{PV}} \tag{3}$$

To determine T_{PV}, we use the next equation [6]:

$$T_{PV} = T_0 + \frac{I\tau\alpha(1 - \eta_{ref})}{O_{loss}} \tag{4}$$

The thermal power can be written as next [8]:

$$Q_{th} = \eta_{op}\ x\ (1 - \eta_{mod}\ x\ \eta_{PV})\ I\ x\ A_{PV}\ x\ Cr \tag{5}$$

The useful power that is taken by the HTF (High Thermal Fluid) can be determined [9]:

$$Q_u = Q_{th} - Q_{loss} \tag{6}$$

$$Q_{loss} = A_{PV}\eta_{PV}\left[h_{cc}(T_{PV} - T_0) + \varepsilon_{PV}\sigma\left(T_{PV}^4 - T_0^4\right)\right] \tag{7}$$

3.2 Trigeneration System Model

The next equations describe the absorption chiller in Steady State [10]:

$$Q_G = \dot{m}_3h_3 + \dot{m}_4h_4 + \dot{m}_7(h_7 - h_6) - \dot{m}_5h_5 \tag{8}$$

$$\dot{m}_9(h_9 - h_5) = \dot{m}_8(h_8 - h_4) \tag{9}$$

The efficiency in the Heat Exchanger is the next [10]:

$$\eta_{HEX} = \frac{h_4 - h_8}{h_4 - h_9} \tag{10}$$

We will consider the Turbine and pumps with isentropic transformation [10]:

$$W_{Turbine} = \dot{m}_3(h_3 - h_{12}) \tag{11}$$

$$Q_{Absorber} = \dot{m}_{12}h_{12} + \dot{m}_{11}h_{11} - \dot{m}_{10}h_{10} \tag{12}$$

$$Q_{evap} = \dot{m}_{15}(h_{16} - h_{15}) \tag{13}$$

$$Q_c = \dot{m}_{19}(h_{19} - h_{20}) + \dot{m}_{21}(h_{21} - h_{20}) \tag{14}$$

To determine the thermal efficiency, we can use the following expression [10]:

$$\eta_{th} = \frac{Q_{evap} + W_{net} + Q_h}{Q_{solar}} \tag{15}$$

$$Q_{solar} = A_{ap} \times G_{CPV} \tag{16}$$

For the equation to calculate the exergy efficiency, the next equation can be used [11]:

$$\eta_{ex} = \frac{E_{uevap} + E_{unet} + E_{uh}}{E_{usolar}} \tag{17}$$

$$E_{u\ solar} = Q_{solar}\left[1 - \frac{4}{3}\left(\frac{T_0}{T_{sun}}\right) + \frac{1}{3}\left(\frac{T_0}{T_{sun}}\right)^4\right] \tag{18}$$

To determine the payback period (PP), we use the next equation [10]:

$$PP = \frac{\text{Total cost of the system}}{W_{net}.\ T\ .0.2\ \$/\text{kWh}} \tag{19}$$

3.3 Green Hydrogen

In this study, the green hydrogen will be produced by alkaline water electrolysis (AWE). By using electricity, we can produce hydrogen and oxygen [12]:

$$H_2O\ (l) \rightarrow H_2(g) + \frac{1}{2}O_2(g) \tag{20}$$

The current flow can be written as [12]:

$$I_{ely} = \eta_{in}.\frac{P_{el}}{N_{cell}U_{cell}} \tag{21}$$

To calculate N_{cell}, the next equation is proposed [12]:

$$N_{cell} = \frac{\eta_{in}P}{1.8x750} \tag{22}$$

The formula to calculate the quantity of hydrogen produced is [12]:

$$m_{H2} = 80.69\eta_F\frac{N_{cell}I_{ely}}{zF}\rho_{H2} \tag{23}$$

The energy consumption gives information about how to the cost of production can be reduced. To calculate the energy consumption ratio, we use the next equation [13]:

$$ECR = \frac{E_{el}}{m_{H2}} \tag{24}$$

E_{el} is the necessary electricity to produce hydrogen (kWh). m_{H2} is the produced quantity of hydrogen (kg). To calculate the efficiency of the hydrogen we use the equation below [14].

$$\eta_{ely} = \frac{m_{H2} LHV_{H2}}{E_{el}} \tag{25}$$

In this paper, economic analysis will be made by using the levelized cost of energy (LCOE) and levelized cost of hydrogen (LCOH) equations below [12]:

$$\mathrm{LCOE} = \frac{\sum_{T=1}^{N} \frac{(\mathrm{Investment}_T + \mathrm{AC}_T)}{(1+r)^T}}{\sum_{T=1}^{N} \frac{(E_{el})_T}{(1+r)^T}} \tag{26}$$

$$\mathrm{LCOH} = \frac{\sum_{T=1}^{N} \frac{(\mathrm{Investment}_T + \mathrm{AC}_T + \mathrm{EUCC} \times P_{ely} + \mathrm{OM}_T)}{(1+r)^T}}{\sum_{T=1}^{N} \frac{(m_{H2})_T}{(1+r)^T}} \tag{27}$$

4 Validation of the Mathematical Models

In this paragraph, the validation of the mathematical models of the trigeneration plant is realized. Table 1 shows the performances between the actual thermal and those of previous studies [10]. Results highlighted that the gap between these two models were 4.5% for the Thermal power, 6% for the rate of heat transfer and 4.8% for the thermal efficiency.

Table 1. Validation of the mathematical model [10]

Parameters	Rate of heat transfer (kW)	Electric power (kW)	Thermal efficiency (%)
Previous work [10]	1003	4.25	40.07
Present model	1050	4.5	42
Difference (%)	4.5	6	4.8

5 Results

In this section, the results obtained in the proposed system will be presented. Table 2 shows the power for each component of the system by considering $T_{col,out}$ equal to 125 °C.

The effect of changing the fixed $T_{col,out}$ is represented in the next table:

In Fig. 2, the impact of the inlet pressure is illustrated. To achieve maximum power generation (9.6 kW), the inlet pressure should be set at 70 kPa. Beyond this point, the production of electricity begins to decline. Although maximizing electrical exergy outlet is highly desirable, it is also important to maintain adequate performance for cooling and heating loads in residential or industrial applications (Table 3).

Figure 3 shows the influence of the temperature variation in the evaporator. Qc remains constant. However, electricity generated by the turbine decreases as the evaporator temperature increases which proves that it is non profitable raising the temperature beyond 10 °C. On the other hand, the cooling load improve with the raise of evaporator temperatures. Setting the evaporator temperature at 8 °C could maximize turbine outlet power, but it would result in a substantial reduction in the cooling load, making this option inefficient.

In Fig. 4, the influence of solar radiation on thermal and exergy efficiency of the trigeneration is presented. The exergy efficiency remains practically constant because as explained in the Eq. (17), when solar radiation Gt increases, the energy provided by the turbine as well as the power of the condenser and the evaporator rises. The same results are obtained with the thermal efficiency.

Figure 5 depicts CPVT surface area impact on thermal and exergy efficiency the graph shows when total surface area increases, the exergy efficiency and thermal efficiency remain virtually constant. The Eqs. (15), (16), (17) and (18) show that an increase of the CPVT area systems leads to increase the collected solar energy, and as a result, the outlet power of the turbine, condenser and evaporator raise maintaining practically the same value.

In this section, the amount of hydrogen produced in steady state will be estimated. In fact, the number of the electrolysis cells is 38 cells. The amount of hydrogen production is approximatively 0.87 kg/h, in full regime. These results are extremely interesting.

Table 2. Parameters of the trigeneration system

Parameters	Result
Q_{evap}	8.4 kW
W_{net}	56.3 kW
Q_h	12.53 kW
η_{th}	42%
η_{ex}	12%

Table 3. $T_{col,out}$ influence in the Trigeneration plant

$T_{col,out}$	105 °C	110 °C	115 °C	120 °C
Q_{evap}	9.18	8.95	8.78	8.62
W_{net}	50.9	52.1	53.8	55.2
Q_h	12.53	12.53	12.53	12.53
η_{th}	47%	45.7%	44.3	43.1

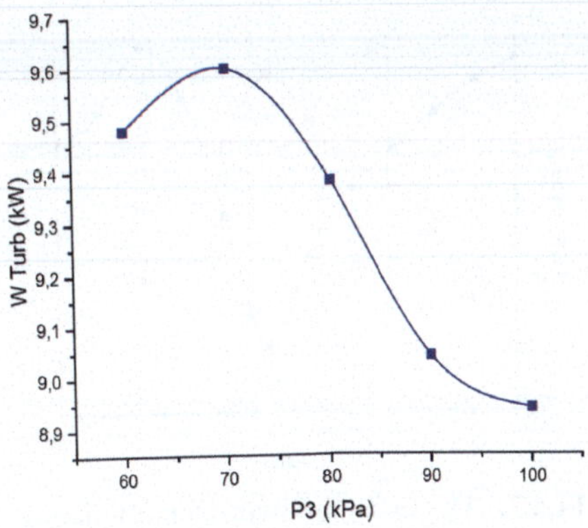

Fig. 2. Effect of the change of the pressure P_3 on the electricity produced

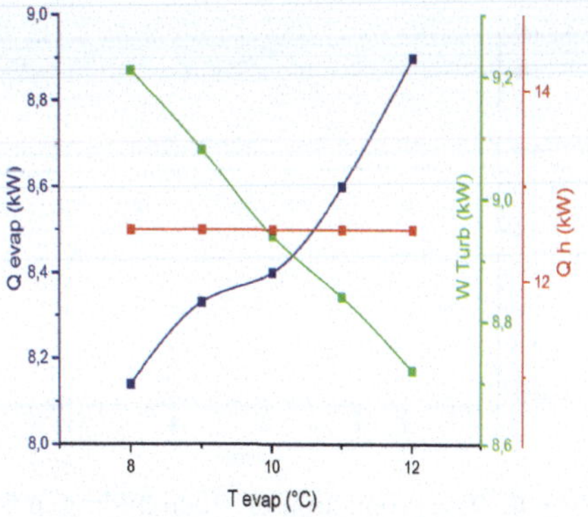

Fig. 3. Temperature influence of the evaporator on power generation, heating and cooling load

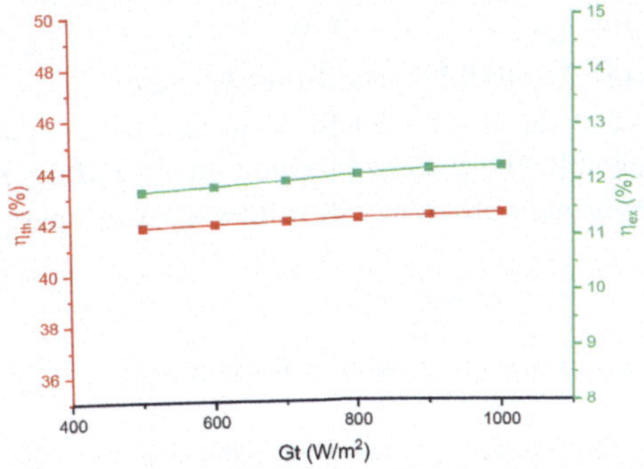

Fig. 4. Effect of the solar radiation in exergy and thermal efficiency

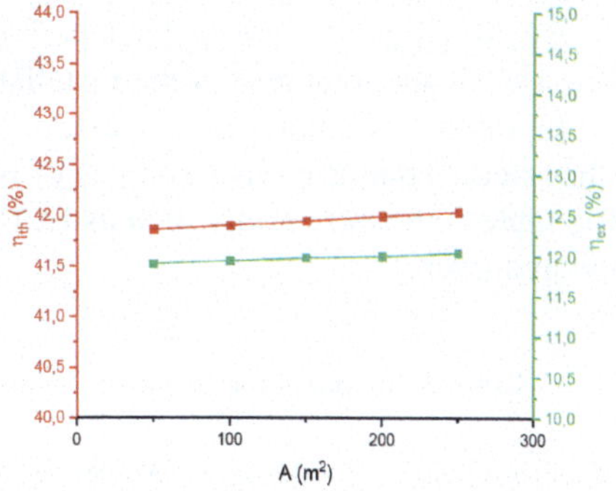

Fig. 5. Area influence of CPVT in thermal and exergy efficiency

Figure 6 shows for each month the production of green hydrogen. The number of hours of sunshine has been shown to have a significant effect on hydrogen production each month. The graph shows that production is minimal in December with 228.78 kg, but from January it begins to increase until reaching its maximum peak with 391.5 kg in June and then decreases. Figure 7 illustrates the monthly variation of ECR and hydrogen efficiency. From January, the ECR begins to decrease until July when it reaches its minimum with 62.5 kWh/kg and increases until reaching the maximum in October with a value of 64.8 kWh/kg. The results show that hydrogen efficiency increases from January to July, peaking at 63%, and then decreases again. The hydrogen efficiency during winter is low due to the reduction of the solar radiation received by the CPVT system.

For the LCOE and LCOH, the total investment is 23000 $ (Turbine cost), AC is equal to 5870.4 $, r (discount rate) is 0.05 and the total lifespan is considered to be 20 years. The EUCC is taken as 750 $/kWh and the OM is 500 $. The total net power is 56.3 kW.

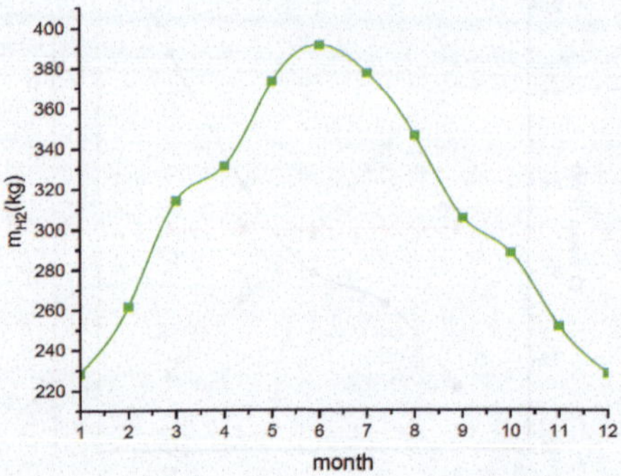

Fig. 6. The production of green hydrogen for each month of the year

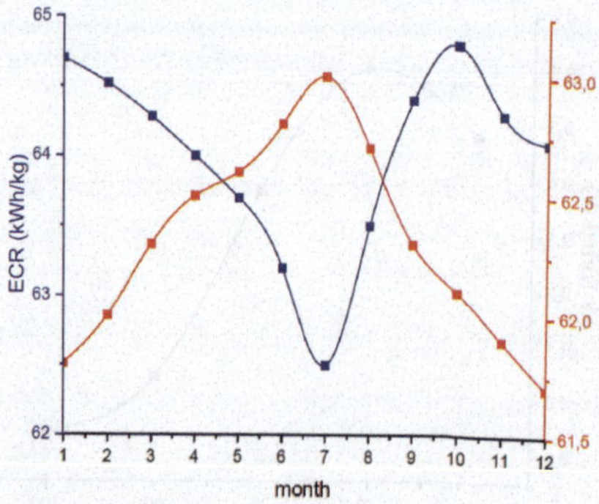

Fig. 7. The ECR and hydrogen efficiency during the year

After performing the simulation in C++, the LCOE value is 0.12 \$/kWh and the LCOH is 18.7 \$/kg.

By using Eq. (19), the payback period (PP) can be calculated. W_{net} is equal 56.3 kW, T = 4380 h and total cost of the installation is 236969.2 \$. The PP is 4.8 years.

In Table 4, a comparison is made between the results obtained in this paper with other studies that use other sources of renewable energies to produce green hydrogen. The table compares results such as ECR, hydrogen efficiency, levelized cost of energy and hydrogen.

Table 4. Comparison between the results obtained in this work with other studies.

Renewable energy	ECR (kWh/kg)	Efficiency (%)	LCOE (\$/kWh)	LCOH (\$/kg)	Reference
CPVT	64.71	62	0.12	18.7	This work
CPVT	64	62	0.01 to 0.07	3 to 6.5	Khouya [12]
Wind energy	–	–	1.278	33.7	Mohamed Awad [15]
Photovoltaic	–	–	0.021 to 0.032	3.59 to 4.34	Youssef Achour [16]

After making the simulation in C + + and comparing the results obtained by alkaline water electrolysis with other technologies of hydrogen production such as proton exchange membrane (PEM), we conclude that the PEM produces more hydrogen (0.92 kg/h) than AWE by using the same input current. Also, PEM electrolysis consumes 4% less energy per kg. The ECR obtained in the optimal month by using AWE is 64.71 kWh/kg and for the PEM is 61.2 kWh/kg. The results obtained in hydrogen efficiency by using AWE was 62%, and for the PEM the result was 63%.

From a point of view, the large-scale adoption of CPVT-powered trigeneration for hydrogen production is hindered by both economic and policy challenges. Economically, the high initial investment in CPVT systems, along with some efficiency limitations, maintenance costs, and energy storage expenses, make it difficult to compete with traditional hydrogen generation methods. Additionally, the levelized cost of hydrogen

(LCOH) remains a concern, as fossil-fuel-based hydrogen continues to be cheaper due to the existence of infrastructure and subsidies. On the policy side, there is a lack of incentives and regulatory frameworks, challenges related to grid integration and slow hydrogen infrastructure development, delay large-scale deployment. Overcoming these barriers requires government support and to increase the investment in hydrogen storage and distribution, making the CPVT-powered trigeneration a more viable and scalable solution.

6 Conclusion

This research introduces an innovative model that integrates concentrated photovoltaic thermal technology with an absorption power cycle and an electrolysis plant for hydrogen production. The key findings of the study are:

- Trigeneration system delivers a heating load of 12.53 kW, a cooling load of 8.4 kW, and a net electrical output of 56.3 kW.
- The exergy and thermal efficiency are 12% and 42% respectively.
- The total cells are 38 and the mass of hydrogen produced is 0.87 kg/h.
- The results of ECR and hydrogen efficiency are 64.71 kWh/kg and 62% respectively.

References

1. Ziyaei, S., Panahi, M., Manzour, D., Karbasi, A.R.: For reducing CO2 emissions, what is the most effective: making power plants more efficient or developing renewable resources? Int. J. Environ. Sci. Tech. **21**, 9007–9030 (2024). https://doi.org/10.1007/s13762-024-05705-6
2. Farajollahi, A., Rostami, M., Feili, M., Ghaebi, H.: Thermodynamic and economic evaluation and optimization of the applicability of integrating an innovative multi-heat recovery with a dual-flash binary geothermal power plant. Clean Technol. Environ. Policy **25**, 1673–1698 (2023). https://doi.org/10.1007/s10098-023-02465-8
3. Bellos, E., Tzivanidis, C., Torosian, K.: Energetic, exergetic and financial evaluation of a solar driven trigeneration system. Thermal Sci. Eng. Prog. **7**, 99–106 (2018). https://doi.org/10.1016/j.tsep.2018.06.001
4. Hosseini, S.E., Wahid, M.A.: Hydrogen from solar energy, a clean energy carrier from a sustainable source of energy. Int. J. Energy Res. **44**, 4110–4131 (2020). https://doi.org/10.1002/er.4930
5. Khouya, A.: Performance assessment of a heat pump and a concentrated photovoltaic thermal system during the wood drying process. Appl. Thermal Eng. **180**, 115923 (2020). https://doi.org/10.1016/j.applthermaleng.2020.115923
6. Khouya, A.: Performance analysis and optimization of a trilateral organic Rankine powered by a concentrated photovoltaic thermal system. Energy **247**, 123439 (2022). https://doi.org/10.1016/j.energy.2022.123439
7. Jaramillo, O.A., Venegas-Reyes, E., Aguilar, J.O., Castrejón-García, R., Sosa-Montemayor, F.: Parabolic trough concentrators for low enthalpy processes. Renew. Energy **60**, 529–539 (2013). https://doi.org/10.1016/j.renene.2013.04.019
8. Mittelman, G., Kribus, A., Dayan, A.: Solar cooling with concentrating photovoltaic/thermal (CPVT) systems. Energy Convers. Manag. **48**, 2481–2490 (2007). https://doi.org/10.1016/j.enconman.2007.04.004

9. Renno, C., Petito, F., D'Agostino, D., Minichiello, F.: Modeling of a CPV/T-orc combined system adopted for an industrial user. Energies **13**, 3476 (2020). https://doi.org/10.3390/en13133476
10. Karim, S.H., Tofiq, T., Shariati, M., Nikafshan Rad, H., Ghasemi, A.: 4E analyses and multi-objective optimization of a solar-based combined cooling, heating, and power system for residential applications. Energy Rep. **7**, 1780–1797 (2021). https://doi.org/10.1016/j.egyr.2021.03.020
11. Bellos, E., Tzivanidis, C., Antonopoulos, K.A.: Exergetic, energetic and financial evaluation of a solar driven absorption cooling system with various collector types. Appl. Thermal Eng. **102**, 749–759 (2016). https://doi.org/10.1016/j.applthermaleng.2016.04.032
12. Khouya, A.: Levelized costs of energy and hydrogen of wind farms and concentrated photovoltaic thermal systems. A case study in Morocco. Int. J. Hydrogen Energy **45**, 31632–31650 (2020). https://doi.org/10.1016/j.ijhydene.2020.08.240
13. Micheli, L., Fernández, E., Almonacid, F., Mallick, T., Smestad, G.: Performance, limits and economic perspectives for passive cooling of high concentrator photovoltaics. Solar Energy Mater. Solar Cells **153**, 164–178 (2016). https://doi.org/10.1016/j.solmat.2016.04.016
14. Wang, F.C., Hsiao, Y.S., Yang, Y.Z.: The optimization of hybrid power systems with renewable energy and hydrogen generation. Energies **11**, 1948 (2018). https://doi.org/10.3390/en11081948
15. Awad, M., et al.: A review of water electrolysis for green hydrogen generation considering PV/wind/hybrid/hydropower/geothermal/tidal and wave/biogas energy systems, economic analysis, and its application. Alexandria Eng. J. **87**, 213–239 (2024). https://doi.org/10.1016/j.aej.2023.12.032
16. Achour, Y., Berrada, A., Arechkik, A., El Mrabet, R.: Techno-economic assessment of hydrogen production from three different solar photovoltaic technologies. Int. J. Hydrogen Energy **48**, 32261–32276 (2023). https://doi.org/10.1016/j.ijhydene.2023.05.017

Smart Cities Infrastructure for Sustainable Urban Development

Energy Transition as a Mechanism for Achieving Urban Sustainability: The Case of Jerada City

Mohammed Ziani(✉)

The Regional Academy of Education and Formation in the Oriental Region, Oujda, Morocco
mohammedziani84@gmail.com

Abstract. In this article, we review the most significant energy transformations/changes that took place in the city of Jerada in the eastern part of Morocco, especially with regard to the issue of electricity production and solar energy, as this transformation has become a trajectory pursued by the local charged authorities, considering this transition to renewable energies, as an alternative option to achieve local development and decrease unemployment. The research aims the move towards comprehensive solutions for sustainable development through renewable energy production, particularly solar energy, in order to achieve sustainable local development that meets all the immediate needs of individuals while preserving intergenerational within clean environmental rights, focusing on projects that It takes as its priority the environmental and social dimensions resulting from this transformation, especially after the closure of coal mines at the beginning of the millennium. The analysis delineates key infrastructural challenges compromising the city and its broader regional context, exploring the potential mechanisms for overcoming structural obstacles impeding energy transformation. Moreover, the paper critically assesses the local community's role in fostering awareness and facilitating the successful implementation of renewable energy initiatives in the urban landscape of the city.

Keywords: renewable energies · sustainable development · urbanization

1 Introduction

The city of Jerada is recognized as a geographical area for the extraction of coal, which contributed significantly to the city's prosperity and development over many years. Coal mining served as the sole pillar of local development. However, with the closure of the mines in recent decades, the city has experienced a sharp decline in both economic and living standards for its residents. This situation has compelled the city to seek effective solutions to the crisis caused by the mine closures, leading to a shift away from reliance on traditional natural resources towards the adoption of renewable energy sources.

In light of the city's new circumstances and within the framework of transitioning towards renewable energy, an electricity production unit was established with the aim of alleviating unemployment. However, it was unable to fully compensate for the repercussions of the mine closure, prompting city officials to consider comprehensive solutions

C. El Mokhi et al. (Eds.): ICASET 2025, SUCI, pp. 53–59, 2025.
https://doi.org/10.1007/978-3-031-98334-4_6

for achieving sustainable development. This approach emphasizes the shift towards utilizing renewable energy, particularly in electricity production, to foster local sustainable development that meets the immediate needs of individuals while safeguarding the rights of future generations within a clean environment.

The purpose of this paper is to explore how Jerada can address the challenges of electricity production in the context of achieving urban sustainability, as well as to examine the opportunities and barriers that the city faces in transitioning to renewable energy amidst its environmental, social, and economic challenges.

2 Conceptual Definitions

Voltaire's assertion, if you want to converse with me, define your terms [1], emphasizes the significance of clearly defining concepts in intellectual discourse. In social sciences, terminology is often multifaceted and influenced by both ideological and subjective factors, which are reflected in the researcher's perspective and interpretation of the concepts used [1]. The process of developing these concepts should be grounded in solid, practical foundations—being observable, measurable, and applicable in the field. To avoid abstraction or idealism, [2] "concepts must include dimensions and indicators that represent the studied reality accurately, as seen through the researcher's" lens. With this in mind, and acknowledging the importance of defining terms in social research, we will proceed to define the key concepts of sustainable development, urbanization, and renewable energy, which are central to this paper.

2.1 Sustainable Development

The concept of sustainable development emerged more with the research of Pakistani researcher Mahbub ul Haq and Indian researcher Amartya Sen during their work for the United Nations Development Program (UNDP). For them, sustainable development is both social and economic development, which makes the human being its starting point and goal. We deal with the human or more specifically social dimensions of development as the dominant element and we consider material capabilities as a condition for achieving sustainable development [3]. Thus, sustainable development represents development based on the logic of equitable distribution of wealth, improvement of services and the climate of freedoms and rights, in full balance with development without compromising natural and political resources in general. In this way, development is directed to the benefit of society in general, as it prioritizes the needs of the current society while taking into account the right of future generations, which exactly means sustainability [4].

2.2 The Concept of Renewable Energy

According to the Intergovernmental Panel on Climate Change (IPCC), renewable energy is "any energy that has a solar, geophysical or biological source that is renewable in nature at a rate equal to or greater than its utilization rates, and is generated from successive and continuous currents in nature such as biomass energy, solar energy, subterranean energy, water movement, tidal energy in the oceans and wind energy, Water movement, tidal energy in the oceans and wind energy, and there are many mechanisms that allow converting these sources into primary energies such as heat and hydroelectricity and into kinetic energy using multiple technologies that allow the provision of energy services such as fuel and electricity [5].

2.3 The Concept of Urbanization

Urbanization is a concept that refers to a state, condition, or "way of life," as expressed by Louis Wirth. It is envisioned as a defining characteristic of cities or urban communities [6]. According to Ibn Khaldun, urbanization represents "an advanced and developed stage in all aspects of human life. It denotes human societies marked by advancements in sciences, industries, commerce, and literary life, reflecting a diversity that impacts the lives of both communities and individuals [7].

3 Transition from Coal to Sustainable Energy: Case of Jerada

Jerada City, located in eastern Morocco, south of Oujda-Angad, covers 9,300 square kilometers. The region is mostly mountainous with a continental climate of cold winters and hot, dry summers. It borders Algeria to the east, Taourirt to the west, Oujda to the north, and Figuig to the south. The population of Jerada province is about 108,727, with 43,490 living in the city center [8]. Jerada's establishment was driven by intensive coal production, with the mining company exploiting poor farmers and shepherds for labor. During the colonial period, the workforce grew to around 6,000 workers by 1947, including 700 Europeans who enjoyed better living conditions. By 1953, the city's coal mines produced about 350,000 tons, making Jerada a growing industrial hub in North Africa. However, the closure of the mines by Charbonnages du Maroc in 2001, following a 1998 labor agreement, marked a turning point. This led to economic decline, environmental damage, and significant social challenges for the city.

Environmental and Social Impacts: The cessation of coal extraction reduced carbon dioxide emissions and improved air quality. However, the region still faces challenges with degraded lands and polluted water from past mining activities.

Socially, the mine closures led to higher unemployment and poverty, forcing many residents to migrate for better opportunities. This economic strain sparked social protests, with citizens demanding economic alternatives and improved living conditions.

4 Renewed Hope with Renewable Energy Investments

Despite challenges after the coal mine closures, the situation prompted investments in renewable energy. The government has initiated projects to tackle unemployment, including an agreement between the National Office of Electricity and the Chinese company SEPCO3 to build a 318 MW coal-powered electricity plant in Jerada. The project, valued at three billion dirhams, aimed to revitalize the city's economy by utilizing its existing coal energy infrastructure.

In addition, Jerada's strategic location supported renewable energy projects, such as solar and wind farms, capitalizing on its vast lands and sunlight. A key example is the Ain Beni Mathar Solar-Gas Power Plant in the province's east, covering 160 hectares. The plant uses concentrated solar power technology in a combined cycle system, generating 472 MW, with 20 MW from solar energy. It produces an estimated 3,538 GWh [9] annually and meets environmental standards, bringing significant socio-economic benefits, including: Promotion of renewable energy sources; Construction of à 5.3 km access road to connect nearby villages to the plant; Creation of small and medium-sized enterprises locally; Job creation during construction and operation phases, amounting to 360,000 workdays; Infrastructure upgrades for neighboring areas, including water exploration equipment; Donation of buses, IT equipment, and school supplies to support education; Distribution of 200 bicycles to students as part of efforts to combat school dropouts; Allocating compensation to the land users whose plots were utilized for the construction of the plant, which spans a total area of 160 hectares".

5 The Role of Local Policies in Supporting the Transition to Clean Energy

Local policies are key to advancing sustainable energy in Morocco, particularly in Jerada. The responsible ministry promotes investments in energy transition and sustainable development, positioning Morocco as a global leader in carbon neutrality. In the 2025 Climate Change Performance Index (CCPI), Morocco ranked 8th, a notable achievement that highlights its commitment to sustainable development and reducing greenhouse gas emissions. The country's goal of reaching 52% renewable energy in its national mix by 2030, along with progress in solar and wind projects, has solidified its leadership in Africa and the Arab world. Morocco's ranking places it ahead of countries like Germany (16th), Norway (9th), and Sweden (11th). Denmark leads the ranking (4th), with the top three spots vacant. Public awareness campaigns are also crucial in garnering support for the transition to clean energy.

The city of Jerada possesses a geographical and environmental structure conducive to transitioning towards renewable energy, serving as a critical pathway to improving local living conditions and achieving sustainable development. In developing countries, sustainable food security (Sustainable Food Security) typically hinges on three key factors: the sustainability of natural resources (land and water) and biodiversity (plant and animal resources). However, these resources must be utilized through well-defined strategies and plans to ensure the success of renewable energy projects and the achievement of their objectives.

To realize these goals, it is essential to engage all stakeholders, including public and private sectors as well as civil society organizations. Such collaboration forms the foundation for advancing this transition, enabling risk-sharing and securing necessary funding from financiers, while private companies contribute by transferring global technology and expertise. Simultaneously, involving local communities through joint initiatives enhances public acceptance of these projects. This principle is reinforced by Morocco's 2011 Constitution, which emphasizes participatory democracy in implementing development projects. Article 12 states: "Associations concerned with public affairs and non-governmental organizations shall, within the framework of participatory democracy, contribute to the preparation, implementation, and evaluation of decisions and projects by elected institutions and public authorities, under conditions defined by law."

6 Urban Sustainability and Local Development

Urban sustainability aims to balance economic, social, and environmental development in cities. Job opportunities in eco-friendly sectors are crucial to urban sustainability, improving quality of life and reducing environmental impacts [10]. Urban planning plays a key role in achieving this balance by adopting a sustainable development approach. This approach preserves natural resources while addressing the needs of future generations. Achieving urban sustainability requires resolving various challenges—physical, social, and economic—ensuring balanced development that meets both present and future needs [11].

This vision has directed stakeholders in Jerada Province toward renewable energy production, which could significantly improve the quality of life. Reducing reliance on fossil fuels would improve air quality, decrease pollution, and create a healthier environment for residents. Jerada, where 40% of former coal mine workers suffer from silicosis [12], would especially benefit from such initiatives. Clean energy can also lower electricity costs, improving access to essential services like lighting and heating. Given Jerada's mountainous terrain and cold winters, this would enhance household comfort. Additionally, developing renewable energy projects would boost the local economy by creating jobs in construction, operation, and maintenance of power plants, along with indirect jobs in manufacturing, services, and distribution, ultimately raising living standards and incomes.

7 Challenges and Opportunities

The transition to clean energy in Jerada faces significant social and cultural challenges, primarily community resistance. In a region historically dependent on coal extraction, residents may fear job losses and economic disruption. A cautious approach is necessary, considering the lessons from the 1998 closure of the Charbonnages du Maroc coal mine, the region's largest employer. This closure left Jerada among Morocco's poorest cities, with unemployment over 30% (compared to the national average of 10%) and a poverty rate of 11%, six points higher than the national average [13].

Coal is deeply embedded in the region's culture and economy, fueling resistance to change. To overcome this, community mobilization and collaboration with local leaders are crucial. Government policies offering financial incentives and social guarantees for coal workers are essential to gaining community support.

Civil society also plays a key role in raising awareness about the benefits of clean energy. Public education and training can highlight the environmental and economic advantages of this transition. With these combined efforts, Jerada can become a model for sustainable energy development and social resilience.

8 Challenges in the Energy Transition of the City

Despite the efforts made by authorities in Jerada to transition towards renewable energy, the city faces several challenges, particularly in financing energy projects, which remains a major obstacle to their implementation. Additionally, infrastructure limitations hinder the effective realization of this transition, as adequate infrastructure is essential for extending renewable energy projects to remote areas and achieving local development.

However, despite these obstacles, Jerada has managed to join the ranks of cities pursuing energy transition, thanks to effective policies and the active involvement of civil society. For instance, several cooperatives have initiated projects to convert coal residues into eco-friendly products. One such initiative by the Renewable Energy Cooperative involves creating an innovative product called "Pavico" (ecological flooring), which has been granted an international patent by the Moroccan Office of Industrial and Commercial Property in Casablanca. This product is made from metallic and plastic waste, combined with a unique biological material developed by the innovator.

9 Following Jerada's Path: Similar Models of Energy Transition

Jerada is not alone in its pursuit of energy transition. Morocco has made significant strides in renewable energy, which accounted for 0.4% of the national energy balance (excluding biomass) and 10% of electricity production in 2007. The country has leveraged its strong hydroelectric and wind energy potential, with current wind energy capacity at 147 megawatts and additional installations underway to produce 975 megawatts. Morocco plans to invest $13 billion to increase wind, solar, and hydroelectric capacity, aiming to generate 42% of its electricity from renewable sources by 2020 [14].

A notable example is the Noor Solar Power Station in Ouarzazate, developed by the Moroccan Agency for Solar Energy (MASEN). It is the largest solar power complex in the world, with a total production capacity of 580 megawatts [15]. This project exemplifies Morocco's commitment to renewable energy and serves as a model for cities like Jerada in their transition efforts.

10 Conclusion

Achieving urban sustainability in the city of Jerada requires adopting a comprehensive approach that integrates environmental, social, and economic dimensions. The production of electricity from renewable energy sources presents an opportunity to redefine the city's development path, enhancing the quality of life for residents by improving environmental conditions and reducing dependence on fossil fuels. This transformation cannot be achieved in isolation from the local community; it requires collective efforts from all stakeholders.

The study demonstrates that transitioning to renewable energy significantly contributes to achieving sustainable urban development. The findings highlight that comprehensive development cannot be realized without shifting towards renewable energy sources. Jerada is rich in renewable energy resources, making it a suitable area for wind and solar energy projects. These initiatives can create new economic opportunities in solar and wind technology, generate permanent jobs, and improve environmental conditions. However, financing remains a critical obstacle to the desired energy transition. To advance the energy transition, several recommendations are proposed. First, it is essential to preserve energy resources and optimize their utilization through efficient and sustainable practices. Second, a legal framework should be established to support and incentivize the energy transition, including policies to attract local and foreign investors. Third, investment in renewable energy should be encouraged by providing grants and funding for scientific research in this field. Finally, Jerada's strategic geographical location should be leveraged to maximize the potential of renewable energy projects. These measures can help position Jerada as a leader in renewable energy development, fostering sustainable growth and improving the quality of life for its residents.

References

1. Abrash, I.: Fundamentalism and secularism: The debate thought and reality (2007)
2. Quivy, R., Campenhondt, L.V.: Manuel de recherches en sciences socials, Edition Dunod, p. 120 (1995)
3. Aouki, A.: The role of the urban environment in achieving sustainable development. J. Soc. Legal Sci. Legal Sci. (3), 786 (2023)
4. Afrah, Z.: The government's strategy in eradicating unemployment and achieving sustainable development. Int. Sci. Sympos. M'sila (2011)
5. Munawwar, A.D., Hammou, M.: Environmantal economics. Dar Al khaldounia for Publishing and Distribution, no. 1st Edition, p. 133 (2010)
6. El-Sayed, A.A.: Urban sociology, Dar Al-Ma'arifa, p. 114 (2008)
7. Bougassas, A.: Rural-Urban models of third world societies in light of the rural urban continuum. D.M.J. p. 103
8. H.-C. F. Planning, Results of the general census of population and housing, (2024)
9. N. o. o. electricity, estern region (2009)
10. U. N. E. Programme, Cities and climate change: integrating sustainability into Urban Planning (2021)
11. Jassiba, A.: Urban Planning and Its Role in Achieving Urban Development Goals. J. Res. Hum. Soc. Sci. **12**, 249
12. M. o. H. Statistics, Health Delegation of Jerada Province (2009)
13. Barada, K.L: Beyond the mines of Jerada: The need for a sustainable approach, Moroccan Institute for Policy Analysis
14. Mosaad, A.: Jerada: Cooperative transforms coal residues into eco-friendly products, https://ar.hibapress.com/details-142368.html. Accessed 04 2025]
15. M. o. E. T. a. S. Development, Energy Transition Sector (2016)

Exploring the Corporate Sustainability Reporting Trends: A Bibliometric Analysis Using Clarivate's WoS Database Between 2014–2025

Hicham Chbihi Kaddouri(✉) and Mohammed Kadous

Moulay Ismail University, Meknes, Morocco
h.chbihikaddouri@edu.umi.ac.ma, m.kadous@umi.ac.ma

Abstract. Corporate sustainability reporting has evolved through time from a nice gesture or philanthropism to a necessity for a successful company. Companies are increasingly under pressure from stakeholders, investors, regulatory bodies, and consumers to adopt sustainable practices. Corporate sustainability reporting is no longer an option but a strategic necessity for companies that wish to thrive in a constantly evolving world. Nowadays, a reel needs for standardized high-quality sustainability language, as robust and widely recognized as language for financial language.

Corporate sustainability reporting has known in the past decades a rise in number and in fame from all parties. However, bibliometric literature review on this topic remains limited.

This paper aims to discuss and answer many questions from the emergence of corporate sustainability reporting to their evolution until today; it advances our understanding of corporate sustainability reporting, clusters and maps academic literature, and attempts to give a roadmap and research agenda for future research.

Using a rigorous search protocol, we gathered 869 articles from the Web of Science (WoS) database between 2014–2025 and conducted a bibliometric analysis via Biblioshiny also known as "Bibliometrix 3.0 package".

This analysis highlights influential authors, institutions, and key trends, offering valuable insights for scholars and practitioners.

Keywords: environmental · social and governance · corporate sustainability reporting · bibliometric analysis · WoS database · Biblioshiny · citation analysis · thematic mapping

1 Introduction

Today, the world is facing a serious threat that continues to grow in scope. Climate change affects every aspect of our lives, including social and economic stability, and one might even say that it puts in danger both our very existence as a human race and entire ecosystems. Responsibility for acting against its severe implications lies with the most

C. El Mokhi et al. (Eds.): ICASET 2025, SUCI, pp. 60–77, 2025.
https://doi.org/10.1007/978-3-031-98334-4_7

powerful instances of society: governments, international non-governmental organizations, and financial market players such as financial institutions and big corporations. We are interested in the contribution of finance and financing while moving towards to a more sustainable economy, one that can satisfy present needs without compromising those of coming generations.

As the globe recovers from the tyranny of the COVID-19 pandemic, Firms are accountable for their actions towards the environment. Their ESG practices and the risks and impacts related to their activities are on the scope of many governmental and non-governmental organizations who seek to increase awareness of people towards Corporates Social Responsibility. Corporate sustainability has become a crucial metrics for assessing organizational resilience and, consequently, a sustainable society. (Krishnamoorthy, 2021). Corporate sustainability appeared and gathered the attention of many in the past several decades. Hence, (Arduino et al., 2024) asserted that "Stakeholder pressure has increased on executives and non-executives' directors to demonstrate ethical operations and incorporate ethical practices into their corporate image".

Socially Responsible Investors are not the only party interested in Responsible companies. In fact, other stakeholders consider companies practices, risks and impacts on Environment, on Society and on Governance. For example, customers' awareness increases regarding these concerns. They consider, increasingly, ESG in their choice. Another example is Governments, who penalize companies for their harmful practices towards ESG, especially those related to Environmental concerns.

ESG factors are non-financial criteria that assess the commitment of companies to achieving (SDGs). They evaluate how firms consider long-term sustainability in their operational activities and their mid to long-term strategies.

ESG is firstly "Environmental". This section E assesses the company regarding its risks and impacts on the environment. For instance, GHG emissions, energy consumption, water and waste management etc. ESG is also "Social". This section S considers how the company manages its human resources and acts in favor of its external stakeholders (e.g. local communities, social actors). Finally, ESG is "Corporate Governance". This section G takes into account how the company's management team operates. For example, Respect of rights of minority shareholders, Board diversity and structure.

The ESG framework has become a key foundation for business models and decision making, helping investors and employees assess a company's enduring sustainability.

The corporate sustainability reporting goes beyond compliance. Companies have a chance to show their dedication to sustainability and gain a competitive edge, interestingly financial benefits align with this approach; it's a win-win scenario, where sustainability and financial interest align for long-term success.

Corporate sustainability reporting offers a full summary review of a of how well a company did across various environmental, social, and governance aspects.

Unlike financial reporting, which must adhere to mandatory and truthful standards, firms have greater flexibility and autonomy in disclosing ESG reports.

(Elkington, 1998) outlined that corporate sustainability is no longer an option but a must for businesses that wish to thrive in a constantly evolving world.

This citation from one of the pioneers of corporate sustainability resonates with particular force in the current context. It highlights an unavoidable reality: companies can no longer be disregarded to ignore Environmental, Social, and Governance (ESG) standards if they want to ensure their sustainability and long- term success.

(Jain et al., 2019) describe a sustainable investment as a plan for financial investments that aims to combine environmental, social, and governance advantages with financial gains. Companies have recently begun integrating this aspect into their business strategy due to its perceived impact on investments and performance.

In a world where the expectations of stakeholders are rapidly evolving, integrating ESG criteria is no longer just a beneficial practice but a strategic requirement. Investors and stakeholders expect companies to disclose ESG reporting, as well as risks and opportunities, with the same rigor and discipline as they do for financial reporting.

Nowadays, financial reporting alone is insufficient to fully capture and reflect the influence of a company's operations on its environment. Studies and academic research show that companies must consider, evaluate, and communicate their stakeholder impacts.

Consequently, many frame references providing indicators and structures for corporate sustainability reporting have emerged in recent years, intending to elevate sustainability reporting to a standard comparable to financial reporting. These standardization initiatives offer different approaches to ESG reporting: (GRI, SASB, CDP, CDSB, TCFD, ISO 26000, IIRC). These studies have also shown that companies engaged in a Corporate Social Responsibility (CSR) approach benefit from more sustained and significant growth than others. The notion of performance must now go beyond financial performance to consider extra-financial performance.

Several names for corporate sustainability reporting have emerged, including ESG (Environnement, Social and Governance) reporting, ESG information disclosure, SRI (Socially Responsible Investment) disclosure, CSR criteria (Corporate Social Responsibility), and even extra-financial analysis criteria. Whatever its name, ESG reporting considers all elements other than the accounting and financial indicators used in traditional financial analysis. It emphasizes the level of the company's dedication to sustainable development. By integrating social, ethical, and environmental criteria, we can thus understand the comprehensive performance of the company.

Corporate sustainability reporting has also become a crucial data point for both shareholders and all stakeholders, complementing financial reporting to offer a thorough perspective of the company.

"It has become an important feature of global capital markets, with companies rated according to ESG criteria now representing 78% of the total global market capitalization". (Boffo and Patalano, 2020). The tremendous rise in ESG reporting studies over the past few years makes it challenging to understand the present trends and trajectories. Additionally, we can use a bibliometric analysis to ascertain and fill the research gaps from previous investigations. Moreover, by using bibliometric tools, researchers can highlight influential authors, institutions, and key trends, offering valuable insights for scholars and practitioners. Using a rigorous search protocol, we gathered 869 articles from the Web of Science (WoS) database between 2014–2025 and conducted a bibliometric analysis via Biblioshiny also known as "Bibliometrix 3.0 package".

The purpose of employing Biblioshiny was to conduct bibliographic coupling and identify the most cited articles, authors, sources, and organizations. It was used to get an annual graph of scientific output, the locations of the top countries, the impact of authors and their sources, and the number of locally referred sources in the field. Subsequently, we performed many keyword studies to identify the primary keywords, including tree-map, word growth, trending topics, thematic map analysis, and cluster analysis.

The research we made will be useful for academic purposes since it completes the work of many previous literatures who studies corporate sustainability reporting in general and adds more details to their works. This paper will also be useful to practice since it displays to companies the importance of ESG practices in facilitating their financial activities and benefiting their operational cycle in general to fit with market requirements.

Given the obvious gap, and the growing interest in the corporate sustainability reporting, our paper seeks to fill the void by providing a comprehensive understanding of this field. Therefore, the objectives of this study are: (1) to systematically explore and critically examine the ongoing knowledge in corporate sustainability reporting; (2)to advance our knowledge of corporate sustainability reporting; (3) to pinpoint knowledge voids as a way to map future directions for this paradigm; (4) to determine the patterns of scientific development, publishing, and citation that have emerged over the course of time in the field of corporate sustainability reporting study; (5) ranking the most fruitful and significant nations, journals, writers, and organizations; finally (6) to illustrate the conceptual framework and most often used keywords of corporate sustainability reporting research.

Our study will revolve around three research questions are:

- **RQ1.** First, regarding publications, references, journals, authors, and linked nations, what are the current trends in the corporate sustainability reporting?
- **RQ2.** Second, regarding this field of research, what are the present advancements? what is their current situation as of today
- **RQ3.** And finally, which are the primary gaps and research questions in the corporate sustainability reporting field that require future research?

To answer those questions, it is rational to adopt in Sect. 2 the PRISMA methodology, in which we thoroughly elucidate the methods and procedures employed for the bibliometric review, Sect. 3 analyzes some of the key observations, trends, and patterns, and endeavors to identify noteworthy aspects of corporate sustainable reporting. In Sect. 4, we discuss the theoretical and practical implications of our study, and finally, in Sect. 5, we conclude by highlighting both future research directions, analyze deficiencies in the extant literature to propose suitable domains that could be further researched and limitations of this literature review.

2 Research Methodology: Procedure and Strategies

To answer our research questions related to corporate sustainability reporting, this study adopts a bibliometric literature review research design. This type of review aims to report what is known and what the literature ignores, rather than producing descriptive reports (Briner et al., 2009). We used the bibliometric analysis for its potential to introduce a transparent, rigorous, objective, and reproducible review process and thus to improve the quality of the review. Recent bibliometric studies (Srivastava and Sivaramakrishnan,

2024; Rafi et al, 2024; Verma and Ghosh, 2024) have embraced the Biblioshiny software and seem to have benefitted from it. For this work, we have selected the PRISMA method, which we deemed most appropriate given our research objectives. For our literature review, an initial dataset of 1452 articles were extracted from Web of Science (WoS) database as our source of articles. Many advantages of using this database for bibliometric analysis have been shown in earlier studies (Adriaanse et al., 2013; Mengist et al., 2020). We retain English articles published between 2014 and 2025, that investigate the search word "ESG reporting". Once we eliminate duplicates and categorize the document type and subject categories, we arrive at a final sample of 869 articles. The content was coded according to the publishing year, journal, research method, and main findings.

The methodology flow is illustrated in Fig. 1.

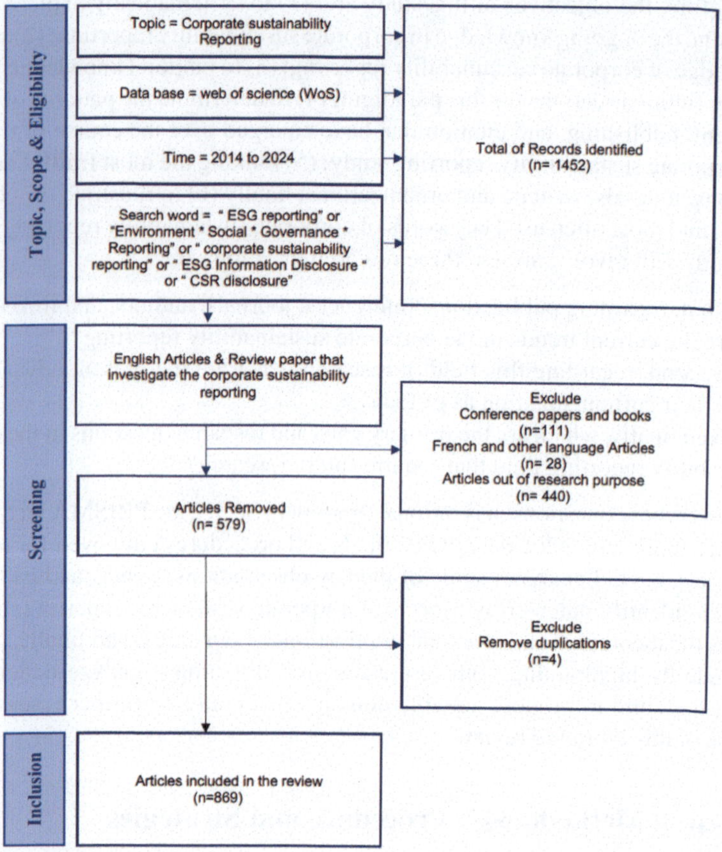

Fig. 1. PRISMA flowchart for Bibliometric review (Source: Created by the authors.)

While there are multiple databases used for bibliometric analysis (e.g., Scopus, Web of Science, PubMed, IEEE Xplore) (Moral-Muñoz et al., 2019), "the first two (WoS and Scopus) encompass multidisciplinary research from business and management journals" (Srivastava, 2021). we specifically chose Clarivate's WoS over Elsevier's Scopus.

We decided to adopt these terms and use them in combination in our keyword search. For instance, we combined "ESG reporting" or "Environment * Social * Governance * Reporting" or "sustainability reporting" or "ESG Information Disclosure" or "CSR disclosure" along with the Boolean operators like "AND," "OR" and "*" and to create the search protocol. The investigation was conducted in the "title, abstracts, and keywords" field. we have chosen to utilize Bibliometrix or Biblioshiny, a popular R package for bibliometric research investigations, to create a bibliographical study of research articles. It offers numerous outcomes for the importation, analysis, and visualization of bibliographic databases. According to (Aria and Cuccurullo, 2017), the software is written in R language and provides a set of instruments for quantitative analysis in bibliometrics.

3 Findings and Interpretation: Bibliometrics and Trends

This study conducts a bibliometric analysis of the current literature regarding corporate sustainability reporting. The analysis seeks to evaluate the current research, identify prominent authors and institutions, and elucidate significant patterns, providing essential insights for researchers and practitioners. It provides a roadmap for future investigations in the field of corporate sustainability reporting.

Table 1 provides a concise overview of the information gathered from the (WoS) database utilizing the search words "ESG reporting" AND "corporate sustainability reporting.". It presents data about authors, their collaborations, documents, and their contents.

Table 1. Main information about data

Description	Results
Timespan	2014:2025
Sources	289
Documents	869
Annual Growth Rate %	42,64
Document Average Age	1,97
Average citations per doc	22,08
References	38027
Keywords Plus (ID)	1126
Author's Keywords (DE)	2410
Authors	2273
Authors of single-authored documents	100

Source: Biblioshiny.

3.1 List of Top Cited Articles and Most Productive Journals on "Corporate Sustainability Reporting"

The number of citations is a representation of the level of interest that researchers have exhibited in a particular scientific study and the potential impact this paper can make.

Tables 2 list the top cited articles focused on ESG reporting based on their number of citations. As revealed, the top influential articles with the largest number of citation count with 391, 261, 254, 225, 194 citations, respectively. The paper published in the review of accounting studies by (Christensen et al., 2021) looked at the possible economic impact of mandatory disclosure and reporting requirements for sustainability and corporate social responsibility (CSR) topics. Similarly (Persakis, 2024) elucidates how ESG performance influences the relationship between climate policy uncertainty and both business performance and carbon dioxide emission performance, published in the environment development and sustainability, which is considered as the top two productive journal in relation to environment, social and governance reporting research (see Table 3). Furthermore, (Sulkowski et al., 2022) identify the key obstacles to global consolidation of ESG reporting. (Jebe, 2022) examined the consequences of Both financial reporting and ESG reporting streams to shape firms' disclosure obligations. (Cao et al., 2019) delineates the current state of Pacific accounting review's topics to identify potential directions for future publications. The results of this study demonstrate the extensive nature of ESG research.

Table 2. Top influential articles

Authors	Article Title	Journal	Year	Total citations
Christensen, HB. et al	Mandatory CSR and sustainability reporting: economic analysis and literature review	Review of Accounting Studies	2021	391
Persakis, A	The impact of climate policy uncertainty on ESG performance, carbon emission intensity and firm performance: evidence from Fortune 1000 firms	Environment Development and Sustainability	2024	261
Sulkowski, A. et al	Evolving ESG Reporting Governance, Regime Theory, and Proactive Law: Predictions and Strategies	American Business Law Journal	2022	254
Jebe, R	The Convergence of Financial and ESG Materiality: Taking Sustainability Mainstream	American Business Law Journal	2019	225
Cao, J. et al	Pacific accounting review in 2013–2023: a bibliometric analysis	Pacific Accounting Review	2024	195
Wen, Q. et al	Does creation-oriented culture promote ESG activities? Evidence from the Chinese market	Global Environmental Change-Human and Policy Dimensions	2024	185

(*continued*)

Table 2. (*continued*)

Authors	Article Title	Journal	Year	Total citations
Almici, A	Does sustainability in executive remuneration matter? The moderating effect of Italian firms' corporate governance characteristics	Meditari Accountancy Research	2023	185
Dilling, PFA et al	The Impact of Corporate Characteristics on Climate Governance Disclosure	Sustainability	2024	179
Chahed, Y	Words and Numbers: Financialization and Accounting Standard Setting in the United Kingdom	Contemporary Accounting Research	2021	177
El-Jourbagy, J. et al	In Space, No One Can Hear You're Green: Standardization of Environmental Reporting, the SEC's Proposed Climate Change Disclosure Rules, and Remote Sensing Technology	American Business Law Journal	2022	176

Source: created by the authors based on WoS data.

Table 3. Journal Rankings in 2024

Journal	Rank	No of Articles and %
Sustainability	1	131 (54,81)
Journal of cleaner production	2	28 (11,71)
Environment development and sustainability	3	14 (5,85)
Journal of environmental management	4	12 (5,02)
Environmental science and pollution research	5	10 (4,18)
Frontiers in environmental science	6	7 (2,92)
Utilities policy	7	6 (2,51)
Clean technologies and environmental policy	8	2 (0,83)
Environmental science \& technology	9	2 (0,83)
Management science	10	2 (0,83)
Polish journal of environmental studies	11	2 (0,83)
Agribusiness	12	1 (0,41)
Atmospheric environment	13	1 (0,41)
Carbon balance and management	14	1 (0,41)
Ecological indicators	15	1 (0,41)
Environmental development	16	1 (0,41)

(*continued*)

Table 3. (*continued*)

Journal	Rank	No of Articles and %
Environmental engineering and management journal	17	1 (0,41)
Environmental pollution	18	1 (0,41)
Environmental progress \& sustainable energy	19	1 (0,41)
Food policy	20	1 (0,41)
Frontiers in marine science	21	1 (0,41)
Global nest journal	22	1 (0,41)
Journal of contaminant hydrology	23	1 (0,41)
Journal of medical ethics	24	1 (0,41)
Resources conservation and recycling	25	1 (0,41)
Sustainable chemistry and pharmacy	26	1 (0,41)
Technovation	27	1 (0,41)

Source: created by the authors based on WoS data.

Table 3 illustrated the most productive journals in 2024 concentrated on corporate sustainability reporting and their total papers count over the full duration. Sustainability and journal of cleaner production are the most active journals with 131 and 28 publications, respectively. These periodicals are essential for disseminating research that investigates corporate sustainability reporting. They address the economic, ethical, and strategic aspects of sustainability, providing a comprehensive view of how financial factors may impact sustainable practices in various sectors.

Stakeholder theory argues that financial performance and environmental performance are positively related since stakeholders are more satisfied when this element is taken into account which improves the company's external reputation and its financial performance consequently (Allouche and Laroche, 2005).

Literature also presents the resource perspective which stipulates that even though investing in socially and environmentally responsible ways of doing business generates costs and requires financial means from the company, this investment may help the company develop more resources in the future both financial and human.

3.2 Research Production Over Years

The study reviewed bibliometric data, showing a significant increase in publications related to corporate sustainability reporting, particularly from 2018 to 2025, highlighting a growing academic interest.

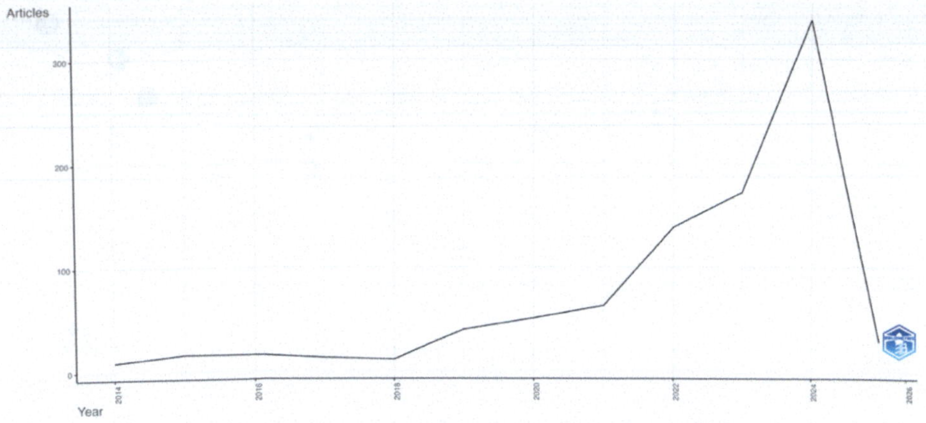

Fig. 2. Annual scientific production (source: Biblioshiny)

Given the limited number of papers published annually, we might consider the period 2014–2018 as an embryonic stage.

The scientific literature has long addressed the topic of corporate sustainability reporting. It's important to note that research in this field has been extensive from 1970 to this day.

The consistent growth in publications throughout this era, reaching 338 papers by 2024, signifies a sustained escalation in research endeavors within this field. The reduced quantity of articles in January 2025 is likely attributable to the recency of the data. Based on these tendencies, we anticipate that the number of articles in 2025 will rise as the year advances, perhaps exceeding the high of 2024.

3.3 Most Popular Keywords

Figure 2 encapsulates the predominant terms employed in the sampled papers. Keyword analysis is regarded as a crucial undertaking as it facilitates the identification of primary subjects and focal regions in contemporary research. It also offers insights into how scholars tackle corporate sustainability reporting.

Primarily, "performance" emerges as the predominant keyword, occurring (231) times over the research period. Examining the correlation between financial performance and ESG characteristics is crucial and has garnered sustained interest. The acronym "CSR" ranks second with a total of 209 appearances. Emphasizing the significance of social and environmental concerns within corporate practices. Other terms include "impact" (200) and "sustainability" (92), highlighting the growing importance of ESG impact in research.

Nowadays, the impact regarding CSR and environmental issues is stronger than it had ever been in the past century. As such, ESG reporting tends to give a general idea for people to whom it matters where they are investing and the impact of their investment.

Being socially responsible has risen in fame as more and more people understood the real impact of decisions made today and their consequences in their future daily life (Fig. 3).

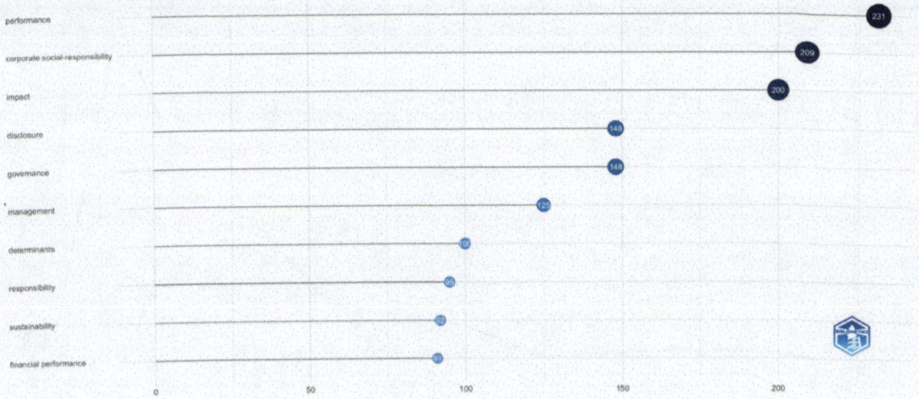

Fig. 3. Most trendy keywords on corporate sustainability reporting (source: Biblioshiny)

According to the statistics, financial performance, corporate social responsibility, sustainability, and other ESG-related topics have recently attracted a lot of academic attention.

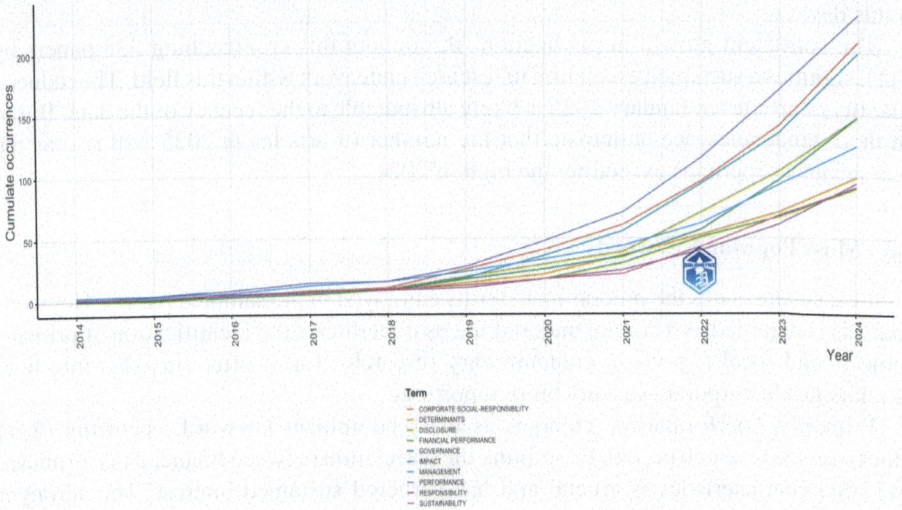

Fig. 4. Word frequency over time related to corporate sustainability reporting (source: Biblioshiny)

The graph reveals that corporate sustainability reporting and financial performance have been the focal points of current studies. In the present day, corporate sustainability reporting and financial performance holds a higher place in society than it ever did and more and more companies big or small, willingly or not contribute to societal actions. Because pressure from NGOs, government and people is stronger than ever in the past decade. Socially Responsible Investors are not the only party interested in corporate sustainability reporting. In fact, other stakeholders consider companies practices, risks and impacts on Environment, on Society and on Governance. For example, customers' awareness increases regarding these concerns. They consider, increasingly, ESG in their choice. Another example is Governments, who penalize companies for their harmful practices towards ESG, especially those related to Environmental concerns.

3.4 Top 10 Contributing Countries

This data pertains to the most often cited nation. Analyzing nations with the most citations in a particular research area is essential for understanding and valuing the global distribution and influence of scientific contributions. Furthermore, recognizing these leading nations allows researchers and institutions worldwide to seek collaborations and partnerships while benchmarking their accomplishments against the highest global standards.

Table 4. Top 10 most contributing countries

Rank	Country	No. of articles and % (N = 1727)	Total citation	Average articles citation
1	USA	340 (19)	2535	25.40
2	CHINA	315 (18)	723	8.80
3	ITALY	227 (13)	1838	25.90
4	UK	172 (9,9)	3700	77.10
5	INDIA	163 (9,43)	760	14.60
6	CANADA	130 (7,52)	551	15.30
7	AUSTRALIA	124 (7,18)	1141	27.20
8	SPAIN	96 (5,55)	280	26.20
9	POLAND	83 (4,80)	787	9.70
10	GERMANY	77 (4,45)	1283	44.20

Source: created by the authors based on WoS data.

As noted in Table 4, the USA ranks first among the top 10 countries, contributing 19% of total publications. This illustrates United State of America's importance in sustainability research and its substantial investments. Followed by China follows with 18% and Italy with 13% of the total contributions. Based on citations average, the UK possesses the highest citation average (3700).

The publishing figures in line with the citation counts of USA and UK show a harmonic degree of output and impact, therefore reflecting their effective efforts in carrying out environmentally friendly research (Fig. 5).

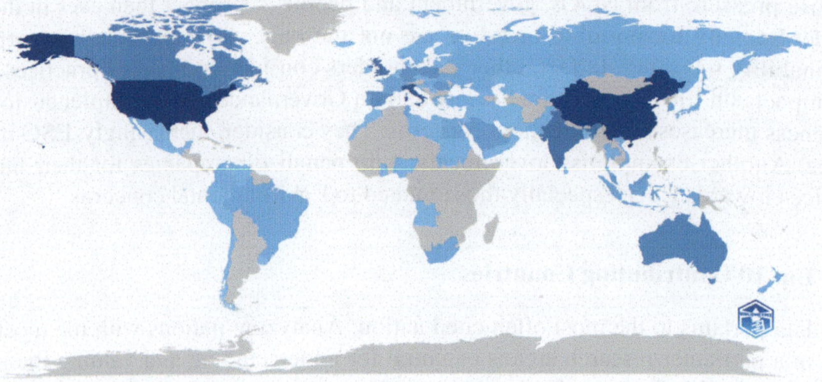

Fig. 5. Countries' Production on corporate sustainability Reporting (source: Biblioshiny)

Figure 4 presents a global map indicating the nations that have published research on corporate sustainability reporting. Colors represent publishing frequency; deeper colors indicate more frequent publication, while lighter colors indicate less frequent publication for the countries.

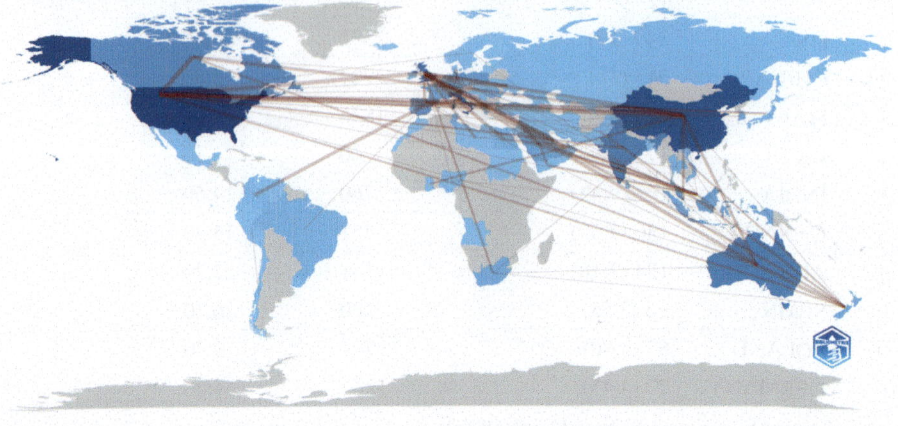

Fig. 6. Countries' Collaboration Map on corporate sustainability Reporting (source: Biblioshiny)

The publishing output of USA is strongly reflected in statistics, underscoring its pre-eminence. Conversely, CHINA, ITALY, and the UK have substantial citation effects, indicating the existence of high-quality research. European countries and Australia have significantly contributed, illustrating a comprehensive global effort to advance sustainable production and consumption habits.

An increasing number of people have been interested in corporate sustainability, which resulted in the rise of fame of ESG reporting. In USA more than elsewhere, being sustainable has become a trend. A trend where surprisingly enterprises are sharing their social. And environmental actions and showing that they are getting more and more involved regarding that (Fig. 6).

United Kingdom, USA, China, Canada, Australia, and Malaysia each own over 30 partnerships. The nations with the fewest connections include France, Morocco, Tunisia and Saudi Arabia, each having less than 10 collaborations.

3.5 Thematic Evolution and Trends Linking

Three clusters are formed from the analysis; Fig. 7 illustrates the co-occurrence network, which examines the publication content.

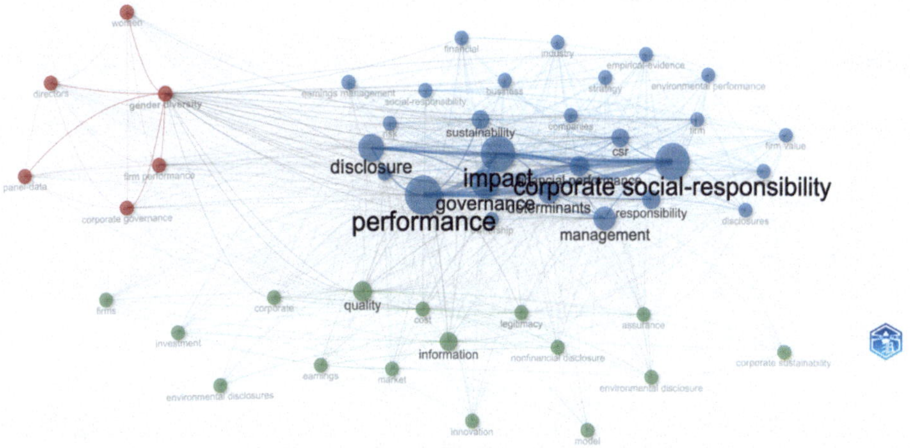

Fig. 7. Co-occurrence analysis (source: Biblioshiny)

The blue, red, and green clusters in the keyword co-occurrence network represent the highly referenced terms in the literature on ESG reporting. One of the most often occurring words is performance (425 times) followed by "Corporate Social responsibility" (304 times), "Sustainability" (253 times), "Disclosure" (207 times), and "mpact" (196 times).

Various existing literatures have studied the effects of ESG reporting on firms' performance (Landi and Sciarelli, 2018; Balatbat et al., 2012; Alareeni and Hamdan, 2020).

(Landi and Sciarelli, 2018) tackled the topic from a market return point of view. They analyzed the impact of sustainability reporting on the abnormal Financial Performance (FP). They studied the case of Italian firms listed on the FTSE Milano Indice di Borsa (MIB) index over a sample period from 2007 to 2015.

(Balatbat et al., 2012) also studied the same relationship for the top 300 firms listed in the Australian Securities Exchange from 2008 to 2010.

They examine the influence of ESG score on Financial Performance using financial profitability ratios (ROA, ROE, ROIC, EBITDA margin).

The same paper goes on in the following section to mentions briefly the issue of access to green information. Financial institutions don't have access to green databases in order to obtain environmental information and evaluate the risks of renewable energy projects. The author suggests as a solution the establishment of a centralized database for the renewable energy industry to help market players have more access to risks that the companies in this industry encounter.

Additionally, (Alareeni and Hamdan, 2020) investigated the impact of ESG on US S&P 505-listed firms' Operational, Financial and Market performance for the period 2009–2018. They base their assessment on ROA for operational performance, ROE for financial performance and price earnings ratio for market performance.

3.6 Thematic Evolution

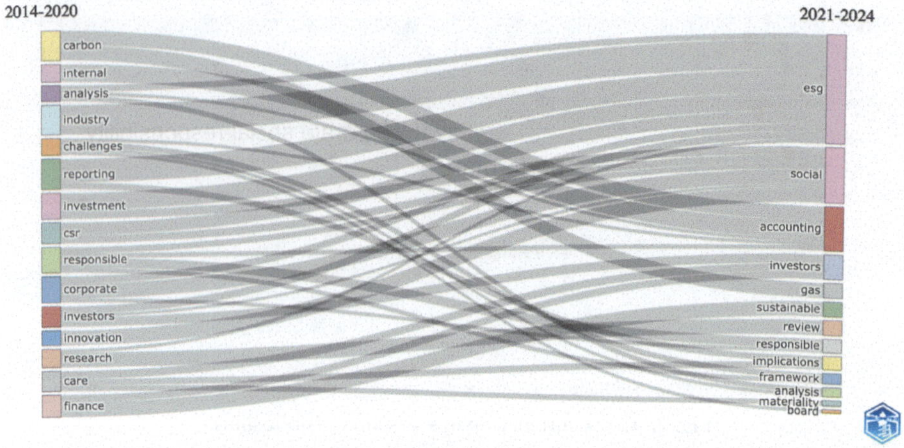

Fig. 8. Thematic evolution (source: Biblioshiny)

Figure 8 reveals thematic evolution in two different stages (2014–2020, 2020–2024).

The first phase of this study (2014–2020) highlighted the main themes such as carbon emission and industry sectors, Corporate Financial Disclosure. Corporate Social Responsibility (CSR). Between 2018 and 2024, the emergence of ideas such as "ESG", "corporate sustainability"," and "green finance" signified a notable shift towards employing financial strategies to promote sustainability. Recently, there has been heightened acknowledgment of the financial sector's role in advancing sustainable development. Furthermore, the emphasis on "renewable energy" and "financial performance" indicates a dual commitment to environmental goals and economic sustainability. The progression of research topics indicates an increasing understanding of how financial management strategies may successfully enhance sustainable production and ESG reporting.

4 Discussion and Implications

Our analysis shows growing academic interest in the interplay between stakeholder engagement practices and corporate sustainability reporting offers a promising avenue for academic research. Moreover, our bibliometric analysis indicates that stakeholder theory is crucial in influencing CSR outcomes, highlighting the necessity for further investigation into how diverse stakeholder engagement practices affect the transparency and credibility of sustainability reporting across various industries.

This bibliometric literature review has significant practical consequences, especially for firms seeking to incorporate sustainability into their fundamental strategy and operations. Our findings underscore the essential importance of stakeholder participation in improving the openness and credibility of sustainability reporting.

Businesses must engage a diverse array of stakeholders in their sustainability projects to cultivate trust and guarantee that their efforts are acknowledged and esteemed by both internal and external audiences. Practitioners have the problem of efficiently adopting these techniques while managing the intricacies of various industrial environments, especially in developing nations. By cultivating a culture of sustainability inside their enterprises and employing sophisticated management methods, firms may effectuate significant changes, improve their corporate reputation, and achieve larger environmental and social goals.

5 Conclusion, Limits and Future Research Directions

Nowadays, the impact of corporate sustainability and ESG issues is stronger than it had ever been in the past century. An increasing number of people have been interested in corporate social involvement, which resulted in the rise of fame of corporate sustainability reporting. In USA and China more than elsewhere, align with global sustainability agendas, and secure long-term value has become a trend. A trend where surprisingly enterprises are sharing their social actions and showing that they are getting more and more involved regarding that. The current paper offers an overview and provides several insights regarding corporate sustainability reporting. According to our findings, the volume of research conducted on the topic of environment, social, governance reporting has attracted significant attention in both academia and industry during the past decades, especially the period 2018–2025, had known a rise in number and in fame from all parties. In addition, sustainability, journal of cleaner production, and environment development and sustainability are the top three productive journals on this topic. Furthermore, USA, China, Italy and UK are the most active countries-researchers in this field. The mapping and clustering research have shown that they interacted extensively both among themselves and with other nations. This paper offers significant implications and delineates persuasive future research avenues for this field. Like any research work, our study is not free from limits, and it should be exposed to guide future research paths. For example, while it provides valuable insights, it has been inherently constrained using the Web-of-Science database. Despite being a widely acknowledged and extensive data source, it may not include all pertinent publications, especially those found in other databases such as Scopus or non-indexed journals and grey literature, potentially leading to the omission of important studies and viewpoints from the analysis. Secondly, the bibliometric techniques utilized, such as Biblioshiny, rely on the algorithms and criteria applied for data processing and display. Despite their potency, these methods may engender biases in the detection and portrayal of relationships among articles. The grouping and mapping of themes may be affected by the specific settings established inside the program, thereby influencing the assessment of the structure and patterns in the existing literature. Future research may build upon this study's findings by utilizing varied analytical tools, refining search tactics, and merging data from other databases. This will enhance our comprehension of the topic and uncover more potential study avenues.

References

Adriaanse, L., Rensleigh, C.: Web of science, scopus and google scholar: a content comprehensiveness comparison. Electron. Libr. **31**, 727–744 (2013). https://doi.org/10.1108/EL-12-2011-0174

Alareeni, B., Hamdan, A.: ESG impact on performance of US S&P 500-listed firms. Corp. Gov. **20**(7), 1409–1428 (2020). https://doi.org/10.1108/CG-06-2020-0258

Almici, A.: Does sustainability in executive remuneration matter? The moderating effect of Italian firms' corporate governance characteristics. Meditari Account. Res. **31**(7), 49–87 (2024). https://doi.org/10.1108/MEDAR-05-2022-1694

Arduino, F.R., Buchetti, B., Harasheh, M.: The veil of secrecy: family firms' approach to ESG transparency and the role of institutional investors. Financ. Res. Lett. **62**, 105243 (2024). https://doi.org/10.1016/j.frl.2024.105243

Aria, M., Cuccurullo, C.: Bibliometrix: an R-tool for comprehensive science mapping analysis. J. Informetr. **11**(4), 959–975 (2017). https://doi.org/10.1016/j.joi.2017.08.007

Balatbat, M., Siew, R., Carmichael, D.: ESG scores and its influence on firm performance: Australian evidence. Aust. Sch. Bus. Sch. Account. **2012**, 1–30 (2012)

Boffo, R., Patalano, R.: ESG investing: practices, progress and challenges. Open Access Libr. J. **11**(1) (2020). https://doi.org/10.1787/5504598c-en

Cao, J., Huang, Z., Kristanto, A.B., Scott, T.: Pacific accounting review in 2013–2023: a bibliometric analysis. Pac. Account. Rev. **36**(3/4), 297–347 (2024). https://doi.org/10.1108/PAR-03-2024-0049

Chahed, Y.: Words and numbers: financialization and accounting standard setting in the United Kingdom. Contemp. Account. Res. **38**(1), 302–337 (2021). https://doi.org/10.1111/1911-3846.12614

Christensen, H.B., Hail, L., Leuz, C.: Mandatory CSR and sustainability reporting: economic analysis and literature review. Rev. Acc. Stud. **26**(3), 1176–1248 (2021). https://doi.org/10.1007/s11142-021-09609-5

Dilling, P.; Harris, P.; Caykoylu, S.: The impact of corporate characteristics on climate governance disclosure. Sustainability **16**(5), (2024). https://doi.org/10.3390/su16051962

El-Jourbagy, J.; Gura, P.P.: In space, no one can hear you're green: standardization of environmental reporting, the SEC's proposed climate change disclosure rules, and remote sensing technology. Am. Bus. Law J.**59**(4), 333–820. (2022). https://doi.org/10.1111/ablj.12214

Elkington, J.: Accounting for the triple bottom line. Meas. Bus. Excell. **2**(3), 18–22 (1998). https://doi.org/10.1108/eb025539

Gundlach, G.T., Achrol, R.S.: Governance in exchange: contract law and its alternatives. J. Public Policy Mark. **12**(2), 141–155 (1993)

Jain, M., Sharma, G.D., Srivastava, M.: Can sustainable investment yield better financial returns: a comparative study of ESG indices and MSCI indices. Risks **7**(15), (2019). https://doi.org/10.3390/risks7010015

Jebe, R.: The convergence of financial and ESG materiality: taking sustainability mainstream. Am. Bus. Law J. **56**(3), 645–702 (2019). https://doi.org/10.1111/ablj.12148

Krishnamoorthy, R.: Environmental, social, and governance (ESG) investing: doing good to do well. Open J. Soc. Sci. **9**, 189–197 (2021). https://doi.org/10.4236/jss.2021.97013

Rafi, A., Rehman, M.A., Musu, A., Rather, R.A., Kautish, P.: Revisiting sustainable tourism research: bibliometric insights and future research agenda. Anatolia. 1–16 (2024). https://doi.org/10.1080/13032917.2024.2411988

Landi, G., Sciarelli, M.: Towards a more ethical market: the impact of ESG rating on corporate financial performance. Soc. Responsib. J. **15**(1), (2018). https://doi.org/10.1108/SRJ-11-2017-0254

Persakis, A.: The impact of climate policy uncertainty on ESG performance, carbon emission intensity, and firm performance: evidence from fortune 1000 firms. Environ. Dev. Sustain. **26**(9), 2403–2408 (2024). https://doi.org/10.1007/s10668-023-03634-x

Mengist, W., Soromessa, T., Legese, G.: Method for conducting systematic literature review and meta-analysis for environmental science research. Methods X **7**(1) (2020). https://doi.org/10.1016/j.mex.2019.100777

Moral-Muñoz, J.A., López-Herrera, A.G., Herrera-Viedma, E., Cobo, M.J.: Science mapping analysis software tools: a review. In: Springer Handbook of Science and Technology Indicators, vol. 3, no. 1, pp. 159–185 (2019)

Srivastava, M., Sivaramakrishnan, S.: Mapping the themes and intellectual structure of customer engagement: a bibliometric analysis. Mark. Intell. Plan. **39**(5), 702–727 (2021). https://doi.org/10.1108/MIP-11-2020-0483

Sulkowski, A., Jebe, R.: Evolving ESG reporting governance, regime theory, and proactive law: predictions and strategies. Am. Bus. Law J. **59**(3), 449–503 (2022). https://doi.org/10.1111/ablj.12210

Verma, P., Ghosh, P.K.: The economics of forest carbon sequestration: a bibliometric analysis. Environ. Dev. Sustain. **26**, 2989–3019 (2023). https://doi.org/10.1007/s10668-023-02922-w

Wen, Q., et al.: Does creation-oriented culture promote ESG activities? Evidence from the Chinese Market. Global Environ. Change – Hum. Policy Dimensions **86**(1), 254–607 (2023). https://doi.org/10.1016/j.gloenvcha.2024.102852

Adaptive Approach to Enhance ITS-G5 Protocol Performance in High-Density Environments Using Artificial Intelligence

Kawtar Jellid(✉) and Tomader Mazri

National School of Applied Sciences, Kenitra, Morocco
kawtarjellid@gmail.com

Abstract. Within environments linked to intelligent transport, the efficiency and reliability of vehicular communication protocols are crucial, especially in dense urban environments. This work proposes an enhancement to the ITS-G5 protocol by integrating adaptive mechanisms based on real-time traffic analysis. The approach relies on intelligent adjustment of communication parameters to optimize data flow, reduce latency, and also avoid congestion-related losses. By continuously analyzing the network's state, the system dynamically adapts to varying traffic conditions, thus improving the scalability and responsiveness of Its-G5. Experimental evaluations carried out in simulated urban scenarios have shown that the proposed solution significantly enhances communication quality and resource management. Compared to the standard protocol, notable improvements were observed in terms of data delivery rate, latency reduction and overall network performance. These results demonstrate the relevance of incorporating intelligent decision-making into existing communication architectures. The proposed method opens promising perspectives for the deployment of more resilient and efficient networks, contributing to safer and more coordinated autonomous vehicle navigation in complex traffic environments.

Keywords: ITS-G5 · Autonomous Vehicles · Vehicle-to-Vehicle · High-Density Environments · Artificial Intelligence

1 Introduction

New challenges emerging from contemporary market and regulatory trends include the adoption of green technologies, intelligent vehicles, In a context marked by the rise of alternative energy solutions, online platforms and smart vehicles, the optimization of engines and response times remains a crucial challenge, yet communication and networking are poised to be crucial components for advancing future transportation systems [1].

C. El Mokhi et al. (Eds.): ICASET 2025, SUCI, pp. 78–87, 2025.
https://doi.org/10.1007/978-3-031-98334-4_8

In the near future, the integration of vehicles equipped with autonomous driving capabilities and roadside infrastructure is set to become more commonplace, aimed at enhancing road safety and optimizing traffic flow to support these advancements, two main communication standards are considered as critical enablers: ITS-G5, based on IEEE 802.11p, and C-V2X, defined by 3GPP [2].

Recent studies [3] have highlighted the potential of Cellular V2X as an alternative to (DSRC) Dedicated Short-Range Communications based on IEEE 802.11p. However, DSRC\ITS- G5 remains the most tested and mature protocol for vehicular communications [4].

In Europe, substantial efforts by ETSI TC ITS have led to the standardization of ITS-G5, a framework for both safety-critical (G5A) and non-safety-critical vehicular communications based on the IEEE 802.11p standard [5].

C-ITS Cooperative Intelligent Transport Systems have been proposed as a solution to enhance road transport safety and efficiency. However, C-ITS is susceptible to misbehaviors that could result in serious consequences, such as accidents and loss of life. Although there are numerous studies on technical aspects of misbehavior detection within C-ITS [7], these often focus on narrow areas and do not provide a comprehensive or unified perspective [7, 8].

Our paper is structured as follows: Sect. 2 we present the fundamental principles and architecture of the ITS-G5 protocol. Section 3 explores its role in autonomous vehicle communication, highlighting its benefits and integration capabilities. Section 4 discusses both the advantages and limitations of ITS-G5, particularly in high-density environments. Section 5 introduces the proposed adaptive AI-based approach to enhance ITS-G5 performance. Finally Sect. 6 concludes the paper and outlines potential future research directions.

2 Exploring the ITS-G5 Protocol: Principles

2.1 ITS-G5

The technology of connected vehicles to the internet represents an important challenge in wireless networking. Two main approaches are being pursued. The first, (WLAN) the wireless local area network method, is based on the IEEE 802.11p standard (in its European version known as ETSI ITS- G5) [9].

C-ITS systems have the fundamental goal of delivering information to drivers through v2x communications. This information can originate from both infrastructure V2I and vehicles themselves V2V. From the infrastructure perspective, services are often provided downstream, where information, authorizations, and warnings are transmitted by Traffic Management Centers (TMCs) to vehicles in specific geographic areas. For example, one such service is the generation of traffic congestion. Warnings [10, 11].

Intelligent Transport Systems (ITS) employ various scientific and technical methodologies to improve the efficiency, safety, and security of transportation systems, facilitated by advanced communication systems [12].

Cellular networks also play an important role in vehicular communication, providing support for road weather and traffic information exchange to enhance traffic safety. The 3rd Generation Partnership Project (3GPP) has developed the V2X (Vehicle-to-Everything) communication standard based on LTE technology [13].

The Finnish Meteorological Institute (FMI) has developed a hybrid vehicular network infrastructure that combines ITS-G5 with 5G cellular technology to improve communication capabilities and vehicle safety.

2.2 ITS-G5 Architecture

This section presents the architecture of the ITS-G5 protocol, an important protocol for ensuring the communication of autonomous vehicles [14].

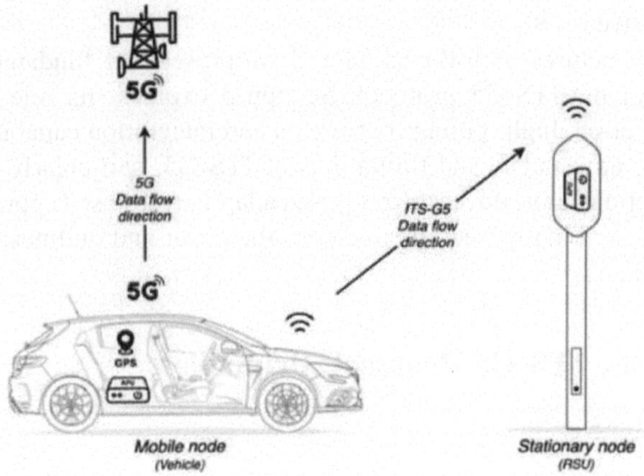

Fig. 1. Overview of ITS-g5 Architecture for V2V and V2I Communication

The ITS-G5 system is dedicated to ensuring transparent and clear communication between autonomous vehicles and road infrastructure based on technologies such as wireless technologies in order to improve at the same time the efficiency and safety of autonomous vehicles and cars [15, 16]. The Key Components of the Architecture are (Fig. 1):

5G Communication Module. This module provides efficient, long-range, and high-speed connectivity, ensuring the efficient transfer and transmission of large amounts of data, such as traffic updates, real-time weather, and road condition information.

RSU - Roadside Unit: Stationary Node. A component that plays a vital in the system, the RSU plays a central role in ensuring the exchange of information between road infrastructure and vehicles. This component is typically placed in strategic locations, such a high-risk areas and intersections. It is typically equipped with the following components:

- ITS-g5 Communication: This component uses short range communication (DSRC) technology to establish direct, low-latency interactions, ensuring fast and reliable data exchange between vehicles and road infrastructure.
- Sensors and Transmission System: The role of these elements is to collect and distribute important data to ensure communication between autonomous vehicles, namely traffic light information, safety alerts, and real time road condition updates, thus improving situational awareness for each autonomous vehicle [17].

5G Antenna. The integration of this element into the system allows autonomous vehicles to maintain continuous connectivity with the cellular network. This step and capability are crucial for high bandwidth applications and services requiring extend long distance coverage, such as real time dynamic mapping and cloud-based vehicle coordination.

Combining direct short-range communication technology with the advantages of cellular networks, the Its-g5 architecture represent an efficient and robust solution to address the communication challenges of autonomous vehicles. This integration guarantees a secure, efficient and reliable interaction transport systems [18].

3 The Importance of the ITS-G5 Protocol in Autonomous Vehicle Communication

The ITS-G5 protocol plays a key role in the development and improvement of intelligent transportation systems, particularly communication between autonomous vehicles [19].

- **Benefits of the ITS-G5 Protocol**

Among the main advantages of the ITS-g5 protocol is that it can operate independently of the cellular infrastructure, it has the ability to operate autonomously, which allows is to adapt to situations with limited network coverage [20].

Several recent studies have demonstrated that the integration of Its-G5 in autonomous vehicles provides a considerable improvement for real-time system reactivation, an important and key factor in accident prevention and thus improving traffic flow [21].

For example, a recent study showed that the use of Its-G5 reduces communication latency to less than 10 milliseconds, a crucial threshold for rapid communication, such as rapid decision-making in emergency situations [22].

- **Integration and Compatibility**

The ITS-G5 protocol is designed to work and interact seamlessly with other communication technologies, including C-v2x (Cellular Vehicle-to-Everything). This compatibility ensures a hybrid communication approach, allowing vehicles to dynamically switch between two technologies based on real-time environmental conditions and network availability thus optimizing reliability and performance [23].

4 Advantages and Disadvantages of ITS-G5 for Autonomous Vehicles: A Focus on High-Density Environments

ITS-G5 offers significant advantages, especially for autonomous vehicles, but it also poses some challenges in densely populated environments. This section examines the strengths and limitations of ITS-G5, focusing on its constraints in heavy traffic and urban environments.

4.1 Advantages of ITS-G5 for Autonomous Vehicles

- Improved Traffic Management: Its-G5 contributes to reducing congestion and optimize traffic flow, and also and enhancing overall transportation efficiency [24].
- Facilitation of Autonomous Decision-Making: The protocol supports real-time data exchange, enabling autonomous vehicles to male informed and timely decisions [25].
- Stability and Long-Term Viability: Designed to adapt to evolving transportation needs; ITS-G5 ensures seamless integration with future mobility technologies [26].

4.2 Disadvantages of ITS-G5 for Autonomous Vehicles

Challenges in High-Density Environments. In densely populated areas, namely high-traffic urban centers, the ITS-g5 presents specific limitations that can impact its overall performance:

Limited Communication Range: The presence of physical obstacles such as buildings, vehicles and other structures significantly reduces the effective communication range between the Route and other structures significantly reduces the effective communication range between the route response Unit (RSU) and vehicles. In addition, interference from multiple communication devices further limits signal and propagation.

Increased Latency: A Large number of connected devices and vehicles leads to network congestion, which presents the challenge of increased latency. This delay in data transfer can compromise the real-time responsive required for autonomous vehicles.

Signal Interference: The simultaneous operation of numerous communication devices within the same frequency bands creates interferences, which can degrade signal quality and reduce the reliability of data exchange between vehicles and infrastructure (Table 1).

Table 1. Comparison of ITS-G5 Performance in Low-Density vs. High-Density Environments

Feature	Low-Density Environments	High-Density Environments
Communication Range	Long-range, minimal interference	Reduced range due to obstacles
Latency	Low latency, fast data exchange	High latency, slower communication
Signal Interference	Minimal interference, stable communication	Increased interference
Network Congestion	Low congestion, stable connections	High congestion, unreliable connections

5 Adaptive Approach to Enhance ITS-G5 Protocol Performance in High-Density Environments Using Artificial Intelligence

The adaptive approach that we propose involves using machine learning models, particularly reinforcement learning (RL), to dynamically adjust the communication strategies employed by the ITS-G5 protocol in high-density environments. This approach considers several critical factors, including signal interference, communication range, and latency. The key advantage of using AI is the ability to continuously learn from the environment, adapting the communication strategy in real time to ensure optimal performance. The following steps outline the key components of our approach:

Real-Time Data Collection: Autonomous vehicles and roadside units (RSUs) will continuously collect data regarding traffic density, signal strength, and communication latency. This data is fed into the AI model to enable real-time decision-making.

- **Pseudo-code Implementation for Reinforcement Learning-Based Adaptive Communication**

To illustrate the implementation of the adaptive approach, the following Python code leverages the Gym library, which is a toolkit for comparing and developing robust learning algorithms. The code simulates an ITS-G5 environment where the communication protocol's parameters are adjusted based on real-time traffic data.

```
# Define a simple environment for ITS-G5 communication optimization
class ITS_G5(Env):
    def __init__(Mine):
        super(ITS_G5, Mine).__init__()
        Mine.action__space = gym.spaces.Discrete(3)  # Adjust
transmission power (Low, Medium, High)
        Mine.observation_space = gym.spaces.Discrete(10)  # Traffic
density levels (0-9)
        Mine.state = 0  # Initial traffic density (low)
        Mine.done = False
        Mine.max_steps = 100

    def step(Mine, action):
        transmission_power = ['Low', 'Medium', 'High'][action]
        traffic__density = Mine.state
        # Reward is based on the combination of transmission power
and traffic density
        if transmission_power == 'High' and traffic__density < 5:
            reward = 10  # High transmission power is optimal for
low density
        elif transmission_power == 'Medium' and traffic__density ==
5:
            reward = 5  # Medium power for moderate density
        elif transmission_power == 'Low' and traffic__density > 5:
            reward = 8  # Low power is optimal for high density
        else:
            reward = -10  # Suboptimal communication configuration
        return reward

# Define a simple Q-learning agent to optimize ITS-G5 communication
class QLearningAgent:
    def __init__(Mine, action_space, observation_space, alpha=0.1,
gamma=0.9, epsilon=0.2):
        Mine.action_space = action_space
        Mine.observation_space = observation_space
        Mine.alpha = alpha  # Learning rate
        Mine.gamma = gamma  # Discount factor
        Mine.epsilon = epsilon  # Exploration-exploitation balance
        Mine.q_table = np.zeros((observation_space, action_space))
# Table initialization

    def select_action(Mine, state):
        if random.uniform(0, 1) < Mine.epsilon:  # Exploration
            return random.choice(range(Mine.action_space))
        else:
            return np.argmax(Mine.q_table[state])  # Choose best
action based on Q-value

    def update_q_table(Mine, state, action, reward, next_state):
        best_next_action = np.argmax(Mine.q_table[next_state])
        Mine.q_table[state, action] += Mine.alpha * (
            reward + Mine.gamma * Mine.q_table[next_state,
best_next_action]
    - Mine.q_table[state, action]
 )
```

6 Conclusion

This study proposed an adaptive reinforcement learning-based mechanism to enhance vehicular communication in densely populated urban environments. By dynamically adjusting critical communication parameters, the approach addresses some of the key limitations of the ITS-G5 protocol, particularly under high traffic loads where conventional static configurations often fail to maintain reliability and performance. Through simulations conducted in OMNeT + +, we validated the effectiveness of the proposed solution, observing significant gains in terms of communication stability, latency reduction, and overall network efficiency. These results demonstrate the relevance of integrating machine learning into vehicular networks to enable more context-aware and responsive communication systems.

Moreover, the proposed model opens new research directions for the future evolution of intelligent transportation systems. It can be extended to hybrid architectures combining multiple communication standards, or integrated into real-time decision-making systems in autonomous vehicles. While the simulation results are promising, further validation through field testing and hardware implementation is essential to assess scalability and robustness in real-world conditions.

In conclusion, this work contributes to ongoing efforts in creating more resilient, adaptive, and intelligent vehicular communication protocols. By anticipating network dynamics and adjusting accordingly, such solutions play a key role in ensuring the safety, efficiency, and coordination of autonomous mobility in increasingly complex traffic ecosystems.

References

1. Sukuvaara, T., Mäenpää, K., Stepanova, D., Siili, T., Karsisto, V.: Combined ITS-G5 and 5G test track for vehicle winter testing and advanced road weather services. In: Proceedings of 8th Transport Research Arena TRA 2020 April 27–30, Helsinki, Finland (2020)
2. Maglogiannis, V., Naudts, D., Hadiwardoyo, S., Van Den Akker, D., Marquez-Barja, J., Moerman, I.: Experimental V2X Evaluation for C-V2X and ITS-G5 Technologies in a real-life highway environment. IEEE Transactions on Network and Service Management **19**(2) (2022)
3. 5GAA V2X Functional and Performance Test Report. Tests Procedures and Results (2018). http://5gaa.org/wp-content/uploads/2018/11/ P-180106-V2X-Functional-and-Performance-Test-Report Final 051118.pdf
4. Amjad, Z., Sikora, A., Hilt, B., Lauffenburger, J.: Low latency v2x applications and network requirements: performance evaluation. In: 2018 IEEE Intelligent Vehicles Symposium (IV), pp. 220– 225 (2018)
5. Masini, B., Bazzi, A., Zanella, A.: A survey on the roadmap to mandate on board connectivity and enable v2v-based vehicular sensor networks. Sensors **18**(7), 2207 (2018)
6. ETSI ES 202 663 v1.1.0 Intelligent Transport Systems (ITS); European profile standard for the physical and medium access control layer of Intelligent transport systems operating in the 5 GHz frequency band. http://www.etsi.org/deliver/ etsi es/202600 202699/202663/01.01.00 50/es 202663v010100m.pdf
7. Amanullah, M.A., Loke, S.W., Baruwal Chhetri, M., Doss, R.: A taxonomy and analysis of misbehaviour detection in cooperative intelligent transport systems: a systematic Review. ACM Comput. Surv. (CSUR) **56**(1) Article No.: 3, 1–38 https://doi.org/10.1145/359659 8,2023

8. Integration of C-ITS in Autonomous Driving, José E. Naranjo, Member, IEEE, Felipe Jiménez, AitorGómez, Alfredo Valle, Jorge González and Alberto Cruz, 2020 IEEE Intelligent Vehicles Symposium (IV) October 20-23. Las Vegas, USA (2020)
9. Dias, D., et al.: A dataset of ITS-G5 and cellular vehicular connectivity in urban environment. Data Brief **52**, 109846 (2024)
10. C-ITS Platform, C-ITS Platform Report Phase I, European Commission (2016)
11. C-ITS Platform, C-ITS Platform Report Phase II, European Commission (2017)
12. Tahir, M.N., Fatima, S., Bashir, N.: Car-to-Car Communication using ITS-G5 & 5G. In: IEEE 5th International Conference on Intelligent Transportation Engineering, Beijing, China (2020)
13. Tahir, M.N., Katz, M.: Heterogeneous (ITS-G5 and 5G) vehicular pilot road weather service platform in a realistic operational environment
14. Chen, M., Zhang, Y.: Next-Generation ITS-G5 for Autonomous Vehicle Communication: Architecture and Performance. In: IEEE Transactions on Intelligent Transportation Systems. This paper explores the architecture and performance metrics of the ITS-G5 system, focusing on its efficiency and scalability in enabling real-time vehicle-to-infrastructure communication (2023)
15. Sharma, R., et al.: 5G-Enabled ITS-G5 Communication for autonomous vehicles: a survey. This article provides a comprehensive survey of 5G integration with ITS-G5, highlighting the impact on transportation systems and safety through enhanced vehicle-to-vehicle and vehicle-to-infrastructure communication. J. Commun. Netw. (2022)
16. Li, X., Yang, Z.: Design and optimization of ITS-G5 architecture for autonomous vehicle networks. Wireless Communications and Mobile Computing. This research examines the technical design and optimization of the ITS-G5 architecture for autonomous vehicles, with an emphasis on the interaction between mobile nodes (vehicles) and roadside units (RSUs) (2022)
17. Khan, Z.S., et al.: Evolution of ITS-G5: enabling safe and efficient autonomous transportation. IEEE Access. This paper delves into the evolution of ITS-G5 technology, discussing its integration with GPS, 5G modules, and its role in improving transportation efficiency and safety through low-latency communication (2023)
18. Bai, H., Wang, L.: Real-Time communication and data sharing in ITS-G5 systems for autonomous vehicles. The study highlights the importance of real-time communication in ITS-G5 systems, especially the interaction between vehicles and RSUs using 5G networks for dynamic data transmission such as traffic updates and road condition reports. Int. J. Comput. Sci. Telecommun. (2023)
19. Bazzi, A., Masini, B.M., Zanella, A., Thibault, I.: On the performance of IEEE 802.11p and LTE-V2V for the cooperative awareness of connected vehicles. IEEE Transa. Vehicular Technol. **67**(4), 3727–3731 (2018)
20. Sun, Y., Chen, Y.: Performance analysis of ITS-G5 communication in autonomous vehicle scenarios. Int. J. Automot. Technol. **21**(3), 515–522 (2020)
21. Kihl, M., Håkansson, M.: Real-time communication in autonomous vehicle networks using ITS-G5. J. Transp. Eng. **145**(6), 06019002 (2019)
22. Matolak, D.W., Sen, I.: Wireless channel characterization for vehicular Ad Hoc networks. IEEE Trans. Veh. Technol. **66**(4), 3403–3415 (2017)
23. Boban, M., Tonguz, O.K.: Hybrid vehicle-to-everything (V2X) communication for connected and autonomous vehicles. IEEE Commun. Mag. **57**(10), 126–132 (2019)

24. Sharma, R., et al.: 5G-Enabled ITS-G5 communication for autonomous vehicles: a survey. J. Commun. Netw. (2022)
25. Li, X., Yang, Z.: Design and optimization of ITS-G5 architecture for autonomous vehicle networks. Wireless Commun. Mobile Comput. (2022)
26. Khan, Z.S., et al.: Evolution of ITS-G5: enabling safe and efficient autonomous transportation. IEEE Access (2023)

E-government, Governance and Sustainable Development: What Synergy for African Countries?

Ahlam Er-rahmani(✉) and Aziz Bensbahou

Economics and Public Policy Laboratory, Faculty of Economics and Management, Ibn Tofail University, Kenitra, Morocco
{ahlam.er-rahmani,bensbahou.aziz}@uit.ac.ma

Abstract. This paper seeks to measure the dynamics of E-government in 51 African countries in comparison with other regions of the world over 2003–2023 and test the effect of its various dimensions in ensuring successful governance and promoting sustainable development. By combining an analytical and econometric approach, this paper attempts to 1) verify the relationship that can link E-government to governance, 2) evaluate the role of governance in the improving or worsening of the level of sustainable development, and, 3) verify the impacts that E-government could have on the level of sustainable development. The results of the econometric estimates confirm the positive dynamics achieved by African countries in the implementation of E-government, with a significant impact on the improvement of sustainable development indicators and those related to the governance indicators. They also emphasize bridging the gaps between countries with geographical proximity, fostering regional cooperation and integration, as well as addressing the similarities and complementarities in their socio-economic and institutional characteristics.

Keywords: E-government · Sustainable Development · Governance · African Countries · Panel Regression

1 Introduction

The proliferation of data networks, combined with the technological has undergone, has contributed to transform the global landscape. In this context, E-governance has emerged as an innovative application playing a pivotal role in delivering efficient public service provision [1] and fostering connections between governments, citizens, businesses, and institutions [2, 3]. It has established a framework for enhancing governance [4] and promoting the social, economic, and environmental dimensions of sustainable development [5, 6].

In this context, the intelligent pathway to achieve good governance is E-governance [7], which is also a critical means to achieve sustainable development. Indeed, this smart pathway promises to make life more sustainable, resilient, and greener

C. El Mokhi et al. (Eds.): ICASET 2025, SUCI, pp. 88–103, 2025.
https://doi.org/10.1007/978-3-031-98334-4_9

[8], particularly by leveraging disruptive technologies. Within this framework, good governance, combined with digital innovation, represents a significant opportunity to support climate transition initiatives and environmental protection.

This study aligns with this framework by evaluating the state of E-government across the African continent, which faces digital and institutional challenges that significantly impact its sustainable development [9]. Consequently, this paper aims to address a gap in the existing empirical research regarding the linkages of E-government for governance and Sustainable Development (SD) for African countries from 2003 to 2023. Specifically, this study bases on E-government Development Index (EDGI). Sustainable Development is measured through dimensions relating to environmental and disaster management as well as economic and social aspects, while Governance Indicators (GI) are evaluated based on six dimensions of the Worldwide Governance Indicators. This paper is structured around four sections:

- Conceptual framework and a literature review of E-government for Sustainable Development (second section);
- Variables Description and Trend Analysis in the African context (third section);
- Model Specification and Empirical Results (fourth section).
- Finally, the last section concludes with key findings and offers derived recommendations (conclusion).

2 E-government for Sustainable Development

E-government (shortly E-gov) become a transformative paradigm based on ICT [10], that holds substantial potential for modernizing public administration [11]. It represents a fundamental pillar for the establishment of inclusive institutions, in alignment with Sustainable Development Goal (SDG) [12, 13]. Despite significant progress made in these fields independently, there is a little research treating the role of E-gov as a lever offering countries, especially Africa, new opportunities to respond to the needs of the current generation without affecting future capacities.

2.1 Conceptual Framework

Two papers serve as the foundation for this study (Fig. 1): the conceptual framework performed by Estevez and Janowski [14], focused on the components of E-gov promoting SD (EGov4SD) and the paper of Dhaoui [15], providing a structured approach for the MENA region to analyze the interaction between (EGov4SD). The conceptual framework refers to case studies and successful country initiatives conducted by Estevez and Janowski, which generated an in-depth conceptual and qualitative analysis, underlining the role the importance of digital advances for economic, social, and environmental sustainability. On the other hand, Dhaoui, building on the work of Lim, J. H. (2015) [5], developed a framework facilitating the transition from qualitative to quantitative analysis and subsequently modeled the relationships between E-gov, SD, and governance. The objective link the trilogy (EGov4SD) using a static panel for 16 distinct models.

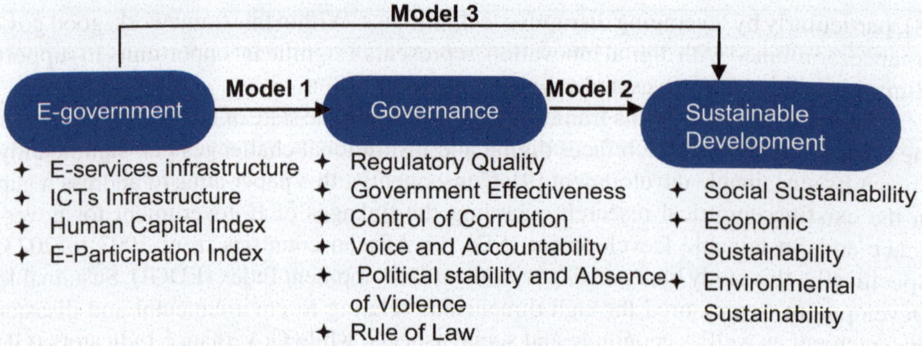

Fig. 1. Conceptual Framework

Good governance is defined as the efforts by international institutions [16] to regulate the institutional environments [17]. According to the World Bank [18], governance is the art of managing public affairs for individual and collective well-being and the SD. For the OECD [19], good governance is linked to the way in which public authorities design, implement, and evaluate their reforms and policies. Heeks characterizes E-gov as the key to achieving better governance through the integration of ICTs [20]. Whereas OECD [21], asserts that e-Gov is more about government than about "e" and it is challenging countries to re-think government organizational structures and processes [22]. Defined as a desirable solution to arrive at the right path [23], the dimensions of sustainable development cover all economic, social and environmental aspects [24]. It's summarized through five Ps: Planet, People, Peace, Partnerships and Prosperity [25]. In this context, the presence of efficient institutions, accompanied by technology, is essential to promote sustainable development, which faces limits related to technology and social organization [26].

2.2 Literature Review

The success of digitalization will only be approved as revolutionary if all populations and regions of the world benefit from it [2]. In this regard, numerous studies suggest that operationalizing and establishing E-gov policies does not automatically lead to the same improvements for countries [27–30]. Countries facing challenges in addressing the basic needs of impoverished populations in developing nations are unlikely to put e-government at the top of their agenda [31]. Following a benchmark of African countries, Heeks [32] notes that they are not the only ones facing E-gov failures, suggesting implementation of e-government projects adapting to local realities and inspired by occidental experiences. In terms of ICT development in Africa, Adam [33] suggests that the priority should be the improvement of ICT infrastructure rather than the general implementation of initiatives simply categorized as E-gov. Although e-government was initially developed in industrialized countries, it can also benefit sub-Saharan African countries [34], though it requires more efforts due to challenges such as limited technological infrastructure and organizational cultural differences.

3 Variables Description and Trend Analysis

3.1 Variables Description

The UN-EDGI measures a country's readiness, capacity, and advancement in using E-gov for public service delivery [35]. Its importance is reflected in its strong correlations with the SDG index scores [12]. It incorporates three key components [12]: The Telecommunication Infrastructure Index (TII), the Human Capital Index (HCI) and the Online Service Index (OSI). The E-Participation Index (EPI), a sub-index of the OSI, captures the electronic participation performance. The dimension that contributes most to the EGI Africa is the HCI, with an average of 0.51 over the period (2003–2023), compared with the lowest contribution from the TII, with an average of 0.15.

The Governance Indicators cover six dimensions and three pillars of governance [36]: economic (Regulatory Quality, Government Effectiveness), political (Voice and Accountability, Political stability and Absence of Violence) and institutional (Control of Corruption, Rule of Law). The first pillar measures the process of selecting, monitoring, and replacing governments. The second pillar quantifies the government's capacity to formulate and implement sound policies, and the third pillar assesses the degree of respect by the state and citizens for the institutions that govern their social interactions.

The dimensions of SD relating to environmental and disaster management encompass several indicators (Fig. 3). Environmental management is measured by Ecosystem Vitality (EV), part of the environmental performance index [37], which gauges the degree to which ecosystems are maintained, conserved and improved. Disaster management is evaluated using Vulnerability to natural risk (VUL) [38], which evaluates exposure and vulnerability to natural disasters. The choice of dimensions of sustainable development selected for the sample depend on the availability of data (Fig. 2).

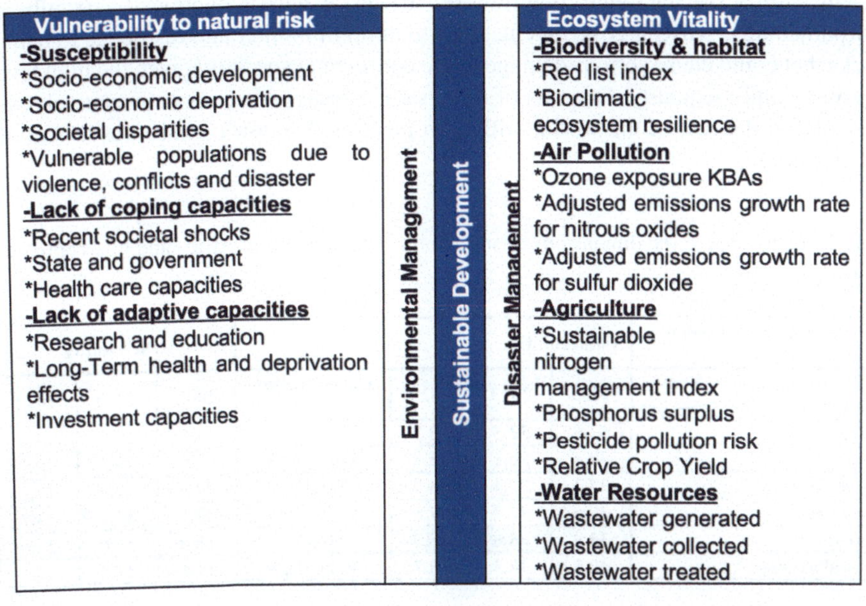

Fig. 2. Sustainable development dimensions used for Africans countries

Figure 3 describes the variables used in the analytical and econometric section and the data sources used to answer the research questions.

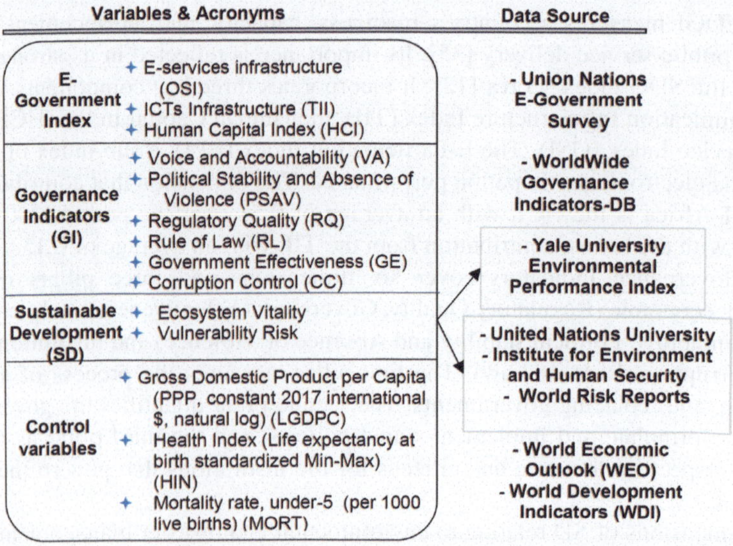

Fig. 3. Variables list and data sources

3.2 Trend Analysis for Africa

An analysis of the dynamics of the dimensions of sustainable development (Table 1) over the last three decades has enabled us to situate Africa's progress in terms of environmental sustainability in relation to other continents. This high level of vulnerability risks having a negative impact on the living conditions of African citizens: economically, socially and environmentally. African governments need to build immunity and resilience to natural shocks that could cause serious damage to the continent, especially given the high levels of poverty and inequality. In terms of ecosystem vitality, Africa scores relatively well compared to the rest of the world, with both positive dynamics and scores around the global average.

Table 1. Dynamics of the dimensions of sustainable development in Africa and intercontinental comparison

Sustainable development dimensions	Vulnerability				Ecosystem Vitality		
Year / Continent	2003	2013	2023	Classification 2023	2003	2013	2023
Africa	53,6	52,3	53,3	Very High	21,3	20,7	23,4
Asia	39,4	36,1	35,7	Very High	24,9	25,2	29,5
Oceania	37,2	37,4	38,0	Very High	24,8	26,9	28,3
America	38,6	37,2	37,5	Very High	33,1	35,3	33,2
Europe	33,9	28,6	28,9	High	18,0	30,8	27,6
World average	40,5	38,3	38,7	Very High	24,4	27,8	28,4

An overview of the EGI at continental level, (Table 2) reveals that:

1) Europe is in first place with a very high EGI, while Africa is in last place with a Middle EGI; 2) in terms of dynamic, all continents have managed to improve their EGI levels, with Europe and Asia in the lead; and 3) Africa and Oceania are the two continents that have reached a level below the world average over the period. Africa's position thus reflects its digital deficiencies. It is true that some countries have managed to change their status (technological progress, digital infrastructure, etc.) over the last two decades, but the average score for African countries is lower than for other continents, indicating the potential that Africa has and that it must be exploited with the means (natural, human, technological, etc.) at its disposal.

Table 2. Dynamics of the e-government indicators in Africa and intercontinental comparison

Continent \ Year	2003	2013	2023	Rating classes 2023	Number of country
Africa	**0,20**	**0,27**	**0,42**	**Middle EGI**	51
Oceania	**0,30**	**0,42**	**0,52**	**High EGI**	14
Asia	**0,35**	**0,50**	**0,67**	**High EGI**	46
America	**0,42**	**0,52**	**0,66**	**High EGI**	35
Europe	**0,54**	**0,70**	**0,84**	**Very high EGI**	44
World average	**0,36**	**0,48**	**0,62**	**High EGI**	**190**

The analysis of the EGI for African countries highlights three key findings: a positive dynamic over the period for all countries; out of 51 countries, only one country has reached a very high EGI (South Africa), 60.8% record a middle EGI, 31% achieve a low EGI and 7.8% have a high EGI; a positive trend in membership of the middle and high EGI groups between 2003 and 2023. This positive dynamic reflects the efforts of countries to improve their EGI and their awareness of its importance for their economy. The Côte d'Ivoire has made the most significant progress over the period, moving up by more than 60 places from 84th in 2003 to 120th in 2023. Since 2012, the country has implemented significant reforms and, in 2021, adopted its National Digital Strategy (NDS) for 2025 [39]. However, most African countries have EGI levels below the global average (0.624 in 2023) (Fig. 4).

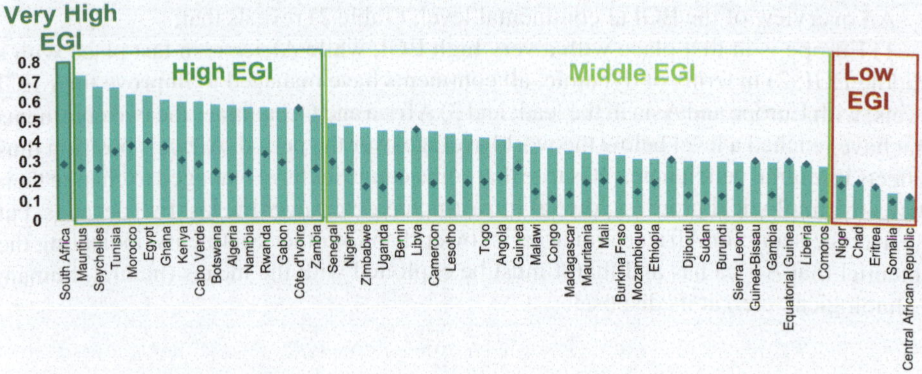

Fig. 4. EGI 2023 level (Histogram) and trends (points) for African Country, 2003–2023

The top five African countries performing at the EGI level are: South Africa, Mauritius, Seychelles, Tunisia and Morocco. The good ranking of these countries is due to their commitment to digital strategies. The South African National E-government Strategy and Roadmap [40] is geared towards an inclusive digital transformation. The Digital Mauritius 2030 Strategic Plan [41] aims to transform the country into an inclusive and high-income economy and fosters a culture of innovation based on emerging technologies such as Blockchain, Fintech and Robotics. Digital technologies are widely used in the Seychelles [42], where the vast majority of government services use at least one digital system to deliver their services. The Digital Tunisia 2020 National Strategic Plan [43] aims to position Tunisia as an international digital reference and to base socio-economic development on ICT. The Digital Morocco 2030 Strategy [44] aims to establish a digital hub to accelerate development.

The analytical mapping of African countries (Fig. 5), which relates the IG (constructed by the principal component analysis method using the six indicators on World Bank governance dimensions) and the EGI, reveals significant differences in the levels recorded by each country and a positive correlation between the two variables. Indeed, the dynamics of the EGI are correlated with the level of development and the presence of institutions [45, 46]. These institutional differences are influenced by contextual specificities, geographic and politico-economic differences, as well as countries' initial conditions [47].

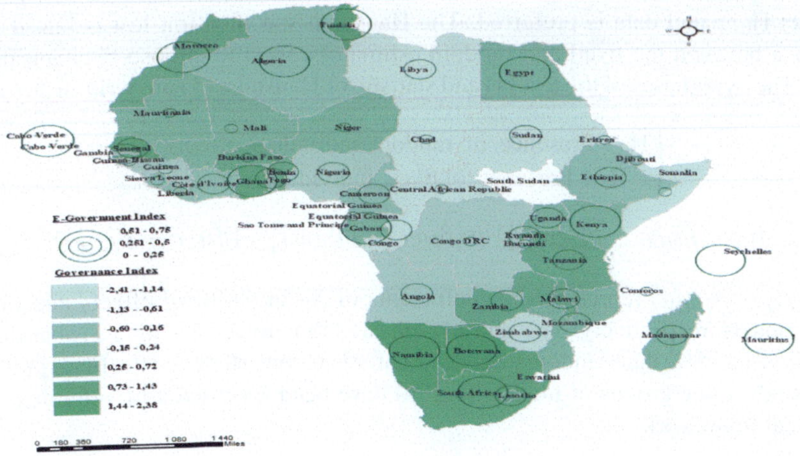

Fig. 5. Geographical distribution of the EGI and GI for African countries, 2003–2023

4 Model Specification and Empirical Results

4.1 Econometric Framework

The literature clearly shows the theoretical link between E-gov, governance and SD. This section attempts to verify the presence of a positive effect of E-gov on governance an SD and of governance on sustainable development, through a series of panel data regressions. Panel data allows to control for individual heterogeneity [48]. Consider the general representation of a panel data model:

$$Y_{it} = \beta_0 + \beta_1 X_{it} + \beta_2 Z_{it} + \mu_{it} \tag{1}$$

where Y: endogenous variable; β_0: the constant term, β_1 and β_2 are coefficients associated with X: independent variable and Z : vector of control variables; i: individual; t: time; $\mu_{it} = (\mu_i + \epsilon_{it})$, includes individual specific effects (μ_i) and idiosyncratic error term (ϵ_{it}). The specific effects component captures the unobservable characteristics specific to each individual. The individual fixed component αi can be decomposed into a specific component α and a global constant γ_i: $\alpha_i = \alpha + \gamma_i$. Before proceeding with model estimation, we tested the heterogeneity of the data using the Fisher test for homogeneity of constants, following Baltagi's approach [52]. The Fisher test is used to decide between a pooled and a specific effects model. The test (2) seeks to test two hypotheses (S 3).

$$\begin{cases} H_0 : \text{Total homogeneity of constants.} \\ H_1 : \text{Partial homogeneity of constants.} \end{cases} \tag{2}$$

$$\begin{cases} H_0 : \text{Absence of individual effects in the data.} \\ H_1 : \text{Presence of individual effects in the data.} \end{cases} \tag{3}$$

If the P-value > 5%, we accept H_0, indicating the presence of a common effect. The estimation method is the ordinary least squares (OLS) method. If the P-value < 5%, we reject H_0, panel data is preferred. The Hausman specification test is based on the difference between the fixed and random estimators, used to choose between the two effects. The hypotheses of the test (4) and statistic of Hausman (5) are written as follows:

$$\begin{cases} H_0 : \text{Presence of random individual effects} \\ H_1 : \text{Presence of fixed individual effects} \end{cases} \quad (4)$$

$$H = \left(\hat{\beta}_{MCG} - \hat{\beta}_{LSDV}\right)'\left[Var\left(\hat{\beta}_{MCG} - \hat{\beta}_{LSDV}\right)\right]^{-1}\left(\hat{\beta}_{MCG} - \hat{\beta}_{LSDV}\right) \quad (5)$$

$(\widehat{\beta}_{MCG} - \widehat{\beta}_{LSDV})$ represent the coefficients of X_{it} in the fixed effects and random effects models, respectively. The null hypothesis H0 is rejected with a Probability (P-Value) less than 5%, then random effect estimator is inconsistent and the FE is preferred. In this study, three groups of models (Fig. 6) have been specified with reference to the theoretical framework.

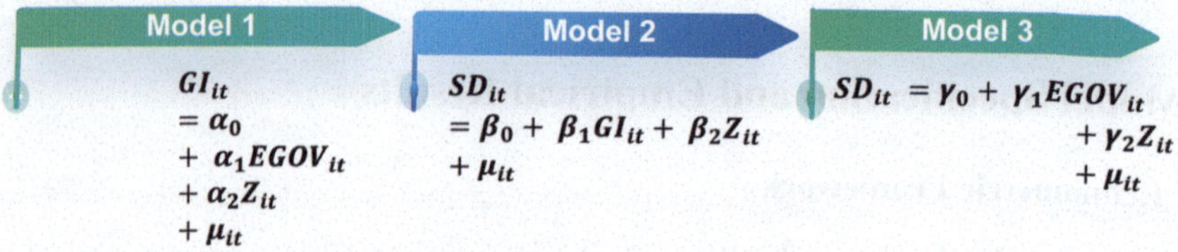

Fig. 6. Design of estimated models

The groups of models were disaggregated into 16 models, which validated the presence of fixed effects (Table 3) as determined by the Hausman test.

Table 3. Hausman test result

Model	Chi2_Hausman	P_value	Choice
1	106	0,0000	Fixed effect
2	47	0,0000	Fixed effect
3	38	0,0000	Fixed effect
4	67	0,0000	Fixed effect
5	39	0,0000	Fixed effect
6	52	0,0000	Fixed effect
7	101	0,0000	Fixed effect
8	679	0,0000	Fixed effect
9	783	0,0000	Fixed effect
10	32	0,0001	Fixed effect

(*continued*)

Table 3. (*continued*)

Model	Chi2_Hausman	P_value	Choice
11	70	0,0000	Fixed effect
12	115	0,0000	Fixed effect
13	626	0,0000	Fixed effect
14	31	0,0000	Fixed effect
15	34	0,0000	Fixed effect
16	150	0,0000	Fixed effect

4.2 Results and Discussion

The results of the study confirm a positive impact of the Human Capital Index and the E-Participation Index on the improvement of all governance indicators (Table 4). For the online services index and the telecommunications infrastructure index, they have a positive impact on government efficiency, control of corruption and voice and accountability, and a negative impact on the rule of law. This nuanced impact of the e-government index on improving accountability calls into question the effectiveness and efficiency of digitization strategies in Africa and calls for greater emphasis on this dimension (the weakest component of the EGI).

Table 4. E-government and good governance

Variables	World Governance Indicators					
	(1)	(2)	(3)	(4)	(5)	(6)
	EG	CC	PSAV	RQ	VA	RL
TII	0.04	0.12*	–0.50***	–0.26***	0.24***	–0.18***
	(0.06)	(0.07)	(0.15)	(0.07)	(0.08)	(0.07)
HCI	0.24***	0.18**	0.57***	0.19***	0.03	0.25***
	(0.07)	(0.08)	(0.16)	(0.07)	(0.08)	(0.07)
OSI	0.30***	0.33***	–0.12	0.35***	–0.05	0.27***
	(0.09)	(0.10)	(0.20)	(0.09)	(0.11)	(0.09)
EPI	0.08	0.13	0.43**	0.02	0.30***	0.17**
	(0.08)	(0.09)	(0.18)	(0.08)	(0.09)	(0.08)
HCIND	–0.11***	–0.10***	–0.11**	–0.06**	0.07**	–0.02
	(0.02)	(0.03)	(0.05)	(0.02)	(0.03)	(0.02)
LPOP	0.01	–0.12	0.36	0.10	-0.83***	–0.02

(*continued*)

Table 4. (*continued*)

Variables	World Governance Indicators					
	(1)	(2)	(3)	(4)	(5)	(6)
	EG	CC	PSAV	RQ	VA	RL
	(0.16)	(0.17)	(0.36)	(0.17)	(0.19)	(0.16)
Constant	−0.99	0.02	−3.29	−1.55	5.03***	−0.75
	(1.07)	(1.19)	(2.46)	(1.14)	(1.32)	(1.09)
Observations	1,071	1,071	1,071	1,071	1,071	1,071
R-squared	0.08	0.07	0.06	0.07	0.07	0.06
Number of countries	51	51	51	51	51	51

Standard errors in parentheses. *** $p < 0.01$, ** $p < 0.05$, * $p < 0.1$

In addition, Table 5 shows that governance indicators have a significant impact on the three dimensions of sustainable development. The positive or negative sign differs according to the dimension (GI and SD) taken into account. The Voice and Accountability is the dimension that negatively affects the majority (number of variables positively influenced) of the SD dimensions, referring to the lack of efficiency of African governments actions. The control of corruption positively affects most of the environmental Sustainability dimensions. In sum, the magnitude of the impact of each variable reflects the nature of the relationship and the quality of governance in Africa. In addition, the economic dimension of sustainability benefits the most from the governance dimensions and the Vulnerability Risk is the most affected (negatively).

Regarding the effect of governance, both negative and positive effects on SD dimensions are observed, depending on the selected indicator (Table 5). The Online Service Index (OSI) is the variable that positively affects all dimensions of SD in Africa, while the Telecommunication Infrastructure Index (TII) positively influences the majority of the SD dimensions. In contrast, the E-Participation Index (EPI) has lesser effects, and the Human Capital Index negatively impacts three of the SD dimensions. This highlights the outcomes of policies adopted in Africa that promote a digital transition. These policies should improve the quality of services provided, enhance the quality of life of its citizens and their well-being, and, to a large extent, spill over into sustainable development (Table 6).

These results confirm those of previous studies [5]. Indeed, the results show that e-government combined with governance can improve environmental, economic and social sustainability in African countries, provided that they are geared towards achieving sustainable socio-economic objectives rather than being an end in themselves. However, more policies are needed to harness the potential of countries and technologies for well-being [2].

Table 5. Governance and sustainable development

Variables	Economic sustainability	Social sustainability		Environmental sustainability	
	(7)	(8)	(9)	(10)	(11)
	LGDPC	MORT	HIN	EV	VUL
CC	0.10***	1.23	–0.11***	1.14***	0.75*
	(0.03)	(1.48)	(0.03)	(0.41)	(0.43)
EG	0.04	7.30***	–0.09**	–0.86*	–1.75***
	(0.03)	(1.66)	(0.04)	(0.46)	(0.48)
RQ	0.07***	–11.79***	0.12***	–0.84**	–1.64***
	(0.03)	(1.46)	(0.03)	(0.40)	(0.42)
VA	–0.05**	–0.09	0.04	–1.26***	–0.12
	(0.02)	(1.10)	(0.02)	(0.30)	(0.32)
PSAV	0.00	0.83	0.00	0.02	–2.20***
	(0.01)	(0.67)	(0.02)	(0.18)	(0.19)
RL	0.13***	–5.80***	0.01	0.23	–0.20
	(0.03)	(1.85)	(0.04)	(0.51)	(0.54)
LPOP	1.14***	–263.11***	4.05***	5.36***	–5.21***
	(0.06)	(3.26)	(0.07)	(0.89)	(0.95)
Constant	0.41	1,905.04***	–27.69***	–24.30***	85.58***
	(0.41)	(22.77)	(0.51)	(6.24)	(6.61)
Observations	1,071	1,071	1,071	1,071	1,071
R-squared	0.35	0.87	0.76	0.06	0.26
Number of countries	51	51	51	51	51

Standard errors in parentheses. *** $p < 0.01$, ** $p < 0.05$, * $p < 0.1$

Table 6. E-government and sustainable development

Variables	Economic sustainability	Social sustainability		Environmental sustainability	
	(–12)	(–13)	(–14)	(–15)	(–16)
	LGDPC	MORT	HIN	EV	VUL
TII	0.13***	16.47***	–0.21***	4.86***	0.96
	–0.05	–2.67	–0.06	–0.7	–0.88
HCI	0.23***	–1.89	–0.23***	0.31	–1.84**
	–0.05	–2.79	–0.06	–0.73	–0.91

(continued)

Table 6. (*continued*)

Variables	Economic sustainability	Social sustainability		Environmental sustainability	
	(–12)	(–13)	(–14)	(–15)	(–16)
	LGDPC	MORT	HIN	EV	VUL
OSI	0.1	11.11***	–0.33***	2.23**	2.46**
	–0.06	–3.61	–0.07	–0.95	–1.18
EPI	0.32***	–13.73***	0.47***	–2.85***	–4.15***
	–0.06	–3.16	–0.07	–0.83	–1.04
HCIND	–0.01	1.34	0.09***	0.91***	0.92***
	–0.02	–0.96	–0.02	–0.25	–0.31
LPOP	0.50***	–289.58***	3.45***	–5.56***	–7.72***
	–0.12	–6.43	–0.13	–1.68	–2.1
Constant	4.46***	2,092.76***	–23.39***	51.58***	107.17***
	–0.79	–44.14	–0.91	–11.56	–14.45
Observations	1071	1071	1071	1071	1071
R-squared	0.34	0.87	0.79	0.11	0.03
Number of countries	51	51	51	51	51

Standard errors in parentheses. *** $p < 0.01$, ** $p < 0.05$, * $p < 0.1$

5 Conclusion

The present paper contributes to enriching the analytical framework of e-government in the African countries and its impact on sustainable development. The analysis revealed an overall positive trend, with notable disparities between countries. Although digitalization policies have shown positive trends over the last two decades, there is a need to improve the effectiveness of this transformation for Africa [49]. In this regard, Digital Transformation Strategies [49], Economic Governance Strategy (EGS) (2021–2025) in Africa and others initiatives aim to promote inclusive governance and foster sustainable development.

A detailed analysis highlights persistent challenges, including weak digital infrastructure, limited connectivity, regulatory frameworks, and insufficient digital skills, make it unlikely that Africa will bridge the digital divide by 2030 [49]. These factors call for a more nuanced approach to defining and measuring governance [50]. In addition, corruption, often considered as the true adversary of digitalization, remains a significant barrier [51], as well as the weakness of regulatory structures, the unnecessary tax incentives that encourage illicit financial flows [52]. This calls for accelerated efforts and innovative solutions to address these gaps [12] and proactively contribute to DTS [53]. To achieve this leadership, African countries must adopt goals that are consistent with the UN's 2030 Agenda and the African UN's 2063 Agenda [54]. In this sense, the adoption of platforms and innovative financing offers sustainable solutions [55].

Furthermore, governance's path to sustainability is not unique [56]. Africa possesses significant potential, yet faces challenges related to resources (natural, human, financial, technological…), which distinguish it from other regions of the world. Countries are distinguished by their specificities, which influence the implementation of strategies and public policies [56]. Even though Africa generates only a tiny proportion (2% to 3%) of global gas emissions, it is including climate change actions in its national plans [57]. Certainly, Africa needs to be supported by the countries involved in climate change (developed polluters), in terms of technological tools to counter these effects [58].

Africa faces a trade-off: either focus on its potential (environmental, technological, institutional) and move forward with e-government and successful governance across all its dimensions to fully benefit from it in favor of the sustainable growth of its countries, or zoom in on the areas of deficit and try to strengthen them in order to drive the desired sustainable growth. The most important thing is to move forward at two speeds, taking into account the resources (strengths and weaknesses) available, with a more accelerated pace (opportunities) to close the sustainability and digitization gaps with other advanced countries.

In summary, this paper provides an analysis of the dynamics of E-gov, governance and sustainable development over the last two decades and how they interact for African countries. The prospects for future research could be expanded in two key directions. First, it could focus on intra-African regions, such as those within the framework of the African Continental Free Trade Area (AfCFTA), to explore into intra-regional disparities and dynamics. Second, incorporating spatial econometrics would allow for a better understanding of how proximity and regional autocorrelation influence governance and sustainable development outcomes. These extensions would offer a more nuanced and comprehensive perspective on the regional and spatial dimensions of e-government for sustainable development.

References

1. Nose, M.: Inclusive govtech: enhancing efficiency and equity through public service digitalization. Social Science Research Network, Rochester (2023)
2. Group, W.B.: World Development Report 2016: Digital Dividends. World Bank Publications (2016)
3. Basu, S.: E-government and developing countries: an overview. Int. Rev. Law Comput. Technol. (2004)
4. Estevez, E., Lopes, N., Janowski, T.: Smart sustainable cities: reconnaissance study. (2016) Accessed 19 Feb 2025
5. Lim, J.: e-Government for Sustainable Development in SIDS. Seoul U. N. Proj. Off. Gov. (2015)
6. Stanimirovic, D.: Conceptualization of an integrated indicator model for the evaluation of e-government policies. Electron. J. E-Gov. **11** (2013)
7. Vinod Kumar, T.M.: E-Governance for Smart Cities. In: E-Governance for Smart Cities, Vinod Kumar, T.M., Ed., Springer Singapore (2015)
8. Mishra, R.K., Kumari, C.L., Chachra, S., Krishna, P.S.J., Dubey, A., Singh, R.B.: Smart cities for sustainable development. In: Advances in Geographical and Environmental Sciences. Singapore: Springer Nature Singapore (2022)
9. Quels sont les défis de DITE en Afrique?. Nations Unies Commission économique pour l'Afrique. Accessed 19 Feb 2025. https://www.uneca.org/fr/dite-pour-l%E2%80%99afrique/quels-sont-les-d%C3%A9fis-de-dite-en-afrique

10. Ojo, A., Shareef, M., Janowski, T., Bardon, S.: La gouvernance électronique en asie: bilan, impact et réduction de la fracture interne. Hermès Rev. **55**(3), 159–167 (2009)
11. Glass, L.-M., Newig, J.: Governance for achieving the sustainable development goals: how important are participation, policy coherence, reflexivity, adaptation and democratic institutions? Earth Syst. Gov. **2**, 100031 (2019)
12. UN, E-Government Survey 2024. Accelerating Digital Transformation for Sustainable Development (2024)
13. Ullah, A., Pinglu, C., Ullah, S., Abbas, H.S.M., Khan, S.: The Role of e-governance in combating COVID-19 and promoting sustainable development: a comparative study of china and pakistan. Chin. Polit. Sci. Rev. **6** (2021)
14. Estevez, E., Janowski, T.: Electronic Governance for Sustainable Development — Conceptual framework and state of research. Gov. Inf. Q. **30** (2013)
15. Dhaoui, I.: E-Government for sustainable development: evidence from MENA countries. J. Knowl. Econ. **13** (2022)
16. Ansell, C., Torfing, J.: Handbook on theories of governance., Edward Elgar Publishing. (2022)
17. UN-ESCAP. “What is Good Governance?” (2009)
18. Lopez-Calva, L.F., Zhou, Y., Al-Dahdah, E., Bulman, D.J., Isser, D.H., Larizza, M., Sharma, S.: World development report 2017: governance and the law (2017)
19. OCDE. Cadre d’action en matière de bonne gouvernance publique: Éléments fondamentaux pour le bon fonctionnement des administrations publiques. OECD (2021)
20. Heeks, R.: Understanding e-governance for development. Soc. Sci. Res. Netw. (2001)
21. OECD. The e-Government Imperative. in OECD e-Government Studies (2003)
22. OECD. E-Government for Better Government. in OECD e-Government Studies (2005)
23. Van Zeijl-Rozema, A., Cörvers, R., Kemp, R., Martens, P.: Governance for sustainable development: a framework. Sustain. Dev. **16**(6), 410–421 (2008)
24. European Commission (2024). https://policy.trade.ec.europa.eu/developmentand-sustainability/sustainable-development_en
25. Sachs, J.D., Lafortune, G., Fuller, G.: The SDGs and the UN summit of the future. Sustainable development report 2024 (2024)
26. Report of the World Commission on Environment and Development: Our Common Future (WCED) (1987)
27. Benjamin, P.: Community development and democratisation through information technology: building the new South Africa. Reinventing government in the information age, 194–210 (1999)
28. Datar, M.: Determining priorities of e-Government: a model building approach. In: 5th International Conference on E-Governance, Hyderabad, India (2007)
29. Heeks, R.: Implementing and Managing eGovernment: An International Text. 1 Oliver’s Yard, 55 City Road, London EC1Y 1SP United Kingdom: SAGE Publications Ltd (2006)
30. Kanungo, S.: Information village: Bridging the digital divide in rural India. The digital challenge: Information technology in the development context, p. 103, 124 (2003)
31. Castro, C., Lopes, C.: Digital government and sustainable development. J. Knowl. Econ. **13**(2), 880–903 (2022)
32. Heeks, R.: E-government in Africa: promise and practice. Inf. Polity **7**(2–3), 97–114 (2003)
33. Adam, I.O.: Examining E-Government development effects on corruption in Africa: the mediating effects of ICT development and institutional quality. Technol. Soc., **61** (2020)
34. Schuppan, T.: E-government in developing countries: experiences from sub-saharan Africa. Gov. Inf. Q. **26**, 118–127 (2009)
35. Nations, U.: E-Government Survey 2024 Technical Appendix. Department of Economic and Social Affairs (2024)
36. Kaufmann, D., Kraay, A., Mastruzzi, M.: The worldwide governance indicators: methodology and analytical issues. Hague J. Rule Law **3** (2011)

37. Block, S., Emerson, J.W., Esty, D.C., de Sherbinin, A., Wendling, Z.A., et al.: 2024 Environmental Performance Index. New Haven, CT: Yale Center for Environmental Law & Policy (2024)
38. World Risk Report 2024. Bündnis Entwicklung Hilft, Ruhr University Bochum – Institute for International Law of Peace and Conflict (2024)
39. Stratégie Nationale de Développement du Numérique en Côte d'Ivoire 2021–2025. Ministère de l'Economie Numérique, des Télécommunications et de l'Innovation
40. Telecommunications and postal services, Republic of South Africa, Electronic Communications and Transactions Act: National e-Government Strategy and Roadmap (2017)
41. Digital Mauritius 2030. Ministry of Technology, Communication and Innovation (2018)
42. World Bank. Seychelles Digital Economy Assessment (2018)
43. Ministère des technologies de la communication, Plan National Stratégique Tunisie Digitale 2020 (2024)
44. National Strategy Digital Morocco 2030. Ministry of digital transition and administration reform (2025)
45. Srivastava, S.K., Panigrahi, P.K.: The impact of e-government and e-business on economic performance: a comparative study of developing and developed countries. J. Contemp. Issues Bus. Gov. **22** (2016)
46. Evans, D., Yen, D.C.: 'E-Government: evolving relationship of citizens and government, domestic, and international development. Gov. Inf. Q. **23**(2), 207–235 (2006)
47. Rodrik, D., Subramanian, A.: The primacy of institutions. Finance Dev. **40**(2), 31–34 (2003)
48. Baltagi, B.H.: Fixed Effects and Random Effects. In: Durlauf, S.N., Blume, L.E. (eds.) Microeconometrics, pp. 59–64. Palgrave Macmillan UK, London (2010)
49. Union Africaine. The Digital Transformation Strategy for Africa (2020–2030) (2020)
50. La mesure de la gouvernance. Centre de développement de l'OCDE, cahier de politique économique, N°39 (2010)
51. Goldmark, L., Denguezli, H., Martin, J.: What African countries can learn from the continent's top performers in e-government. APRI-Africa Policy Research Private Institute gUG (haftungsbeschränkt) (2024)
52. The Africa Governance Report: Promoting African Union Shared Values, African Union
53. Digital Government Africa, Africa's Digital Ministerial Summit (2024). https://africa-digital.com/2024/e-government-for-africa/
54. African Development Bank Group and African Development Bank, Eds., Driving Africa's transformation: the reform of the global financial architecture in African economic outlook (2024)
55. Strategie de la Gouvernance Economique en Afrique (SEGA) 2021–2025. Groupe de la Banque Africaine de Développement (2020)
56. Kemp, R., Parto, S., Gibson, R.B.: Governance for sustainable development: moving from theory to practice. Int. J. Sustain. Dev. (2005)
57. World Meteorological Organization. State of the Climate in Africa (2021)
58. Towards a climate resilient Africa. A wwf africa position paper for the African climate week and the African climate action summit (2023)

Developing Sustainable AI: Policies, Strategies, and Practical Applications

Zoubida Benmamoun(✉), Mahmoud Mistarihi, Raisah Al Ali, Abdulla Al Hammadi, and Eissa Al Hammadi

Faculty of Engineering and Computing, Liwa University, Abu Dhabi, UAE
zoubida.benmamoun@lc.ac.ae

Abstract. In recent times, artificial Intelligence has grown extensively to deliver numerous advantages in almost all industries. Such growth has raised several impacts on the environment challenges. The paper discusses how technological growth can be matched with environmental sustainability. This research combines data from secondary sources of articles and primary research based on a survey from AI professionals to understand environmentally friendly ways of developing AI tools. The findings indicated high level of carbon footprint and energy consumption involved in conventional models of AI system. Certain key strategies identified include developing algorithms that are energy-efficient, employing data centers powered by renewable energy, and full life cycle analysis of AI systems. The study further illustrates some practical use of sustainable AI in resource management and climate change modelling illustrating potential of AI tools. This paper contributed to the literature by covering significant knowledge gap in environmental sustainability of AI tools and provided practical recommendations for policy development and future research. These findings are important for policymakers, researchers, and industry leaders.

Keywords: Environmental impact · environmental responsibility · energy efficiency · sustainable AI · technological advancement

1 Introduction

In the current digital era, AI has become essential across various aspect like finance, medical, health, sports, business. Such development increases cost of natural environment by depleting resources [1]. There is need to input a lot of resources, energy, and time to train and customize big AI models. This training process has also contributed significantly towards carbon footprint and environmental pollution. Therefore, it is important to find sustainable solutions to make use of AI environmental-friendly and greener [2]. Since there is significant increase in the demand of AI tools thus it will lead to more use of energy, resources, and contribute to environmental pollution. Therefore, it has become urgent to look for alternative strategies to minimize environmental footprint of AI tools [3]. In recent climate threat, there has been great scrutiny on application of non-greener technologies throughout the world, particularly in the United Arab Emirates (UAE). Furthermore, it is corporate social responsibility to minimize its carbon footprint and make earth sustainable [4].

C. El Mokhi et al. (Eds.): ICASET 2025, SUCI, pp. 104–114, 2025.
https://doi.org/10.1007/978-3-031-98334-4_10

Even though there has been great stress on the AI tools and associated challenges, but limited research has been conducted to comprehensively look for sustainable options for tackling such challenges. Therefore, it is important to assess the carbon footprint and energy consumption driven by AI tools. Since most of AI models involve great amount of model simulations which are energy-consuming, computation-intensive, and expensive. Therefore, this paper is focused on finding the suitable alternative solutions to make AI systems energy efficient and environmentally friendly without reducing performance. First, the algorithm's optimization requires testing and construction by reducing computational demand and minimize the carbon footprint. Secondly, it is important to conduct a comprehensive life-cycle assessment of AI tools to comprehend carbon footprint creation in every stage of development. Third, engage with key stakeholders to evaluate efficiency and performance of AI tools in transparent way. Therefore, it is important to involve stakeholders like AI professionals and experts for collecting data through survey.

2 Literature Review

2.1 Energy-Efficient Models for Greener Models

The theory of "Green AI" has recently obtained great concerns from researchers, with focus on the growth of environmentally friendly and energy-efficient AI systems. Recently, research has provided an important insight into the optimization of models and efficient algorithms [5]. Various hardware alternatives like FPGAs and TPU have been evaluated for energy efficiency and provided strategies for energy-efficient operations and design at greater level [6]. Therefore, it is important to understand various case studies of Green AI approaches for computer vision applications and NLP. Such applications can deliver a better balance for energy-efficient and performance-based AI technologies with the objective of sustainability and low carbon footprint [7]. It is essential for decreasing the environmental pollution induced by AI approaches and ensure the progress of technology without compromising natural environment [8].

2.2 Green AI: Comprehensive Environmental Considerations

Significant improvements in technology have come from the quick development of artificial intelligence, but there are also worries about how these technologies may affect the environment. The environmental impact of Artificial intelligence in no longer minor as the use of artificial intelligence-based technologies increases [9]. As a result, the discipline of "Green AI", which tries to create and use AI technology in a way that is beneficial to the environment, has grown up [10]. The understanding that AI effects on the environment extend beyond energy use is one of the basic principles of green AI [11]. AI may affect the environment in several ways, such as through the consumption of the resources the production of wastes, and greenhouse gas emissions [12]. As a result, the study of the green AI has to consider these different and the complete AI ecosystem [13].

A number of ways that artificial intelligence can support environmental sustainability have been identified by researchers. For example, decrease resource waste, increase energy efficiency, and optimize production processes, which lowers carbon emissions and makes the industrial sector more sustainable [14]. AI may also be used to address environmental issues including climate change pollution reduction, animal preservation, and sustainable city planning [15]. There are several difficulties in combining AI with environmental sustainability. AI possible reverse effects where efficiency improvements might result in higher resource use and damage to the environment [16]. The creation and implementation of AI systems itself may have a substantial negative influence on the environment, as evidenced by the high in energy training of substantial machine learning models [17].

Researchers and industry professionals have proposed a number of approaches to creating and using AI in a more sustainable way in order to overcome these obstacles. These includes the use of cloud computing and green data centers. The employment of effective hardware and algorithms and the incorporation of environmental factors into architecture and creation of AI systems [18]. An increasing focus is also being placed on the necessity of responsible and ethical AI practices, which may help guarantee that the advantages of AI be realized in a manner consistent with social and environmental principles [19]. Combining AI with environmental sustainability is a difficult task with many facts. A comprehensive strategy that takes into account the whole AI ecosystem and its effects on the environment is necessary for establishing a balance between environmental responsibility and technical developments. Scientists and developers can work towards creating and implementing AI technologies that promote innovation and advancements while also protecting that planet's future by adopting the principles of Green AI.

2.3 Assessment of AI Models and Systems

For sustainability perspective, it is important to conduct a life cycle assessment of AI models. Such assessment includes the examination of negative impacts associated with use of AI technologies during development, simulation, and deployment till disposal [20]. However, such life cycle assessment should also focus on social justice and ecological integrity associated with AI tools and models [21]. Such approach can equip us with great technological advancements with socially responsible and sustainable options. Furthermore, life cycle assessment can optimize the waste production and resource consumption during every phase of AI development that can also contribute towards sustainability. Furthermore, this can also deliver a transparent perspective of the environmental effects associated with AI tools. Such reports then can be shared with key stakeholders to have complete accountability and motivates companies to go for greener options. Without life cycle assessment, it would be hard to have transparency, accountability, energy efficiency, and reduction of carbon footprint associated with AI models in the world [22].

2.4 Sustainability and Technological Innovation

Intersection ground between sustainability and technological innovations is an area very essential during research. Various studies reveal while some technological advancements may drive greater gains; they equally come at serious environmental costs. Huang et al. discussed several issues and associated research needs pertinent in balancing technological innovation with sustainable engineering [23]. The study calls for the need for sustainability in both the development and application process of new technologies, so that these technologies contribute to reducing, rather than exacerbating, environmental problems. It thus provides a platform where detailed economic, environmental, and social analyses relevant to technological innovation, including nanomaterials and biofuels, will highlight both positive and negative implications on human health and the environment. The environmental impact of AI is also reduced with sustainable data centers running on renewable energy and featuring advanced cooling technologies. Carrying out an LCA for AI tools can establish environmental hotspots and further improvement during the entire lifecycle, and hence, ensure that advancement in AI will not challenge environmental health. In all, balancing progress with responsibility will require ethical sourcing of data and long-term ecological stewardship-reducing electronic waste and recycling AI hardware. By incorporating sustainability into the development and deployment of AI, we can harness technological innovation to create a greener, more sustainable future. This approach not only helps immediately address environmental concerns but also ensures that AI continues to evolve responsibly, contributing to the long-term health of our planet [24, 25]. The sustainable AI literature indicates that solutions should be balanced, guaranteeing.

3 Methodology

The objective of this questionnaire is to outline the current state of practice on sustainable AI, challenges, and perceived knowledge about it [26, 27]. The data will be collected by surveying targeted groups involving AI researchers, sustainability experts, developers, and professionals from industry. The questionnaire will be sent via e-mails, through LinkedIn, and professional network contacts, besides distribution over the relevant online forums. These are target participants for the research identified to have both diverse and representative samples, covering different geographical regions and sectors. The questionnaire used during the survey consisted of open-ended to close-ended questions. The close-ended question helped quantify the data while the open-ended question gave qualitative insight. The questions were designed to cover areas of current practices in AI development and their environmental impact, awareness and implementation of sustainable AI practices, challenges faced while integrating sustainability into AI projects, and perception of the importance of sustainability in AI. This survey was administered online with the help of some survey platform like SurveyMonkey or Google Forms. This study has invited responses through emails and professional networking sites. The survey will be open to respond to for four weeks. Analyses of data include both quantitative and qualitative. Quantitative responses, through the closed-ended questions, are analyzed using statistical software in search of trends and patterns. Descriptive statistics in mean, median, and standard deviation show summaries of data. A thematic analysis is conducted regarding qualitative responses from open-ended questions to draw common themes and insights. Use of the survey method in this research was based on its potential to collect plenty of information from a myriad of subjects within a quite short period of time. It allowed for the collection of quantitative and qualitative data that would give a holistic view of what is going on in sustainable AI practices today. Further, the online distribution ascertained the convenience and accessibility of the survey, which may ensure a better response rate. This methodology was followed to gain valuable insights about practices in place, various challenges, and perceptions related to sustainable AI. Results have helped formulate strategies and guidelines to support the sustainability of AI.

4 Case Study: Artificial Intelligence-Enhanced Energy Efficiency in Building Management

Making energy efficient is one of the key focuses of international efforts for sustainable development and artificial intelligence (AI) aids in that cause. Stressing on HVAC (heating, ventilation, and air conditioning) systems which contributes major share to building energy consumption part offer detailed case study of artificial intelligence application in improving energy management of urban structures.

4.1 Context and Application

Manhattan's tall structural office skyscraper kicked off an elaborate plan to integrate artificial intelligence with Heating Ventilation and Air Conditioning systems. It collected variables such as temperature, humidity, occupancy schedule, indoor air quality, and external weather, amongst others, all in real-time through a network of IoT sensors and highly developed machine learning techniques. Thanks to such detailed data which the sensors provided, artificial intelligence adjusted heating and cooling systems for comfort, and minimized energy waste.

Furthermore, the used of the artificial intelligence system was helpful in reduction of energy utilization for power demand during peak hours as well as in off-peak hours by using the analysis of historical data. For example, if the AI predicted that the temperature would be lower the following day, it cooled the facility in the early hours of the morning just before demand was high and costs were higher. This level of control was perhaps made possible by gradual model improvements and continuous feedback to one.

Further, the use of artificial intelligence system helped to offer sensible recommendation to the management of the building. A study of seasonal variations established concrete possibilities of energy conservation, of better insulation during the winter. The system discovered that some of the mechanisms failed to meet their expected productivity levels, which made it possible to repair them promptly without additional costs. Through integration of Predictive maintenance into its processes resulted into a reduction in HVAC related downtime, hence increasing overall efficiency.

4.2 Discoveries and Results

The execution where the product was created yielded wonderful outcomes. In twelve months, the energy consumption of the HVAC systems was reduced by 15.8%. In addition to a few thousand dollars being saved, it also greatly reduced the amounts of greenhouse gases being released into the atmosphere. With improved HVAC running anomalies, decreased downtime, and increased equipment lifetime, the AI system improved Predictive Maintenance (Fig. 1).

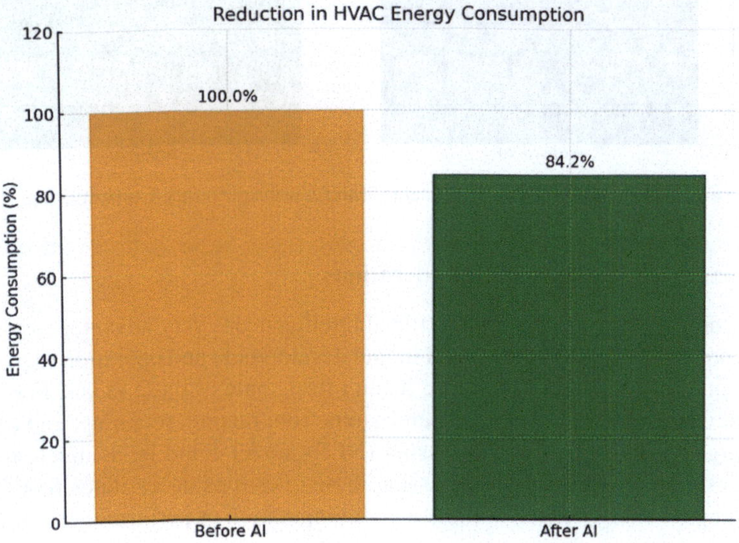

Fig. 1. Reduction in HVAC Energy Consumption

For example, DeepMind engaged data centers of the corporation to optimize cooling systems. Google saved 40% of the energy that had previously been used for cooling through a type of artificial intelligence that can predict the cooling needs based on server workloads and adapt the cooling to. These achievements, therefore, demonstrate how well AI can be scaled in many various disciplines.

All the above-mentioned methods apart from reducing running expenses meant attaining more general sustainability goals. Organizations employing AI solutions engage in international environmental standards at times, including the Paris Agreement's emissions cut targets. These carbon reduction targets and their measures enhance the business Sustainability accounting and reporting, as it increases credibility and readability (Fig. 2).

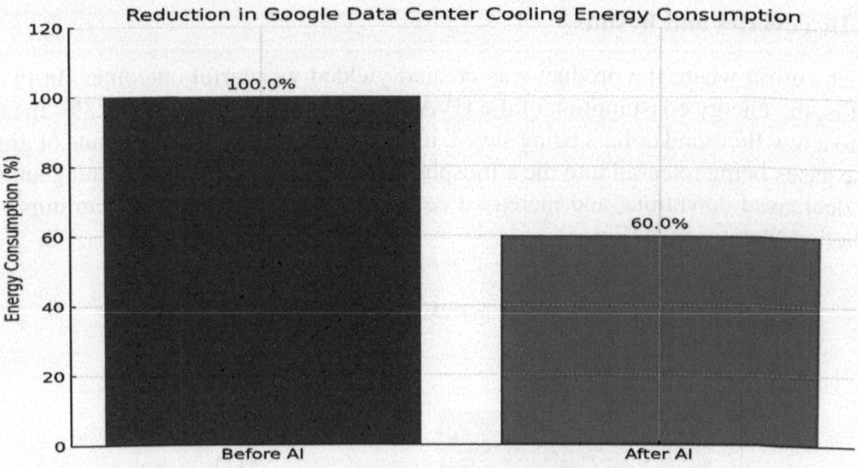

Fig. 2. Reduction in Google Data Center Cooling Energy Consumption

4.3 Influential Elements and Considerations

Although energy efficiency through artificial intelligence is very advantageous, it is disadvantageous in the following ways. AI model development and operation requires complex computational resources therefore it can offset some of these ideals. For instance, a 2019 study suggests that the carbon emissions from raising, retaining, and deploying a single large natural language processing (NLP) model could be as much as the five car lifetimes carbon emissions. Addressing these questions necessitates focused effort in developing AI energy-efficient models for computation or performing computational tasks in green energy.

The first and perhaps the most significant challenge to the AI system is often because of its high initial deployment costs. Fixed costs closely related to the first stages of AI infrastructure and IoT devices acquisition may eventually become problematic for SMEs. Governments and financial institutions must step in with subsidies or low interest loans, or tax incentives that will go a long way in addressing this problem.

The integration of artificial intelligence in energy management systems shows how promising it is to improve sustainability and emphasizes the need of using sensible solutions that minimize its environmental effect. The following suggestions are meant to help to strike this harmony:

Economical artificial intelligence underscores achieving efficient, cost-effective innovative algorithms and algorithms that are not built at the cost of their efficacy. The call for energy efficiency can be addressed through knowledge distillation, quantization, and model pruning both in deployment and in training. These methods save the truly revolutionary potential of Artificial Intelligence and correlate with environmental aims. Also, researchers should study more on federated learning systems wherein the computation is performed into partitions across the separate devices thus offloading the burden on the central server.

AI businesses and data centers want to be powered by green energy like solar and wind energy. Other players in the energy industry need to follow the footsteps of the giants such as Microsoft and Amazon, which aim for 100% dependency on renewable energy within the next one-decade, showing it isn't a hopeless case for major companies. The governments can put conditions that before anybody with high energy consumption can carry on, they must use renewable energy, so the fight against this effect can be hastened.

GPUs and TPUs and other related hardware that are used in the artificial intelligence systems have to be made to work for energy efficiency. That is why, in energy-intensive facilities including data centers, heat recovery solutions as well as liquid cooling would have an impact on business sustainability. To reduce electrical waste hardware makers might research recycling plans or even research biodegradable material to use in the production of their devices.

Managing the sustainability issues provided by artificial intelligence requires collaboration between the government, business and academia. Organizations, such as Green Software Foundation and Partnership on AI are important for having spaces to build up a set of guidelines and to inspire more comprehensive ideas. It is possible for pilot sustainability solutions powered by AI to be piloted at a mass scale with the support of public-private collaborations that will highlight the opportunities of returns on investment through corporate funding.

Evaluation of the environmental effects of artificial intelligence systems depends on standardized measures. These indicators should assess general efficiency, carbon emissions, and energy consumption, thereby providing useful information for enhancing AI's environmental projects. Policymakers have to interact with business leaders to create clear guidelines that direct artificial intelligence growth and implementation in ecologically friendly manner.

Workshop and training courses should stress the need of sustainable artificial intelligence approaches. Fostering a culture of environmental stewardship depends on developers, legislators, and end users all learning more. Universities may add sustainability components into computer science and artificial intelligence courses, so arming next generations of professionals to give environmental issues top priority in their work.

Many argue that the development and implementation of artificial intelligence is positive for the achievement of environmental goals as policymakers need to design and offer incentives. For example, this might entail providing people with bonuses for paying less taxes if they are environmentally friendly, investing in artificial intelligence that is friendly to the environment and penalizing people who use lots of energy. Unfortunately, AI cannot be employed for the benefit of all people if the models that are being developed are not designed with society in mind.

To be long term useful artificial intelligence research have to invent new designs and methods that would require less power consumption. New technologies such as neuromorphic computing, modelled on how well the human brain computes, is one of those things that have some very interesting possibilities for mitigating some of the harm AI does to the world. Research to integrate models of artificial intelligence that leverage the concepts of both quantum and classical computing can open Continued details about scalability and performance that has never been experienced before.

Perhaps the rules and the environment of the AI systems must be similarly speculated to be made to change with the growing advancement. Feedback and dynamic change systems ensure it is possible to guarantee that the change always operates optimally and sustainably with regards to the natural environment. AI models could be trained with such conditions and updated based on these parameters in order to adapt to the emerging conditions of weather changes and other energy requirements.

Through innovations that go beyond simple energy efficiency, able to transform sectors and address world problems, artificial intelligence improves sustainability. Realizing its full potential calls for a coordinated, responsible, progressive plan. Following the above advised suggestions will help stakeholders to maximize the transforming power of artificial intelligence by stressing environmental preservation in its use and development.

5 Conclusion

The fast advancement of artificial intelligence in recent years has changed a wide range of industries, including banking and health. These changes still have an effect on the environment, and give artificial intelligence quick development, energy use and carbon emissions need to be controlled. So, the acceptance of artificial intelligence methods that support establishing a balance between advancement in technology and environmental responsibilities is the main focus of this study.

The negative impact of artificial intelligence on the environment can be reduced by working on energy-efficient algorithms, hardware such as GPUs, TPUs, FPGAs, and renewable energy sources. Artificial intelligence technologies can be assured to be in effective connection with sustainability objectives through lifecycle evaluation and ethical deployment practices. Government cooperation and assistance are also required to advance sustainable Artificial Intelligence Technologies. The development of artificial intelligence will need to overcome certain obstacles, such as a lack of awareness and skill gaps, in order to succeed. Every obstacle at the same time provides a chance for more study in the future to improve artificial intelligence effectiveness and environmental sustainability. Encouraging sustainable Ai not only cares for the environment but also sets the course for responsible technological advancement in the future, improving every member of the next generation.it is practical to accomplish the goals of preserving the environment and technology progress for a better future for all by working together.

References

1. Adewale, et al.: Application of artificial intelligence (AI) in sustainable building lifecycle; a systematic literature review (2024) https://doi.org/10.20944/preprints202405. 2113.v1
2. Benmamoun, Zoubida, Fethallah, W., Raouf, Y.: New green lean assessment framework for enhancing enterprise performance. Benchmarking Int. J. (2025)
3. Saeri, O'Connor: Applying AI to Sustainability Policy Challenges: A Practical Playbook (2023) https://doi.org/10.31219/osf.io/y75rq
4. Umar: Artificial Intelligence in Agricultural Extension for Sustainable Development (2023) https://doi.org/10.59890/ijasr.v1i3.740
5. Advances in Sustainable Irrigation using AI International research journal of modernization in engineering technology and science (2024) https://doi.org/10.56726/irjmets50346

6. Regona, et al.: Artificial intelligent technologies for the construction industry: how are they perceived and utilized in Australia? J. Open Innovation Technol. Market Complexity (2022) https://doi.org/10.3390/joitmc8010016
7. Nuryana, et al.: Exploring the frontiers of school environmental health and artificial intelligence for sustainability. In: E3s web of Conferences (2024) https://doi.org/10.1051/e3sconf/202450101014
8. Yiğitcanlar, et al.: Green artificial intelligence: towards an efficient, sustainable and equitable technology for smart cities and futures. Sustainability (2021) https://doi.org/10.3390/su13168952
9. Cossu, et al.: Sustainable Artificial Intelligence through Continual Learning (2021) https://doi.org/10.48550/arxiv.2111.09437
10. Bolte, et al.: From an ethics of carefulness to an ethics of desirability: going beyond current ethics approaches to sustainable AI. Sustainability (2022) https://doi.org/10.3390/su14084472
11. Integrating Artificial Intelligence for Enhanced Grid Stability and Renewable Energy Management in France. Int. J. Multidis. Res. (2024) https://doi.org/10.36948/ijfmr.2024.v06i03.22279
12. Naeeni: Sustainability and AI: prioritizing environmental considerations in tech advancements (2023) https://doi.org/10.61838/kman.aitech.1.3.1
13. Adanma, Ogunbiyi: Artificial intelligence in environmental conservation: evaluating cyber risks and opportunities for sustainable practices. Comput. Sci. Res. J. (2024) https://doi.org/10.51594/csitrj.v5i5.1156
14. Oyewole, et al.: Promoting sustainability in finance with AI: a review of current practices and future potential. World J. Adv. Res. Rev. (2024) https://doi.org/10.30574/wjarr.2024.21.3.0691
15. Khlie, K., et al.: Generative AI for enhanced operations and supply chain management. J. Infrastruct. Policy Dev. **8**(10), 6637 (2024)
16. Kazeem, et al.: Roles of artificial intelligence and machine learning in enhancing construction processes and sustainable communities. Buildings (2023) https://doi.org/10.3390/buildings13082061
17. Wynsberghe: Sustainable AI: AI for sustainability and the sustainability of AI. AI and ethics (2021) https://doi.org/10.1007/s43681-021-00043-6
18. Kindylidi, Cabral: sustainability of AI: the case of provision of information to consumers. Sustainability (2021) https://doi.org/10.3390/su132112064
19. Jun, et al.: Can artificial intelligence improve the energy efficiency of manufacturing companies? evidence from China. Int. J. Environ. Res. Public Health (2022) https://doi.org/10.3390/ijerph19042091
20. Bachina: Sustainable materials for artificial intelligence (AI) technology adoption for energy-efficient patient-centric healthcare solutions. J. Educ. Health Promot. (2025) https://doi.org/10.4103/jehp.jehp_527_24
21. Sætra: AI in context and the sustainable development goals: factoring in the unsustainability of the sociotechnical system. Sustainability (2021) https://doi.org/10.3390/su13041738
22. Jankovic, Curovic: Strategic Integration of artificial intelligence for sustainable businesses: implications for data management and human user engagement in the digital era. Sustainability (2023) https://doi.org/10.3390/su152115208
23. Xiao: Neuroscience-inspired continuous learning: a sustainable approach to AI energy challenge (2023) https://doi.org/10.31219/osf.io/twn9q
24. Tabbakh, A., Al Amin, L., Islam, M., Mahmud, G.I., Chowdhury, I.K., Mukta, M.S.H.: Towards sustainable AI: a comprehensive framework for Green AI. Discover Sustain. **5**(1), 408 (2024)

25. Van Wynsberghe, A.: Sustainable AI: AI for sustainability and the sustainability of AI. AI and Ethics **1**(3), 213–218 (2021)
26. El Mokhi, C., Addaim, A.: Near-Optimal Cable Layout Design of a Wind Farm Using Genetic Algorithm. In: Peng, S.-L., Hao, R.-X., Pal, S. (eds.) Proceedings of First International Conference on Mathematical Modeling and Computational Science. AISC, vol. 1292, pp. 47–56. Springer, Singapore (2021). https://doi.org/10.1007/978-981-33-4389-4_5
27. Satri, J., El Mokhi, C., Hachimi, H.: Artificial intelligence and machine learning for a better decision making in the public sector. In: Presented at the 8th International Conference on Optimization and Applications, ICOA 2022 - Proceedings (2022)

Assessing the Cooling Impact of Urban Blue Infrastructure on Land Surface Temperature Using Local Climatic Zones from 2014 to 2024 in Varanasi, India

Avinash Yadav, Vishal Chettry(✉), and Amit Maurya

Department of Architecture, Planning and Design, IIT(BHU) Varanasi, Varanasi, India
vishal.apd@iitbhu.ac.in

Abstract. The cooling effects of green infrastructure compared to blue infrastructure in urban areas are well explored in the existing literature. However, the effect of urban blue infrastructure on the surface temperature of urban areas is less explored. Therefore, this study aims to investigate the extent of cooling effects provided by water bodies on the nearby built-up areas. The five largest water bodies in Varanasi, India, were selected to assess the cooling influence on Local Climate Zones (LCZs) within a 200-m radius around the water bodies. The Land Surface Temperature (LST) maps were prepared using geoinformatics. Buffer zone analysis was used to calculate cooling factors, including cooling indices and LCZ indicators. The results exhibited that the average LST of the water bodies and surrounding LCZs increased significantly from 2014 to 2024, with a peak of 43.53 °C in June 2024 and a low of 29.78 °C in April 2014. Built-feature LCZs (3 and 6) exhibited higher and concentrated LST values, while LCZ 9 showed scattered patterns. The LST variation curve near water bodies revealed an initial increase before stabilizing, indicating their cooling effect. These findings will assist policymakers in preparing climate-sensitive urban planning for water bodies.

Keywords: Water bodies · Local Climatic Zones · Land Surface Temperature · Cooling effect · Varanasi · India

1 Introduction

Urban blue infrastructure, including lakes, ponds, and other water bodies, helps mitigate surface temperatures, resulting in significant cooling effects [1]. Water bodies in urban areas effectively maintain a cool environment similar to the urban green infrastructure due to the emission of longwave radiation and consumption of shortwave radiation through evaporation [2]. Urban green infrastructure and its impact on urban areas are well-explored in the literature. However, despite having a similar cooling mechanism, urban blue infrastructure has received less attention. Hence, analyzing the cooling impact of urban blue infrastructure on adjacent built environment is crucial.

Geoinformatics has been extensively utilized to examine urban environments and assess heat island effects [3]. The cooling impact of water bodies is assessed using

C. El Mokhi et al. (Eds.): ICASET 2025, SUCI, pp. 115–123, 2025.
https://doi.org/10.1007/978-3-031-98334-4_11

Waterbody Cooling Distance (WCD), i.e., horizontal cooling distance in meters and Waterbody Cooling Range (WCR), i.e., cooling effect's intensity in degrees Celsius (°C) [4]. Additionally, the Waterbody Cooling Gradient (WCG) is used to evaluate the effectiveness of cooling. Most research on cooling effects primarily examines the impact on surrounding areas. [4]. However, urban composite areas consist of diverse underlying surfaces with distinct morphological features [5]. Therefore, utilizing LCZ theory proposed by Stewart and Oke (2012) may be helpful in understanding the cooling effect on urban composite areas [6].

Gupta et al. (2019), offered a theoretical foundation for studying temperature variations in the surrounding land use cover by analyzing the cooling effect of water bodies on LST [1]. Moreover, studies have examined the influence of waterbody morphological parameters on cooling effects, but their impact on the surrounding environment remains underexplored. Although water bodies play a crucial role in urban cooling, studies on their impact on the adjacent built environment are limited. Therefore, applying LCZ theory significantly improves the understanding of how urban composite areas influence cooling across different LCZs surrounding a water body, as illustrated in Fig. 1 [6].

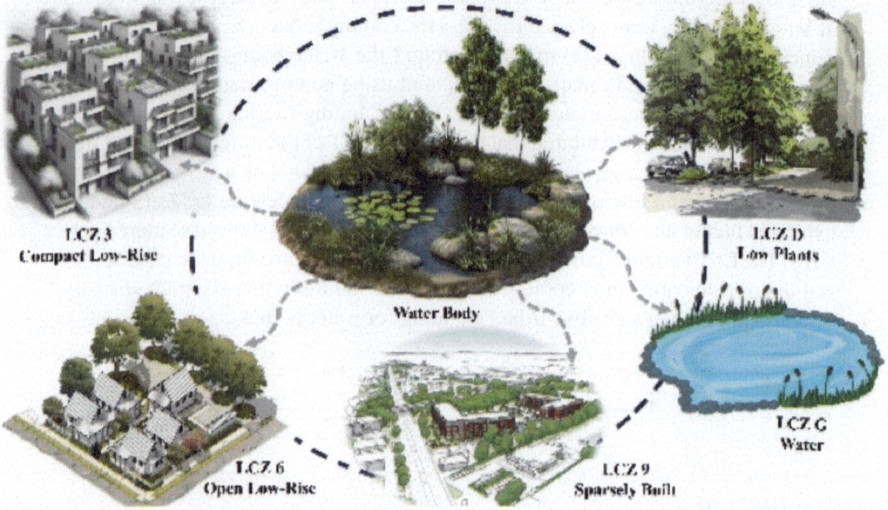

Fig. 1. Selected distinct LCZs around the water body

In this context, this study examined the five largest water bodies located in the composite climate region of Varanasi City, India. Their cooling effects were analyzed using LST maps from the summer months of 2014 and 2024. Further, the influence of water bodies on regulating the thermal environment in nearby composite areas was analyzed. Such studies would provide valuable insights for urban planners in enhancing thermal comfort and emphasizing the importance of water body conservation.

2 Study Area

Varanasi, one of the oldest cities in India, is located in the Gangetic plains of eastern Uttar Pradesh. Based on the administrative boundaries of Varanasi city, 5 largest water bodies were selected for the research study (Fig. 2). These water bodies named Sarang Talab, Pisach Mochan Tal, Chancha Tal, Lahartara Talaab, and Pandepur Pokhra, have surface areas of 16,761.42 m^2, 25,620.56 m^2, 45,908.84 m^2, 20,804.32 m^2, and 25,208.97 m^2, respectively.

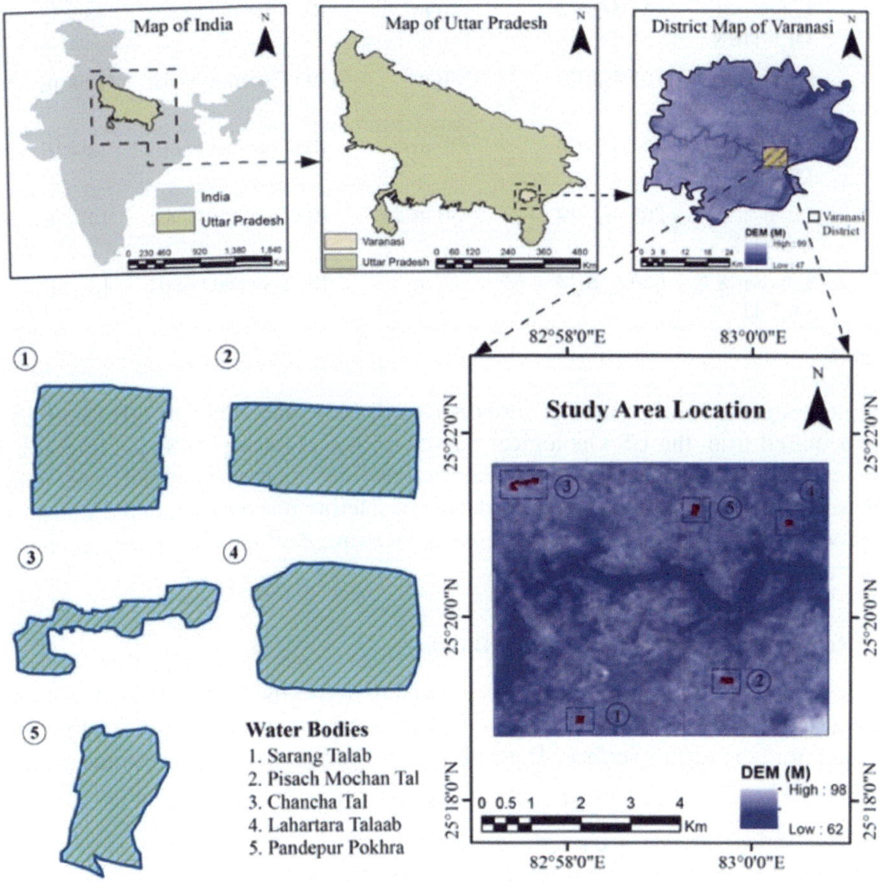

Fig. 2. Location map of the water bodies selected for the study

3 Materials and Methods

Table. 1. Landsat datasets used in this study

S. No	Satellite / Sensors	Acquisition Date	Pixel size	Band used	Path/ Row
1	Landsat-8 / OLI-TIRS	29/04/2024	30 m	10 (Thermal band)	149/36
2	Landsat-8 / OLI-TIRS	15/05/2024	30 m	10 (Thermal band)	149/36
3	Landsat-8 / OLI-TIRS	16/06/2024	30 m	10 (Thermal band)	149/36
4	Landsat-8 / OLI-TIRS	02/04/2014	30 m	10 (Thermal band)	149/36
5	Landsat-8 / OLI-TIRS	20/05/2014	30 m	10 (Thermal band)	149/36
6	Landsat-8 / OLI-TIRS	05/06/2014	30 m	10 (Thermal band)	149/36

Satellite images from the summer months (April, May, and June) of 2014 and 2024 were acquired from the US Geological Survey (USGS) to analyze the cooling effects of water bodies on the surrounding LCZs. The satellite images with cloud cover below 10% were utilized. Table 1 exhibits the details of satellite images used in the study. The pre-processing of satellite images was conducted using ArcMap 10.7.1, i.e., geometric corrections and study area clipping.

3.1 Retrieval of LST from Thermal Bands

The thermal band of satellite image was utilized to derive the LST (Eq. 1). The digital number (DN) of the thermal band was converted into brightness temperature (BT) (Eq. 2) reflected from the earth's surface. The emissivity(e) corrected LST was estimated using a formula [1].

$$\mathbf{LST} = \frac{BT}{1 + \left(\lambda \times \frac{BT}{1.4388}\right)} \times \ln(e) \quad (1)$$

where, $\boldsymbol{\lambda} = 10.8\boldsymbol{\mu m}$ is the wavelength of emitted radiance.

$$\mathbf{BT} = \frac{k_2}{\ln\left(\frac{k_1}{TOA}\right) + 1} - 273.15 \quad (2)$$

where $K_2 = 1321.0789$ and $K_1 = 774.8853$ are the band-specific thermal conversion constant.

Top of the atmosphere (TOA) radiance $= 0.0003342 \times$ thermal band $+ 0.1–0.29$.

3.2 Cooling Effect Indices

For analyzing the cooling effects of water bodies on their surroundings, WCD and WCR have been utilized [7]. While these two cooling indices indicate the impact of water bodies on LST, they do not precisely measure cooling efficiency. Therefore, to quantify the effectiveness of water bodies in cooling, this study introduces the Waterbody Cooling Gradient (WCG) indices, which rely on WCD and WCR.

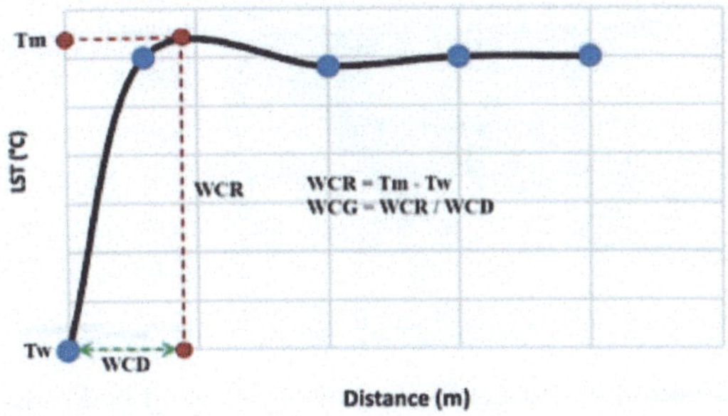

Fig. 3. Representation of cooling effect indices [7]

Figure 3 illustrates the cooling effect indices, depicting the average change in LST between water bodies and their surrounding LCZs. The initial LST of the water bodies, denoted as Tw, acts as the baseline for the curve. There is an increase in LST until it peaks at Tm, which denotes a shift to a stable phase. The distance from the water body's starting point to this phase is termed WCD, while the difference between Tm and Tw is defined as WCR. Additionally, WCG, calculated as the ratio of WCR to WCD, represents the rate at which water bodies cool the surrounding LCZs.

4 Results and Discussion

4.1 LCZ Around the Water Bodies

This study used field surveys and Google Earth imagery to delineate various LCZs at a micro-scale within a 200 m radius around the selected five water bodies. Figure 4 shows the LCZ division around the water bodies. LCZ was classified through a field survey, including measurements of building heights and land cover characteristics. Five LCZ types were identified around water bodies wherein three LCZs, i.e., LCZ 3, LCZ 6, and LCZ 9 included built-up while LCZ D and LCZ G included natural land covers. Similar LCZ patterns were observed for all the five waterbodies selected for the study.

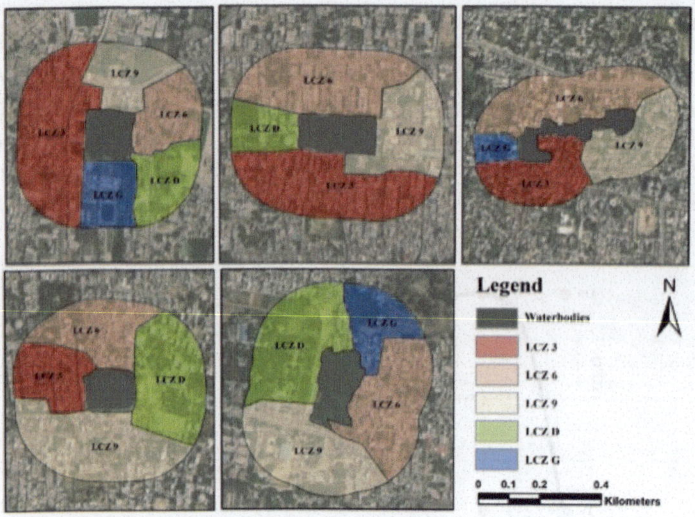

Fig. 4. Distribution of distinct LCZs around the water bodies in this study

4.2 Spatial Distribution of LST on LCZs

The LST distribution across LCZs surrounding water bodies was analyzed using 20 m equidistant buffer rings in ArcGIS software. The built feature LCZs consistently exhibited higher LST values compared to land-cover LCZs.

Figure 5 illustrates the LST distributions for all the five waterbodies and its surrounding areas. The distribution of LST across various LCZs surrounding water bodies was notably non-uniform pattern. Water bodies consistently exhibited lower average LST values compared to surrounding LCZs. The range of LST values increased from 29.05 °C to 41.94 °C during 2014 to 34.59 °C to 43.53 °C in 2024. Comparatively, the built feature LCZs, particularly in LCZ 3 and LCZ 6 showed higher LST values with dense distributions while LCZ 9 displayed scattered LST patterns in 2014 and 2024. Natural land cover LCZs, such as LCZ D and LCZ G, had lower LST values in the summer months of 2014 compared to the elevated LST patterns observed in 2024. Figure 5 highlights LST variations across April, May, and June for both 2014 and 2024, using a color gradient from green (cooler) to red (warmer) around water bodies. In 2014, April LSTs ranged from 29.05 °C to 34.40 °C, May from 34.04 °C to 40.08 °C, and June from 35.75 °C to 41.94 °C. By 2024, April LSTs rose to 34.45 °C–38.43 °C, May to 35.16 °C–40.68 °C, and June to 36.12 °C–43.53 °C. The highest LSTs and concentrated heat patterns were observed in June, both around water bodies and in adjacent LCZs. A noticeable rise in LST values between 2014 and 2024 signifies significant thermal intensification across all LCZs. Additionally, it is worth noting that the same water body i.e., WB 1 exhibited higher LST values in 2014 compared to 2024, influenced by the spatial arrangement of surrounding LCZs. The built feature LCZs (LCZ 3 and LCZ 6) became denser and more compact over the study timeline. This trend reflects an exacerbation of seasonal heating patterns and the Urban Heat Island (UHI) effect, primarily driven by increased dense and compact built features. However, microclimatic factors of water bodies including evaporation rates, local wind patterns, and humidity levels can diminish the overall cooling effect in surrounding areas with stable land cover. Additionally, water bodies are shrinking due to encroachment, and declining groundwater levels reduce evaporation and moisture availability, thereby weakening the natural cooling effect on surrounding LCZs [2].

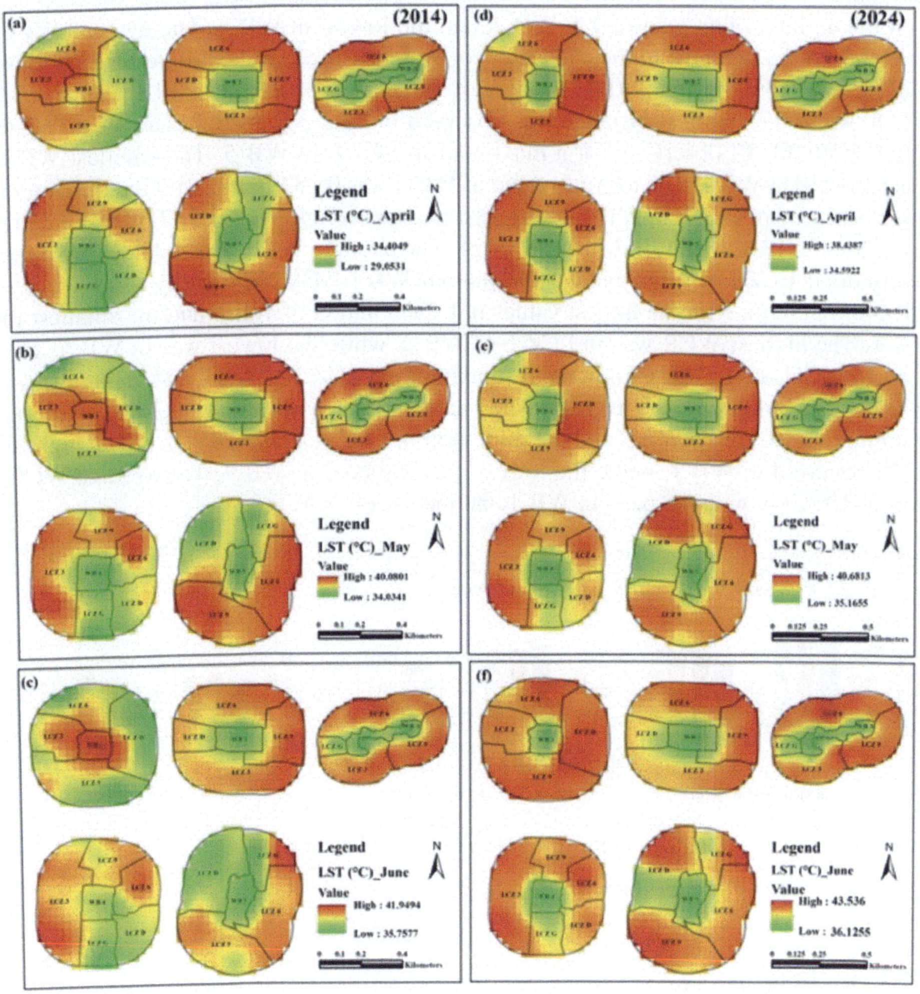

Fig. 5. LST maps of summer months for the decadal years 2014 to 2024

4.3 Pattern of Cooling Indices

The cooling effect indices, including WCD, WCR, and WCG, were computed to quantify the cooling impact of water bodies on surrounding LCZs. Figure 6 shows the summer monthly variations of cooling indices over a considered period. In April 2014, the average WCD, WCR, and WCG values were 60 m, 5 °C, and 0.08 °C/m, respectively. The highest WCD was detected in LCZ 6 of WB 4 whereas the lowest occurred in LCZ 3 of WB 1. The maximum WCR was recorded in LCZ 3 around WB 3, and the smallest (1.0 °C) in WB 1. LCZ 3 in WB 3 exhibited the highest WCG, while WB 1 had the lowest. In May 2014, the largest WCD values were observed in LCZ 6 and G around WB 4, while the smallest WCD was found in LCZ 3 of WB 2. The average WCR value was 6.05 °C, with the highest value recorded in WB 2, and the lowest in WB 4. The average WCG was 0.086 °C/m, with the highest in LCZ 3 of WB 1, while WB 4 exhibited the smallest cooling gradient. For June 2014, the largest WCD was observed in all LCZs around WB 5, while the smallest distance was observed in LCZ 3 in WB 1. The mean WCR value was 6.30 °C, observing for a maximum in LCZ 6 in WB 5. The smallest WCR was recorded at WB 1. The average value of WCG was 0.087 °C/m, with the maximum value observed in LCZ 3 in WB 1. Figure 6 shows an average WCD of 57.32 m, peaking at 68.26 m in LCZ 3 of WB 3 in April 2024. The average WCR was 4.25 °C, with the maximum in LCZ 3 of WB 4 and the minimum in WB 1. May 2024, indicates an average WCD of 63.52 m, with the largest values in LCZ 3 and G of WB 3 and the smallest in WB 1. The highest WCR was in LCZ 6 of WB 2, while the lowest was in WB 5. The average WCG was 0.079 °C/m, with the maximum in LCZ 3 of WB 1 and the minimum in WB 3. In June 2024, the average WCD across water bodies was 66.24 m, with the maximum in LCZ 3 of WB 2 and the minimum in LCZ 3 and D of WB 1. The highest WCR occurred in WB 1, while the lowest was observed in WB 3. The average WCG was 0.010 °C/m, with the peak in WB 1 and the lowest in WB 3.

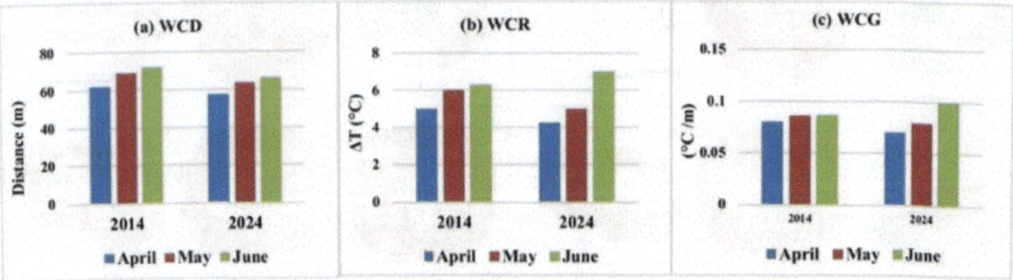

Fig. 6. Summer monthly variations of coding indices

5 Conclusion

This case study investigates the cooling effect of the five largest water bodies located in a composite area of Varanasi City, India during the summer months of 2014 and 2024. The cooling effect indices were used to assess surface temperature changes during the summer months over the study timeline. The study reveals a steady increase in the average LST of the water bodies and surrounding LCZs over the study timeline. The highest LST, 43.53 °C, was recorded in June 2024, while the lowest, 29.78 °C, was observed in April 2014. A marked rise in LST values between 2014 and 2024 reflects significant thermal intensification across all LCZs. The LCZs of built features, such as LCZ 3 and LCZ 6, exhibited higher LST values wherein it initially increased and later got stable, indicating the moderating influence of cooling around the water bodies. Monthly averages of WCD in different LCZs decreased along with a decline in WCR and WCG during study timelines. A limitation of the study is the absence of air temperature data for a detailed comparison of LST changes in the surrounding LCZs. The cooling impact of water bodies depends on microclimatic factors like evaporation rates, local wind patterns, and humidity levels. However, a comprehensive analysis of these dynamics requires air temperature measurements. Despite this limitation, the study provides valuable insights into the cooling effects of urban blue infrastructure.

References

1. Gupta, N., Mathew, A., Khandelwal, S.: Analysis of cooling effect of water bodies on land surface temperature in nearby region: a case study of ahmedabad and chandigarh cities in India. Egypt. J. Remote Sens. Sp. Sci. **22**(1), 81–93 (2019). https://doi.org/10.1016/j.ejrs.2018.03.007
2. Pritipadmaja, R.D. Garg, Sharma, A.K.: Assessing the cooling effect of blue-green spaces: implications for urban heat island mitigation. Water (Switzerland) **15**(16) (2023). https://doi.org/10.3390/w15162983
3. Kumar, A., Chettry, V.: Land cover change and urban heat island trends in varanasi, india from 1991–2021 using geoinformatics. SUPTM 2024 2nd Conf. Futur. Challenges Sustain. Urban Plan. Territ. Manag. Proc. 4–7 (2024)
4. Rahaman, S., Jahangir, S., Haque, M.S., Chen, R., Kumar, P.: Spatio-temporal changes of green spaces and their impact on urban environment of Mumbai, India. Environ. Dev. Sustain. (2020). https://doi.org/10.1007/s10668-020-00882-z
5. Mumtaz, F., et al.: Impacts of Green Fraction Changes on Surface Temperature and Carbon Emissions: Comparison under Forestation and Urbanization Reshaping Scenarios. Remote Sens. **15**(3) (2023). https://doi.org/10.3390/rs15030859
6. Kotharkar, R., Dongarsane, P.: Investigating outdoor thermal comfort variations across Local Climate Zones in Nagpur, India, using ENVI-met. Build. Environ. **249** (2024). https://doi.org/10.1016/j.buildenv.2023.111122
7. Ghosh, S., Das, A.: Modelling urban cooling island impact of green space and water bodies on surface urban heat island in a continuously developing urban area. Model. Earth Syst. Environ. **4**(2), 501–515 (2018). https://doi.org/10.1007/s40808-018-0456-7

Modeling the Relationships Between Green Supply Chain Practices and the Financial and Environmental Performance of the Company

Abdessamad Diani(✉) and Laila Aligod

ERMOT Laboratory, Faculty of Legal, Economic and Social Sciences, Sidi Mohamed Ben Abdellah University, Fez, Morocco
{abdessamad.diani,laila.aligod}@usmba.ac.ma

Abstract. Faced with growing environmental pressures and sustainability requirements, companies are adopting responsible green supply chain practices such as eco-design, reverse logistics, and green purchasing. This study aims to measure the impact of these practices on supply chain performance, as well as on the company's financial and environmental performance. To answer the question: "How do green supply chain (GSC) practices affect supply chain performance (SCP), along with a company's financial and environmental performance?", we adopted a two-phase explanatory sequential methodology. First, we conducted a quantitative study using structural equation modeling (SEM). Data were collected via a questionnaire distributed to Moroccan companies at the Logismed trade fair and processed using IBM SPSS AMOS 26 software. We then enriched the initial findings with a case study of Equatorial Coca-Cola Bottling. Semi-structured interviews with logistics managers confirm that green sourcing, eco-design, and reverse logistics initiatives have a positive impact on supply chain performance (SCP), with significant financial and environmental benefits.

Keywords: Green supply chain · Supply chain performance · Financial performance · Environmental performance · Green procurement · Eco-design · Reverse logistics

1 Introduction

Fast evolution of supply chain management (SCM) research with the capacity for addressing the complexities of current challenges. One of the biggest challenges at this moment we have is the influence of climate change to the business performance. Its far-reaching effects demand attention from all sectors, leading to the emergence of a new approach: green supply chain management (GSCM). This concept encourages the creation of value while respecting the environment, from the selection of suppliers to the collection of used products by customers. The application of these practices places the company face to face with two approaches: the reactive approach and the proactive

C. El Mokhi et al. (Eds.): ICASET 2025, SUCI, pp. 124–134, 2025.
https://doi.org/10.1007/978-3-031-98334-4_12

approach. GSCM focuses on proactively addressing environmental issues by implementing preventive measures starting at the product and process design stages. This consists of eco-friendly sourcing, green manufacturing, sustainable design principles, ecolattice packaging, and its distribution as well reverse logistics [1].

This paper seeks to answer the research question: "How do green supply chain (GSC) practices affect supply chain performance (SCP), along with a company's financial and environmental performance?" The article is organized into three primary sections. The first section, dedicated to the literature review, defines the core concepts of our research and discusses the theoretical contributions regarding the relationships between the variables in our conceptual model. The second section outlines the research methodology used in this study, detailed exposition about methods employed for data feature extraction, mining and processing. Finally, the third section presents a thorough analysis of the results, comparing them with the theoretical contributions discussed in the earlier sections, to effectively position our research within the context of existing studies.

2 Literature Review

2.1 Green Supply Chain Management

The supply chain (SC) encompasses all stakeholders and all stages in a company's procurement process, from suppliers to end consumers. It can be assimilated to a model of activities organized around a network of companies aimed at delivering a product or service to the customer under optimum conditions of cost, lead time and quality [2]. SCM involves identifying, coordinating, and rationalizing physical, human, monetary and, above all, informational flows to ensure smooth operation within the SC. The aim is to ensure efficient production and distribution of goods, minimizing costs while maximizing value for the end consumer [3]. The green supply chain is a two-way SC:

- **Direct logistics** includes eco-design, green purchasing (supplier certification, purchasing of eco-friendly materials/products), green manufacturing, and green distribution.
- **Reverse logistics** manages return flows through the "4Rs" (reduction, reuse, recycling, recovery), thus forming a closed loop [4].

A Green Supply Chain Management (GSCM) is planning, executing and control logistics system using green technologies and environmentally sustainable management principles. It aims to reduce and optimize upstream resource consumption, minimize waste and mitigate environmental pollution during the execution of logistics activities [4]. Throughout the entire supply chain process from raw material extraction to disposal or recycling, the supply chain considers the environment every step of the way, in order to efficiently manage its environmental impact [5]. GSCM aims to ensure the environmental sustainability of the SC network, from suppliers upstream to customers downstream. By integrating environmental considerations into strategic planning, operational practices and policymaking, organizations can minimize their environmental footprint and contribute to sustainable development efforts. The figure below illustrates the various activities of the GSC (Fig. 1):

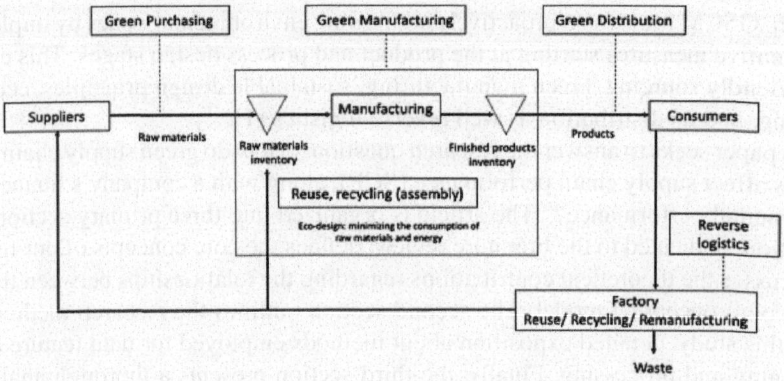

Fig. 1. The activities of a GSC (created by the authors, inspired by [1])

2.2 The Impact of Green Supply Chain Practices on Business Company Performance

Green Purchasing (GP) and Supply Chain Performance. GP consists of making eco-conscious decisions at every stage of the procurement process, from product and process design to product disposal [6]. To ensure ethical material sourcing, the company must choose responsible suppliers who are dedicated to implementing sustainable practices in sourcing raw materials or in the development of eco-friendly products and components [7]. Suppliers who share the same objectives and commitments to respect for the environment make for strong partnerships. Such collaborations help lay out the roadmap to sustainable products and practices, which ultimately enable eco-friendly buying aims to be delivered [8]. Green purchasing encourages adherence to stringent environmental regulations and helps alleviate pressure from stakeholders committed to environmental responsibility. The integration of green purchasing into SCM simultaneously achieves sustainability objectives, is environmentally friendly and is a powerful lever for improving a company's economic performance and operational efficiency.

H1: Green purchasing has a positive impact on SCP.

Eco-design (ED) and Supply Chain Performance. ED as defined by ISO 14062, involves incorporating environmental factors into the design and development of products to enhance their environmental performance across their entire life cycle. This involves designing products with low material and energy consumption, using sustainable materials, and ensuring that products can be reused, remanufactured, and recycled after use to preserve their value and maximize their lifespan [9]. Diabat et al. [10] demonstrate that integrating eco-design principles significantly enhances the efficiency and sustainability of green supply chains.

H2: Eco-design has a positive impact on SCP.

Reverse Logistics (RL) and Supply Chain Performance. Reverse Logistics (RL) refers to the movement of materials in the opposite direction of the conventional SC, traveling upstream in the value chain [11]. Reverse logistics practices include several steps among them: product returns, remanufacturing, recovery, reuse and recycling of any waste resulting from consumption or production [12]. Govindan et al. [11] Stress the strategic importance of reverse logistics for operations management and its effect on profit improvement and the success of organization.

SCP and Financial Performance (FP). In SCM sector, financial performance is a key factor in judging overall system efficiency and a company's competitiveness. Various indicators, such as profitability, cost reduction and operational efficiency, are used to measure this performance. SCM goes beyond simply reducing costs or maximizing production; it also includes strategic aspects such as cooperation between players in the chain, informational openness and well-coordinated processes [10]. Effective information sharing, transparency and coordinated/interlinked participation in the value chain by all stakeholders is the only way a company can develop and grow its relationship with the stakeholder, all of which contribute to improving the company's financial performance [13]. Carter et al. [14] noted that financial performance improved because of greater concentration and integration within the supply chain. By adopting a strategic and proactive approach to SCM, companies can turn these practices into an important competitive asset, capable of generating financial benefits in a dynamic business environment [15].

H4: SCP has a positive impact on financial performance.

SCP and Environmental Performance. Logistics efficiency, often measured by indicators that focus on the financial dimension such as cost reduction, improved delivery times and inventory optimization, helps to mitigate negative environmental impacts by minimizing waste and reducing energy consumption [16]. The adoption of proactive practices helps strengthen adherence to stringent environmental regulations, fulfill the increasing eco-conscious demands of consumers and other interested parties, and enhance the company's brand image in the global marketplace [8]. Moving to the adoption of eco-innovations in supply chain management such as green transportation, ethical sourcing and sustainable technologies is increasingly important for an organization to have maximum environmentally friendly performance [17].

H5: SCP has a positive impact on the company's environmental performance.

This literature review has led to the following conceptual model (Fig. 2):

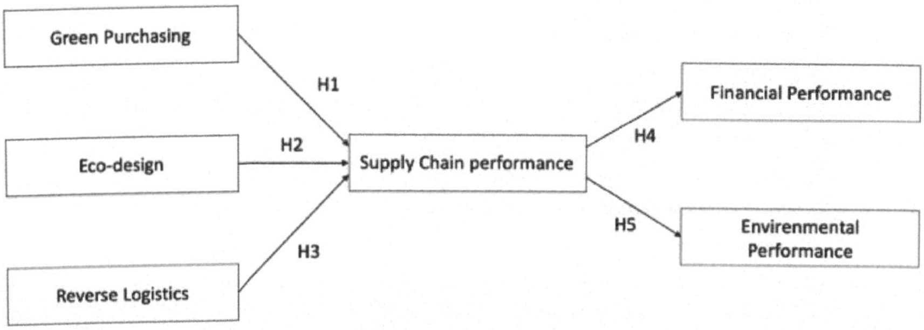

Fig. 2. The conceptual research models

3 Research Methodology

This research project adopts a sequential explanatory design in two phases. This mixed-methods approach involves collecting data through a questionnaire survey during the first phase (quantitative study), followed by data analysis and then conducting interviews (qualitative study) to clarify, elaborate, and explain the initial results obtained from the survey [18].

A quantitative study using SEM was carried out to evaluate and analyze the direct and indirect effects of GSC practices on a company's logistics, environmental, and financial performance. Data collection was conducted using a questionnaire created on Google Forms. The five hypotheses of our theoretical model link six latent variables, with the identification of measurement items based on rigorous research in Scopus- and Web of Science-indexed journals (see table below) (Table 1):

Table 1. Items code, description, source of question

Items code	Description	Source of question
GP1	We evaluate the environmental practices of secondary suppliers	[19]
GP2	Environmental design requirements with our suppliers for items they are going to deliver we shared	
GP3	We collaborate with suppliers to achieve environmental goals	
GP4	We consider suppliers' ISO 14001 certification	
GP5	We conduct environmental audits on suppliers' internal management	
RL1	We gather used items from customers for recycling	[20]
RL2	We offer after-sales repair and maintenance services	[21]
RL3	We reclaim value by remanufacturing and responsibly disposing of returned products	
RL4	We send products back to suppliers for recycling	[21]
RL5	We work alongside operations and reverse logistics teams to develop reverse logistics programs	
ED1	Our products, intended to reduce the material and energy use	[22]
ED2	Our products can emphasize material and component re-use, re-cycling or recovery	
ED3	Our products are crafted to minimize or completely eliminate the reliance on hazardous materials	[19]
ED4	We design processes to minimize waste	
ED5	We design products with life cycle assessment in mind	
SCP1	Our costing is more precise	[23]
SCP2	We have enhanced coordination between departments	
SCP3	We have improved coordination with suppliers	
SCP4	We have improved coordination with consumers	
SCP5	We experienced an increase in sales	
EP1	Minimized emissions of hazardous substances or waste?	[24]
EP2	Minimized direct or indirect consumption of materials?	
EP3	Has the consumption of hazardous materials been minimized?	

(continued)

Table 1. (*continued*)

Items code	Description	Source of question
EP4	Has the overall environmental situation improved?	
EP5	Is there evidence of better adherence to environmental regulations and standards?	
FP1	Have the market share and growth rate shown an increase?	
FP2	Has the profit margin improved?	
FP3	Has sales growth increased?	
FP4	Has there been a reduction in production costs or cost per unit?	
FP5	Has there been an increase in customer satisfaction?	

The questionnaire was administered to a sample of around forty Moroccan companies, conveniently obtained during the 11th edition of the Logismed Exhibition, which took place from May 7 to 9 at the Hyatt Regency Casablanca. The Logismed Exhibition is one of the major events dedicated to logistics in Morocco. It brings together exhibitors, visitors, and experts in the field to discuss the current and future challenges of urban logistics. The processing and analysis of the collected data were carried out using IBM SPSS AMOS 26 software. Following the quantitative study, this study used a qualitative approach of semi-structured interviews with logistics managers at Equatorial Coca-Cola bottling logistics managers in order. An in-depth examination of the relationships between variables in our conceptual model, with a subtle context following the scatter-shot interview survey, contributed further to findings.

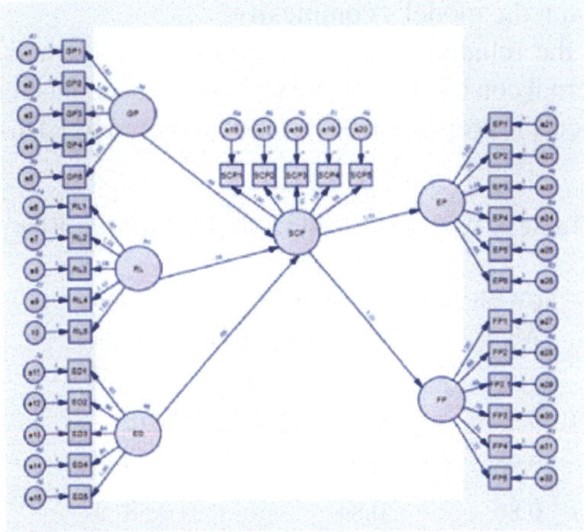

Fig. 3. The conceptual research model (Generated by Amos SPSS Software)

4 Results and Discussion

Significant insights into the CPMS influence on company financial and environmental consequences can be found in the SEM analysis. Below are the outcomes derived from data processing using the AMOS software (Fig. 3).

To assess the model's robustness, we present its fit indices (Table 2) along with the correlation matrix of the study's key variables (Table 4). The fit indices of the SEM model, detailed in Table 2, help determine the relevance and strength of the proposed relationships.

Table 2. Model fit indices

Index	Value	Acceptable threshold
Goodness-of-Fit Index (GFI)	0.92	> 0.90
Comparative Fit Index (CFI)	0.95	> 0.95
Tucker-Lewis Index (TLI)	0.93	> 0.95
Root Mean Square Error of Approximation (RMSEA)	0.05	< 0.06

The GFI of 0.92 surpasses the 0.90 threshold, indicating a strong overall model fit. The CFI stands at 0.95, reflecting a good fit compared to an independent model. Although the TLI is slightly below the recommended 0.95 threshold at 0.93, it remains within an acceptable range. Additionally, the RMSEA of 0.05, being below the 0.06 threshold, signifies a good fit given the model's complexity.

Table 2 assesses the reliability and validity of the constructs. Cronbach's Alpha (>0.70) confirms internal consistency, while CR > 0.70 accounts for item contributions. The AVE > 0.50 reflects the proportion of variance captured by each construct, and factor

Table 3. Statistical analysis of construct precision

Constructs	Cronbach's Alpha	Composite Reliability (CR)	Average Variance Extracted (AVE)	Factor Loadings
Green Purchasing (GP)	0.94	0.93	0.78	0.91 – 0.90
Eco-Design (ED)	0.96	0.95	0.85	0.92 – 0.94
Reverse Logistics (RL)	0.86	0.84	0.58	0.69 – 0.87
Supply Chain Performance (SCP)	0.89	0.88	0.72	0.80 – 0.93
Environmental Performance (EP)	0.95	0.94	0.76	0.79 – 0.92
Financial Performance (FP)	0.88	0.87	0.71	0.82 – 0.90

loadings (>0.70) demonstrate item relevance. The results indicate that most constructs are reliable; however, certain items related to reverse logistics (RL) exhibit weaker correlations. While still within acceptable limits, RL may require further refinement to enhance the model's validity and reliability. Table 3 test accuracy of constructs on reliability and validity.

Table 4. The correlation matrix between constructs

Constructs	GP	ED	RL	SCP	FP	EP
GP	1.000	0.849*	0.731*	0.535*	-	0.732*
ED	0.849*	1.000	0.739*	0.544*	-	0.657*
RL	0.731*	0.739*	1.000	0.743*	-	0.722*
SCP	0.535*	0.544*	0.743*	1.000	-	0.665*
FP	-	-	-	-	1.000	-
EP	0.732*	0.657*	0.722*	0.665*	-	1.000

The correlations between GP and ED (0.849*) and between RL and SCP (0.743*) are among the highest, indicating strong relationships between these constructs.GP, ED, and RL all show positive and significant correlations with EP, suggesting that these responsible practices strongly contribute to the company's environmental performance. The connections between these practices and SCP are moderate yet significant, highlighting the overall influence of GSC practices on SCP (Table 5).

Table 5. Path analysis

Path	Assumption	Path coefficient	Estimate	P-value	Validated
GP → SCP	H1	0.39	0.39	< 0.001	Yes
ED → SCP	H2	0.38	0.38	< 0.001	Yes
RL → SCP	H3	0.83	0.83	< 0.001	Yes
SCP → FP	H4	1.33	1.33	< 0.001	Yes
SCP → EP	H5	1.20	1.20	< 0.001	Yes

The path coefficient indicates its strength and direction on a relationship between two variables in the model. For example, a coefficient of 0.39 between Green Purchasing (GP) and SCP a one-unit rise in green purchasing practices results in a 0.39-unit enhancement in SCP. The p-value determines the robustness of the associations between variables, where a threshold of 0.05 signifies their importance. In this study, all p-values are below 0.001, demonstrating that every connection in the model is highly significant.

As for the first hypothesis, the results display a coefficient of 0.39 and the p. value turned out to be less than 0.001 for green purchasing with SCP so it is a very likely positive relationship. On the other HYPOTHESIS with regard to eco-design (ED) as a driver of SCP, the results for standard coefficient = 0.38, and p-value < 001 suggest eco-design positively have a significant relationship with SCP. Hypothesis 3-shows a coefficient of reverse logistics standardized = 0.83 and p-value < 0.001, indicating that reverse logistics is critical in SCP improvement Hypothesis 4: analyzes the relationship between SCP and financial performance with a standardized coefficient of 1.33 and p < 0.001 showing that is SCP not only has a strong correlation but also is a positive impact on financial performance. Finally,Hypothesis 5 look at the SCP and environmental performance the standardized coeff of 1.20 with p. < 0.001 indicating that the advances in SCP with green practices result into higher environmental performance.

The study concludes, based on these findings, that adopting environmentally friendly practices not only boosts SCP but also leads to improved financial and environmental outcomes for the company.

Our qualitative study focused on the case of the Equatorial Coca-Cola Bottling Company (ECCBC). The study aims to explore in depth SC practices, sustainability initiatives and their impact on SCP, as well as on the company's environmental and financial performance. The SC begins with the acquisition of essential raw materials, such as water, sugar, flavor concentrates and packaging materials. These raw materials are then sent to factories where they are transformed into finished products. For distribution, everything is carefully planned to optimize routes and reduce delivery times, ensuring that beverages arrive at retailers, wholesalers, and distributors in perfect condition. Finally, the SC also considers the management of returns of faulty or unsold products, as well as packaging recycling initiatives, with the aim of limiting environmental impact and supporting sustainable practices.

The study indicates that Implementation of GSCM will have a impact with the performance of Equatorial Coca-Cola Bottling. In terms of sustainable purchasing, Equatorial Coca-Cola Bottling selects suppliers who meet strict environmental criteria for the acquisition of recyclable and sustainable materials, so eco-design, a key element of the company's product development strategy, aims to reduce the environmental footprint right from the design phase. Reverse logistics initiatives encompass managing product returns and the reuse and recycling of materials. From an operational perspective, these initiatives have contributed to the optimization of material and energy flows, a reduction in operating costs, the strengthening of relationships with network partners, and the fulfillment of stakeholder expectations. The results reveal the GSC practices not only improve logistics efficiency but also contribute to a better financial performance and environmental sustainability of a company.

5 Conclusion

The study highlights the value of incorporating preventative environmental measures to bolster supply chain performance, focusing on strategies such as responsible sourcing, eco-design and reverse logistics. These approaches are fundamental to reducing costs, optimizing processes, ensuring compliance with environmental regulations and improving overall company performance by increasing operational efficiency. Companies need to incorporate green supply chain management (GSCM) into their core strategy in order to remain competitive in a market where sustainable development is a top priority.

This research also opens further investigations, especially toward improving GSCM performance through the extension of advanced technologies (i.e. blockchain, Internet of Things (IoT), and artificial intelligence). These technologies are enabling the emergence of innovative methods for managing resources, and are helping to strengthen the sustainability of supply chains. In this way, it's clear that implementing GSCM methods isn't just about regulatory compliance; this global strategy aims to optimize financial and environmental performance in an international context increasingly focused on ecological impact.

References

1. Jermsittiparsert, K., Namdej, P., Somjai, S.: Green supply chain practices and sustainable performance: moderating role of total quality management practices in electronic industry of Thailand. Int. J. Supply Chain Manag. **8**, 33–46 (2019)
2. Mentzer, J.T., et al.: Defining supply chain management. J. Bus. Logist. **22**, 1–25 (2001). https://doi.org/10.1002/j.2158-1592.2001.tb00001.x
3. Srivastava, S.K.: Green supply-chain management: a state-of-the-art literature review. Int. J. Manag. Rev. **9**, 53–80 (2007). https://doi.org/10.1111/j.1468-2370.2007.00202.x
4. Hervani, A.A., Helms, M.M., Sarkis, J.: Performance measurement for green supply chain management. Benchmarking Int. J. **12**, 330–353 (2005)
5. Rauer, J., Kaufmann, L.: Mitigating external barriers to implementing green supply chain management: a grounded theory investigation of green-tech companies' rare earth metals supply chains. J. Supply Chain Manag. **51**, 65–88 (2015). https://doi.org/10.1111/jscm.12063
6. Yook, K.H., Choi, J.H., Suresh, N.C.: Linking green purchasing capabilities to environmental and economic performance: the moderating role of firm size. J. Purch. Supply Manag. **24**, 326–337 (2018)
7. Seuring, S., Müller, M.: From a literature review to a conceptual framework for sustainable supply chain management. J. Clean. Prod. **16**, 1699–1710 (2008)
8. Carter, C.R., Rogers, D.S.: A framework of sustainable supply chain management: moving toward new theory. Int. J. Phys. Distrib. Logist. Manag. **38**, 360–387 (2008)
9. Zhu, Q., Sarkis, J., Lai, K.: Green supply chain management: pressures, practices and performance within the Chinese automobile industry. J. Clean. Prod. **15**, 1041–1052 (2007)
10. Diabat, A., Govindan, K.: An analysis of the drivers affecting the implementation of green supply chain management. Resour. Conserv. Recycl. **55**, 659–667 (2011)
11. Govindan, K., Soleimani, H., Kannan, D.: Reverse logistics and closed-loop supply chain: a comprehensive review to explore the future. Eur. J. Oper. Res. **240**, 603–626 (2015)
12. Jayaraman, V., Luo, Y.: Creating competitive advantages through new value creation: a reverse logistics perspective. Acad. Manag. Perspect. **21**, 56–73 (2007). https://doi.org/10.5465/amp.2007.25356512
13. Hsu, C.-C., Kannan, V.R., Tan, K.-C., Keong Leong, G.: Information sharing, buyer-supplier relationships, and firm performance: a multi-region analysis. Int. J. Phys. Distrib. Logist. Manag. **38**, 296–310 (2008)
14. Carter, C.R., Ellram, L.M.: Thirty-rve years of the journal of supply chain management: where have we been and where are we going? J. Supply Chain Manag. **39**, 27–39 (2003). https://doi.org/10.1111/j.1745-493X.2003.tb00152.x
15. Olšanová, K., Cook, G., Král, P., Vilikus, O., Zlatić, M.: Exploration of opportunities for development of CEE domestic luxury brands: case of the Czech republic. Cent. Eur. Bus. Rev. **9**, 59 (2020)

16. Stentoft Arlbjørn, J., Lüthje, T.: Global operations and their interaction with supply chain performance. Ind. Manag. Data Syst. **112**, 1044–1064 (2012)
17. Ahi, P., Searcy, C.: A comparative literature analysis of definitions for green and sustainable supply chain management. J. Clean. Prod. **52**, 329–341 (2013)
18. Creswell, J.W., Creswell, J.D.: Research design: qualitative, quantitative, and mixed methods approaches. Sage publications (2017)
19. Zhu, Q., Sarkis, J., Lai, K.: Confirmation of a measurement model for green supply chain management practices implementation. Int. J. Prod. Econ. **111**, 261–273 (2008)
20. Zailani, S., Govindan, K., Iranmanesh, M., Shaharudin, M.R., Chong, Y.S.: Green innovation adoption in automotive supply chain: the Malaysian case. J. Clean. Prod. **108**, 1115–1122 (2015)
21. Carter, C.R., Ellram, L.M.: Reverse logistics: a review of the literature and framework for future investigation. J. Bus. Logist. **19** (1998)
22. Matos, S., Hall, J.: Integrating sustainable development in the supply chain: the case of life cycle assessment in oil and gas and agricultural biotechnology. J. Oper. Manag. **25**, 1083–1102 (2007)
23. Lenny Koh, S.C., Demirbag, M., Bayraktar, E., Tatoglu, E., Zaim, S.: The impact of supply chain management practices on performance of SMEs. Ind. Manag. Data Syst. **107**, 103–124 (2007)
24. Zailani, S., Jeyaraman, K., Vengadasan, G., Premkumar, R.: Sustainable supply chain management (SSCM) in Malaysia: a survey. Int. J. Prod. Econ. **140**, 330–340 (2012)

Adaptive Multi-Stage Transfer Learning Approach for Electronic Component Detection in Waste Printed Circuit Boards

M. Mohsin[1(✉)], S. Rovetta[1,2], F. Masulli[1,2], A. Cabri[2,3], and M. G. Abbas Malik[4]

[1] DIBRIS, University of Genoa, 16146 Genoa, Italy
muhammad.mohsin@edu.unige.it
[2] Vega Research Laboratories s.r.l., 16121 Genoa, Italy
[3] Department of Computer Science, University of Milan, 20133 Milan, Italy
[4] CCIS, Prince Sultan University, Riyadh, Saudi Arabia

Abstract. The fast growing trend of electronic waste (e-waste), especially from printed circuit boards (PCBs), has created significant challenges for the efficient recovering of critical raw materials (CRMs). The recovery of valuable materials from these components is crucial for sustaining a circular economy. This paper proposes a adaptive multi-stage transfer learning strategy to improve the automatic detection and localization of selective electronic components on WPCBs containing different distributions of CRMs. We propose an adaptive multi-stage transfer learning strategy using YOLOv11 model that not only improves the detection accuracy, but also addresses the challenges of class imbalance in the WPCBs dataset. The proposed approach improves the detection accuracy by focusing selective electronic components use for CRMs recovery in production processes. This automatic detection and localization system reduces the labor costs, minimizes waste, and supports the principles of sustainability and resource efficiency. Thus, this work presents an effective solution for automated e-waste recycling, contributing to both environmental sustainability and economic growth.

Keywords: Computer vision · Critical raw materials · Waste printed circuit boards · Electronic components · Transfer learning

1 Introduction

Waste from electrical and electronic equipment (WEEE) is growing exponentially, while the rate of recycling and recovery is very low, particularly for waste printed circuit boards (WPCBs). One of the major reason is that conventional manual disassembly processes are tedious and inefficient, creating a need for automated systems that can quickly and accurately identify components for disassembly and sorting [1]. Improving critical raw materials recovery is important for the sustainable production of electronic equipment [2]. The detection and

C. El Mokhi et al. (Eds.): ICASET 2025, SUCI, pp. 135–142, 2025.
https://doi.org/10.1007/978-3-031-98334-4_13

classification of electronic components present on the surface of WPCBs is challenging due to the complex structure and high variation in size and shape. Many new studies have been carried out for the development of automated systems that can efficiently detect and classify these components [3]. The aim of this study is to optimize the detection and localization precision of selective electronic components using a YOLOv11-based object detection model.

This study proposes adaptive multi-stage transfer learning strategy with in domain specific knowledge of WPCBs to improve the detection and localization of electronic component accuracy. To achieve this goal, we use YOLOv11 model specifically YOLOv11x model with our custom dataset named V-PCBs dataset. Below are the main contribution of the paper:

1. We present a multi-stage transfer learning approach for WPCBs electronic component detection, starting with frozen layers and then fine-tuning them for further adaptation.
2. We focus on improving poorly performing classes for instance capacitor, resistor and diodes through selective retraining, improving performance without affecting well-performing classes.
3. The study optimizes the YOLO11x for WPCBs electronic components detection, using pre-trained weights to improve accuracy and training efficiency.
4. Our work contributes to automated WPCBs disassembly and e-waste recycling, with potential applications in sustainable electronics recovery of high density CRMs.

The rest of the paper is organized as follows: Sect. 2 provides a detailed overview of related works. Section 3 presents materials and methods employed, including a proposed adaptive multi-stage transfer learning strategy in detail. Section 4 explains the achieved experimental results and discussions. Section 5 draws the conclusion.

2 State of the Art

Recent applications of computer vision have significantly improved the disassembly and recycling of WPCBs. Specifically they focused on automating the identification and separation of electronic components to improve recycling efficiency. In the study presented in [4], the researchers used a pre-trained YOLOv5 model to analyze a locally developed dataset, achieving high precision and recall, with the ability to optimize maximum electronic components. Silva et al. [5] proposes a method to assess the economic feasibility of recycling integrated circuits from WPCBs, using computer vision techniques for accurate detection.

In another study [6], authors used Real-Time DEtection TRansformer (RT-DETR) model, which outperforms traditional models like YOLOv8 and YOLOv9 in electronic components detection. In 2021, Fridman et al. [7] introduced, an unsupervised detection system for PCB defects detection, which uses computer vision and unsupervised learning to identify defects such as soldering problems and misaligned components, thereby improving quality control in PCB manufacturing. Kuo et al. [8] proposed a data-efficient graph embedding learning method for PCB component detection, which leverages the appearance of individual components and structural information of the PCB to recognize and localize various components, addressing challenges such as domain shift and unbalanced data distribution. A recent studies [1], explored the use of edge computing and Internet of Things to optimize the recycling process, focusing on the automated disassembly of WPCBs.

3 Experimental Methodology

3.1 Dataset

The dataset used in this study consists of high resolution images of WPCBs named V-PCBs dataset, containing a variety of electronic components such as capacitors, resistors, diodes, transformers, electrolytic capacitors and integrated circuits (ICs). We split the V-PCBs dataset into 80% training, 10% validation, and 10% test sets, with a total of 747 WPCBs images. However, the total number of classes of each components is not uniform across the dataset. For instance, diodes, capacitors (non electrolytic) and resistors present in a very low quantity. Therefore, we propose the adaptive multi-stage transfer learning approach to improve the class imbalance problem as well as real time accuracy of the system. We resized the WPCBs images into three different resolutions 640×640, 1280×1280, and 1920×1920 to explore the effect of image resolution on detection accuracy. Preprocessing steps include normalization, data augmentation (random rotations, flipping, and color jittering), and rescaling to standardize the input and improve model generalization.

3.2 Model Architecture

The object detection model used in this work is based on the YOLOv11 [9] architecture, which is known for its high accuracy and real-time detection capabilities. We begin by using pre-trained YOLOv11x weights on our custom V-PCBs dataset. YOLOv11 includes feature extraction, feature aggregation, and final predictions. In the first phase of training, we freeze the initial 15 layers, allowing the model to retain general features learned during pre-training while training the deeper layers to adapt to WPCB electronic components detection. Figure 1 shows the detailed adaptive multi-stage strategy of transfer learning.

3.3 Multi-stage Transfer Learning

Our methodology follows a four-phases training strategy:

- **Phase 1:** In this phase, we freeze the first 15 layers and train the remaining layers on the WPCBs dataset. This helps the model retain general feature extraction capabilities while adapting the deeper layers to the WPCBs components.

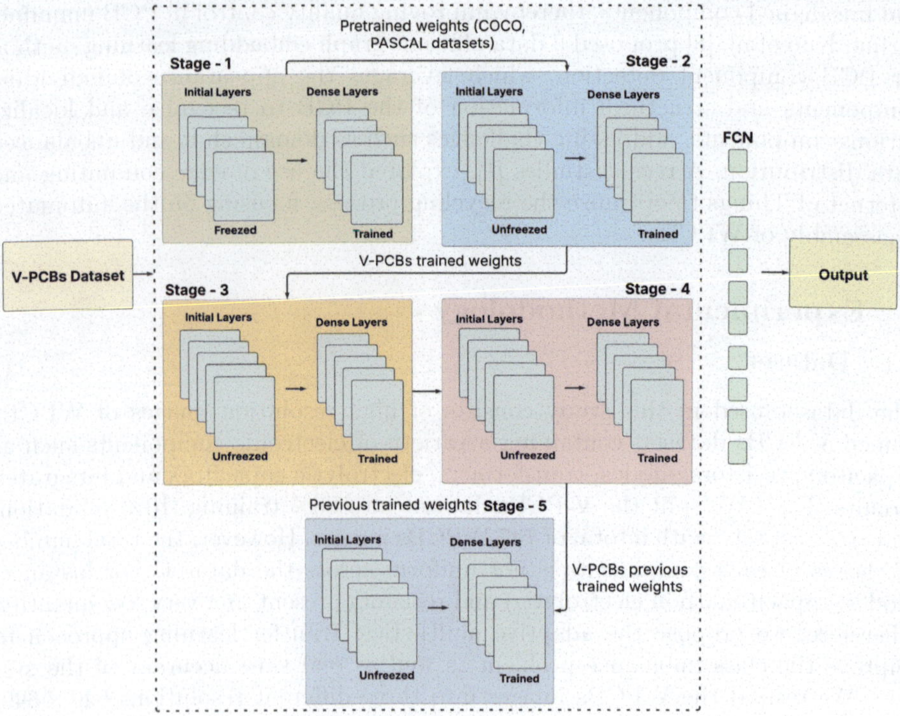

Fig. 1. Multi-stage transfer learning approach.

- **Phase 2:** After phase 1, we unfreeze the layers and retrain the model using the weights from the previous phase. This enables the model to fine-tune its feature extraction and improve the detection of WPCB components.
- **Phase 3:** We observed that certain classes (e.g., resistors and diodes) had low detection accuracy. Thus, we retrain the model only on these low-accuracy classes, using weights from the previous phase to avoid affecting the well-performing classes.
- **Phase 4:** After improving performance on selective classes, the model is retrained on the full dataset. This final phase ensures that the model generalizes well across all classes.

During training, hyperparameters like learning rate, batch size, and epochs were tuned for each phase.

3.4 Experimental Setup

We developed the software using Python 3.11 and PyTorch 2.5.0 with CUDA 12.4. on the CINECA Leonardo supercomputer [10] powered by the NVIDIA A100 GPUs with 64 GB of memory and high-speed connections. We used a

Table 1. Model performance across multi-stage transfer learning at various resolutions (AP_{50}= IoU:0.50 and AP_{75}= IoU:0.75)

Model Details			640 × 640		1280 × 1280		1920 × 1920	
Stages	State	Classes	AP_{50}	AP_{75}	AP_{50}	AP_{75}	AP_{50}	AP_{75}
V-PCBs Dataset								
Stage -1	Freeze	8	0.95	0.94	0.56	0.84	0.86	0.85
Stage -2	Non-Freeze	8	0.97	0.96	0.82	0.91	0.92	0.91
Stage -3	Non-Freeze	3	0.93	0.92	0.84	0.89	0.91	0.92
Stage -4	Non-Freeze	8	0.96	0.95	0.90	0.94	0.95	0.95
Robo Dataset								
Stage -1	Freeze	23	0.93	0.91	0.60	0.85	0.89	0.87
Stage -2	Non-Freeze	23	0.94	0.92	0.84	0.90	0.91	0.90
Stage -3	Non-Freeze	23	0.91	0.89	0.80	0.86	0.90	0.91
Stage -4	Non-Freeze	23	0.95	0.94	0.88	0.93	0.94	0.93

batch size of 8, and 150 epochs of training in each stage. To enhance model generalization, we applied also several data augmentation strategies, including random rotations and flipping.

4 Experimental Results and Discussion

We evaluated the model performance using popular metrics [11] such as Precision (P), Recall, F1 Score, and Mean Average Precision (mAP). The results obtained using our approach are summarized in the Table 1. The average precision (AP) of the model increases from stage 1 to stage 4 using different Intersection over Union (IoU) thresholds. For example, at an IoU of 0.50, AP increased from 0.95 to 0.96 in the final stage. Similarly, at an IoU of 0.75, AP increased from 0.94 to 0.95 in stage 4. Using the same strategy, AP increased from 0.56 in stage 1 to 0.90 in stage 4 and from 0.84 to 0.94 at an IoU of 0.50 and 0.75 respectively. At 1920 × 1920 resolution, AP at IoU 0.50 improved from 0.86 in stage 1 to 0.95 in stage 4, and at IoU 0.75 it increased from 0.85 to 0.95. In particular, the most significant improvements occur between stage 1 and stage 2, suggesting that unfreezing layers and retraining on the full dataset significantly improves detection accuracy. Further more, to test the generalization of the YOLOv11 model in electronic components detection, we applied our proposed strategy on roboflow dataset [12] and achieved significant results.

Table 2. Comparison of model performance across multi-stage transfer learning with other current models (AP_{50}= IoU:0.50 and AP_{75}= IoU:0.75), frames per seconds (FPS)

Model	Stages	Classes	Dataset	AP_{50}	AP_{75}	FPS
YOLOv3	Single	6	FICS-PCBs [13]	N/A	0.79	N/A
RatinaNet	Single	4	PCB-METAL [14]	N/A	0.83	N/A
YOLOv8	Single	8	V-PCBs [15]	0.90	0.82	26
YOLOv9	Single	8	V-PCBs [15]	0.91	0.85	21
RT-DETR	Single	8	V-PCBs [15]	0.99	0.84	20
Proposed (YOLOv11)	**Multi**	**8**	**V-PCBs** [15]	**0.95**	**0.95**	**22**

Table 2 shows the detailed comparison of achieved results with other current object detection models trained on similar task of electronic waste. The results shows that adaptive mutli-stage transfer learning with in similar domain achieve better results in real time as compared to train the model from scratch on custom dataset or retrained the whole model directly using pre-trained weights. Further, the multi-stage strategy also improve the accuracy of limited class present in the dataset. Figure 2 shows the performance of YOLOv11x model using adaptive multi-stage transfer learning approach.

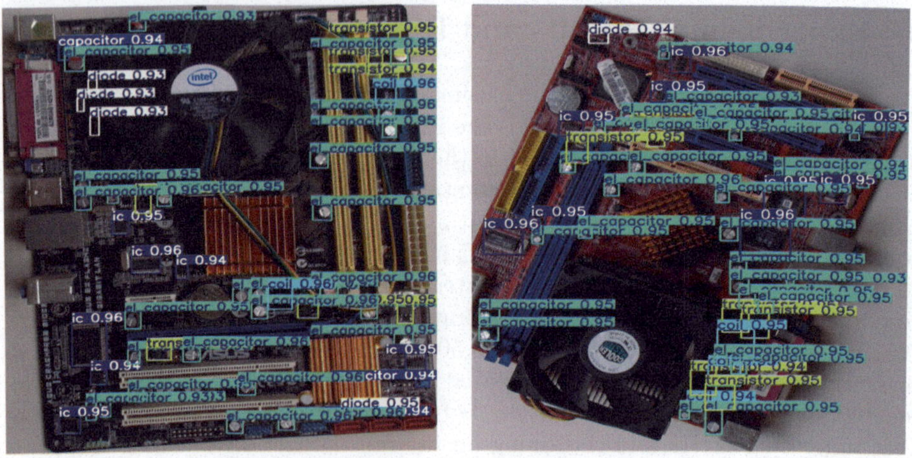

(a) (b)

Fig. 2. Experimental results of YOLOv11x achieved using multi-stage transfer learning approach at 1920 × 1920.

5 Conclusion

This paper presents a adaptive multi-stage transfer learning strategy to improve the automatic detection and localization of electronic components on WPCBs. The model, based on YOLOv11, was trained in a series of four stages, starting with frozen layers, followed by unfreezing layers, selective retraining on low accuracy classes, and full dataset retraining. Our results show that this staged approach significantly improved detection performance across different components. The mAP increased by approximately 11% across different resolutions from stage 1 to stage 4, indicating that the model became more effective at identifying and localizing components after each stage. In particular, we observed significant improvements in the detection of challenging components such as resistors and diodes, which were addressed during the selective retraining phase. This multi-stage transfer learning approach not only automates the dismantling process of WPCBs, but also promotes the sustainable solution towards circular economy principles to increase the recovery of high-density CRMs.

Acknowledgment. We acknowledge ISCRA for awarding this project access to the LEONARDO supercomputer, owned by the EuroHPC Joint Undertaking, hosted by CINECA (Italy).

References

1. Mohsin, M., Rovetta, S., Masulli, F., Cabri, A.: The role of edge computing and iot. MDPI Computer, Automated disassembly of waste printed circuit boards (2025)
2. Cabri, A., Masulli, F., Rovetta, S., Mohsin, M.: Recovering critical raw materials from WEEE using artificial intelligence. In: The 21st International Conference on Modelling and Applied Simulation (2022)
3. Mohsin, M., Zeng, X., Rovetta, S., Masulli, F.: Measuring the recyclability of electronic components to assist automatic disassembly and sorting waste printed circuit boards. arXiv preprint arXiv:2406.16593 (2024)
4. Mohsin, M., Rovetta, S., Masulli, F., Cabri, A.: Virtual mines–component-level recycling of printed circuit boards using deep learning. arXiv preprint arXiv:2406.17162 (2024)
5. Silva, L., Júnior, A.A., Azevedo, G.O., Oliveira, S.C., Fernandes, B.J.: Estimating recycling return of integrated circuits using computer vision on printed circuit boards. Appl. Sci. **11**(6), 2808 (2021)
6. Mohsin, M., Rovetta, S., Masulli, F., Cabri, A.: Real-time detection of electronic components in waste printed circuit boards: a transformer-based approach. arXiv preprint arXiv:2409.16496 (2024)
7. Fridman, Y., Rusanovsky, M., Oren, G.: Changechip: a reference-based unsupervised change detection for PCB defect detection. In: 2021 IEEE Physical Assurance and Inspection of Electronics (PAINE), pp. 1–8. IEEE (2021)
8. Kuo, C.W., Ashmore, J.D., Huggins, D., Kira, Z.: Data-efficient graph embedding learning for PCB component detection. In: 2019 IEEE Winter Conference on Applications of Computer Vision (WACV), pp. 551–560. IEEE (2019)
9. Jocher, G., Qiu, J.: Ultralytics yolo11 (2024)

10. Turisini, M., Amati, G., Cestari, M, Leonardo: a pan-european pre-exascale supercomputer for HPC and AI applications. arXiv preprint arXiv:2307.16885 (2023)
11. Everingham, M., Van Gool, L., Williams, C.K., Winn, J., Zisserman, A.: The pascal visual object classes (VOC) challenge. Int. J. Comput. Vis. **88**, 303–338 (2010)
12. KZLHFS. components-pcb dataset (2023). https://universe.roboflow.com/kzlhfs-s6yfj/components-pcb. Accessed 15 Jan 2025
13. Lu, H., Mehta, D., Paradis, O., Asadizanjani, N., Tehranipoor, M., Woodard, D.L.: Fics-pcb: a multi-modal image dataset for automated printed circuit board visual inspection. Cryptology ePrint Archive (2020)
14. Mahalingam, G., Gay, K.M., Ricanek, K.: PCB-metal: a PCB image dataset for advanced computer vision machine learning component analysis. In: 2019 16th International Conference on Machine Vision Applications (MVA), pp. 1–5. IEEE (2019)
15. Mohsin, M., Rovetta, S., Masulli, F., Greco, D., Cabri, A.: Deep learning-powered computer vision system for selective disassembly of waste printed circuit boards. In: 2024 IEEE 8th Forum on Research and Technologies for Society and Industry Innovation (RTSI), pp. 115–119. IEEE (2024)

Artificial Intelligence, Decision Making and Financial Performance: Evidence from Moroccan Industrial Companies

Abdel-Ylah Amazzal(✉), Mohamed El Ghadouia, Mustapha Boumeska, and Mustapha Bengrich

Faculty of Law, Economics and Social Sciences, Ibn Zohr University Agadir, Agadir, Morocco
abdelylahamazzal@gmail.com

Abstract. Our paper aims to empirically investigate the impact of artificial intelligence components on decision making and to examine its effect on financial performance of Moroccan industrial companies. Empirical primary data were collected from managers using a survey of ninety-three companies. Structural equation modelling is used to test these relationships. Our findings show that machine learning positively and significantly affects decision making. Moreover, they illustrate that computer vision positively and significantly impacts the decision making. Nevertheless, the hypothesis concerning the effect of natural neural networks on decision-making has not been confirmed. In addition, decision making positively and significantly affects financial performance. Besides, our findings indicate that decision making mediates the link between machine learning and financial performance. Decision making mediates also the connectedness between computer vision and financial performance. Nevertheless, hypothesis relating to mediating the relation between natural neural networks and financial performance is rejected. Our manuscript provides some new managerial contributions for managers of industrial organizations and explores the main components of AI and their impacts on decision making and on financial performance. Therefore, our findings will help managers in Morocco and in other countries to introduce artificial intelligence as a secure mechanism for enhance financial performance and mitigate risk.

Keywords: Financial performance · Artificial intelligence · Decision making

1 Introduction

In many areas the decision making is considered as crucial stage and is linked to a high level of risk. However, artificial intelligence does not. As a result, AI provides insights and algorithms that endow decisions makers with the ability to consider decision risk. Besides, speed and quality of decision making are critical to [1]. In this context, many companies adopt emerging technologies to improving their performance and gain a competitive advantage. Among these advanced technologies, Artificial Intelligence (AI) holds a crucial place and attracts increasing interest in both academical and industrial field [2].

C. El Mokhi et al. (Eds.): ICASET 2025, SUCI, pp. 143–153, 2025.
https://doi.org/10.1007/978-3-031-98334-4_14

Moreover, several researchers have attempted to define AI, such as [3]. They describe it as: "a system's ability to perceive its environment and make decisions that maximize its success in that environment". Furthermore, the literature documents that the AI continue to grow and replace traditional practices through innovative technologies to meet the growing complex needs and challenges of globalization. Many empirical studies proved the importance of artificial intelligence in optimizing decision-making process (see for example, [3–4-5–6, 7]. In addition, Other researchers, such as [8–9-10, 11] have confirmed that improving the decision-making process contributes directly to improve company's performance and especially financial performance.

Less works has been conducted in this context. We tried to contribute to fill this gap by investigating this relationship. Our study differs from the others, we attempt to examine the links between the major AI components and financial performance and we consider decision making as a mediator variable. The rest of the text is divided into three sections. The first is a literature review. The second is dedicated to the methodology of the research. The third section is dedicated to the discussion of the results.

2 Literature Review and Hypotheses Development

2.1 Artificial Intelligence: Concepts, Components and Risks

Nilsson [12] argued that artificial intelligence is a branch of computer science. It seeks to create a system that simulates human behavior described as intelligent. AI has several definitions that have been proposed over time. A broad sense, AI is considered as an able system that perceive its environment and make decisions that maximize its success in that environment [3].

In general, Legg & Hutter [13] argued that AI is an able system that interpret external data, to learn from it in order to achieve specific goals and complete tasks in adaptive manner. In addition, AI consists of several sub-fields, each focusing on different aspects of the ability of machines to replicate human cognitive functions. The main components are machine learning, natural neural network and computer vision.

Regarding, the machine learning, it is a crucial field of AI that seeks to empower machines to learn from data, in order to make inferences, predictions, and identify associations that can guide decisions. This area focuses on the development of methods that enable computers to learn from input data and recognized patterns using algorithms to identify patterns and process large volumes of big data. Through this process, machine learning creates a self-learning environment that becomes progressively more intelligent, automating repetitive tasks based on AI learning and repeating them as needed [14]. The field of machine learning has been enhanced by a number of advancements like deep learning and reinforcement learning. These subsets systems of machine learning are made up of neural networks to show superhuman performance by excelling human capabilities [15].

Inspired by the human brains works, neural networks are a cornerstone in the development of AI and machine learning. The introduction of back propagation by [16] has marked a decisive turning point, enabling neural networks to adjust their internal weights efficiently to minimize prediction error, making it possible to learn complex functions through training examples [17].

The third component is computer vision. It is an area of AI that enables computers and systems to understand and interpret visual content (images and videos) in the same way that humans do. It also includes common applications such as facial recognition, image analysis and autonomous driving [18].

In addition, the adoption of AI for decision-making gives rise to risks that need to be addressed. Indeed, increased reliance on AI could introduce new vulnerabilities, such as algorithmic bias or data safe, requiring vigilance and risk management [19].

2.2 Decision-Making Frame

In all areas, decision-making is the process of choosing between several possible actions to solve a problem. This process includes several stages: defining the problem, researching and analyzing relevant information, developing and evaluating hypotheses for solutions based on previous knowledge or experience, selecting an option and implementing it [20]. According to Dean & Sharfman [21], strategic decisions involve the allocation of significant resources, set precedents and have repercussions on smaller decisions [22]. These decisions are characterized by their lack of structure, their non-routine nature and their complexity [23]. They are generally of substantial importance, because they involve major issues for the organization, and also tackle unusual problems that are out of the ordinary and require a particular approach. Moreover, these decisions are pervasive, which means that they affect different aspects and levels of the organization [24].

Strategic decisions concern fundamental issues related to the organization's sustainability and survival, and generally involve a significant proportion of its resources [25]. Strategic decisions focus primarily on problems that are unusual for the organization, rather than on problems that can be solved by routine decisions [25].

2.3 The Concept of Financial Performance

Financial performance of companies has been an important topic of academic study for a long time. Previous literature has examined several explanatory variables that contribute to improve the financial performance in companies. Furthermore, Al-Matari, Al-Swidi & Fadzil [26] argue that financial performance is the firm's capability to achieve its financial objectives. In addition, Baba & Nasieku [27] state that financial performance is a generated income by the company using its assets. It can thus guide stakeholders in their decision-making. More specifically, financial performance examines the factors that directly influence a company's financial position [28]. It is also used as the main evaluation tool by external stakeholders [29]. Malesev & Cherry [30] state that financial performance is a result of various activities carried out within the available financial resources. Thus, financial performance is observed by analyzing financial statements or by analyzing financial ratios. Also, financial performance is a concept or aspect that describes financial information of companies [31]. We can conclude that financial performance is a result achieved in a period that describes the level of excellence of the company.

2.4 The Relationships Between Artificial Intelligence, Decision-Making and Financial Performance

Recently, artificial intelligence has become an essential lever for company's economy. Therefore, examining its effect on the decision-making process and on financial performance has been an unavoidable topic. Many researchers and practitioners have highlighted the importance of artificial intelligence in improving the strategic decision-making process. Russell and Norvig [3] demonstrated the importance of artificial intelligence in optimizing decision-making process. Based on Simon's research work, Jirásek [4] explored that cognitive limitations, insufficient information, and time pressure are the key factors shaping human decision-making. On the other hand, Shrestha et al. [5] analyzed how organizational decision-making structures are influenced by AI integration.

Many studies also confirmed the connectedness between artificial intelligence components and decision-making. Mullainathan and Spiess [6] showed that machine learning is being applied to improve financial decision-making, mainly in financial markets. It allows to predict shares prices and enable effective risk management. Brynjolfsson and McElheran [32] argued also that machine learning is able of becoming an important input for optimized decision-making in modern organizations. Besides, Worasucheep [33] argued that using machine learning algorithms to discover better concepts and analyze data efficiently, provides insights for better financial decisions.

Other researchers, such as Collins, Ghosh, & Scofield [7], Dutta & Shekhar [34], Fletcher & Goss [35], Krizhevsky et al. [36], indicated that natural neural networks have a positive influence on decision-making in companies. They also applied neural networks in decision-making process. Other researchers suggest training these techniques on large databases to make accurate decisions. These techniques are based also on visual information. This will enable innovations [37]. In their study, Li et al. [38] indicated that neural networks positively and significantly affect decision-making. They concluded that this component is capable of evaluating complex information. It also allows to predict market trends and support algorithmic assessment [38].

With regard to computer vision, practitioners have claimed that this technique enables you to make better decisions, faster and with more confidence [39]. In addition, visualization tools can make a better match between the task and environment. This will improve the decision quality [40]. Several researchers confirmed the link between decision-making and financial performance. Haveman [8] showed that when a company adjusts its strategy to respond to rapid changes in the environment, its financial performance improves. Fuller and Stopford [9] argued that companies which have adopted innovative strategies in mature sectors are winners. On the other hand, the strategic decision is a tool that allows changes to be made, allowing threats to be progressively eliminated and opportunities presented to the firm to be seized, thereby improving performance, particularly financial performance [10, 11].

Finally, and in line with this theoretical framework, we have formulated the following hypotheses:

H1: Machine Learning positively and significantly affects decision-making
H2: Natural neural networks positively and significantly impact decision-making
H 3: Computer vision contributes to improving decision-making

H4: Improved decision-making increases the financial performance of companies
H5: Decision-making can mediate the link between AI and financial performance
H5.1: Decision-making can mediate the connectedness between machine learning and financial performance
H5.2: Decision-making can mediate the link between natural neural networks and financial performance
H5.3: Decision-making can mediate the connectedness between computer vision and financial performance

3 Methods

In testing our research hypotheses, used primary data collected by structured questionnaire from ninety-three Moroccan industrial companies. On the basis of a literature review, the sixteen most important items were constructed. Indeed, the scale of Machine Learning (ML) was taken from the previous study by [41] who measured this scale by three items. The Natural Neural Network (NNN) scale was taken from the work of [17] who measured this scale with three items. The Computer Vision (CPV) scale was taken from the work of [18] who used two items. The scale of Decision making (DM) was taken from the study of [20] who used four items. The scale of financial performance (FP) was taken from the work of [42, 43] who measured this scale with four items. The survey has been reviewed by two experts in the field of linguistics and five managers of Moroccan industrial companies. We also applied the tenfold rule proposed by [44], which is largely used within the PLS-SEM [45]. In addition, we measured the indicators above by using the Likert scales, ranging from one to five [46].

4 Results and Discussion

To test constructs reliability and their measurement indicators, we applied principal component analysis (PCA). PCA results show that all items have factor coefficients more than 0.5 and communality more than 0.4 [47]. PCA findings show also that all constructs have more than 60% of total variance explained, which shows the validity our constructs and items. Confirmatory analysis was performed using Smart-PLS3 to test reliability, convergent validity, and discriminant validity. The results are presented in Table 1.

Table 1 shows that all constructs have a Cronbach's alpha more than the minimum of 0.7 [48, 49]. They also have a CR more than (0.7). This finding demonstrates that all constructs are reliable and that there is internal consistency for each construct. In addition, Table 1 shows that constructs AVE coefficients are more than 0.5. All constructs meet therefore the required convergent validity criteria [50].

Table 1. Descriptive statistics, Cronbach's alpha (α), Composite Reliability (CR) and Average variance extracted (AVE)

Constructs	Items	Mean	SD	Alpha	CR	AVE
ML	ML1	3.92	0.23	0.920	0.950	0.863
	ML2	4.16	1.78			
	ML3	3.30	0.19			
NNN	NNN1	2.55	2.02	0.731	0.851	0.659
	NNN2	3.33	1.01			
	NNN3	3.89	0.98			
CPV	CPV1	4.67	1.03	0.813	0.914	0.842
	CPV2	2.78	1.02			
DM	DM1	3.89	0.41	0.875	0.914	0.727
	DM2	4.42	0.99			
	DM3	3.22	0.45			
	DM4	4.12	0.21			
FP	FP1	4.11	1.78	0.947	0.961	0.862
	FP2	3.28	2.05			
	FP3	3.89	1.74			
	FP4	4.51	0.47			

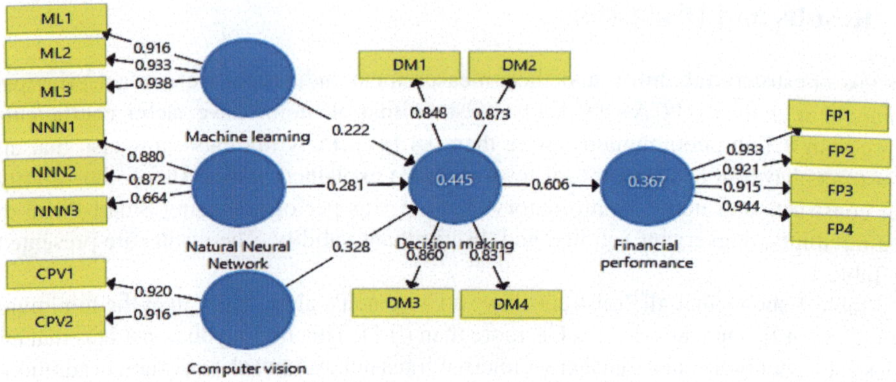

Fig. 1. Results of the research model

Figure 1 illustrates measurement and structural models. In particular, it shows that all loadings coefficients of all items are more than 0.7, which is meaningful [47]. All the measurement indicators have been retained therefore for further analysis. In addition, to test discriminant validity of all constructs, inter and intra-construct correlations were carried out. Table 2 shows that discriminant validity is ensured. Thus, we can confirm that the quality of all the measurement indicators is satisfactory and acceptable. In fact, the reliability, convergent and discriminant validities have significant results that confirm those of PCA.

Table 2. Correlation among constructs and square root of the AVE (Fornell-Larcker).

Constructs	CPV	DM	FP	ML	NNN
CPV	0.918				
DM	0.587	0.853			
FP	0.653	0.606	0.928		
ML	0.680	0.533	0.641	0.929	
NNN	0.386	0.478	0.490	0.317	0.812

Moreover, the findings show the coefficients of determination (R^2-adjusted = 0.427 for decision making and R^2-adjusted = 0.361 for financial performance) generated by the PLS algorithm. It means that our model has moderate explanatory ability [49] and all variables of our model contribute to explain 42.70% of the decision-making, which in turn explains 36.10% of financial performance. Furthermore, redundancy index Q^2 is more than 0. Then, the model has a significant predicting ability.

After that, we applied bootstrap method to test hypotheses.

Table 3. The results of hypotheses testing

Hypotheses	β	T-value	P-value	Results
H1: ML -> DM	0.222	2.850	0.005	Accepted
H2: NNN -> DM	0.281	1.798	0.065	Rejected
H3: CPV -> DM	0.328	2.908	0.004	Accepted
H4: DM -> FP	0.606	8.101	0.000	Accepted

Table 3 shows that Machine Learning positively and significantly affects Decision Making. This result support hypothesis H1 (p-value < 0.05). Our results are concordant with those of previous studies [7, 32] which indicated that machine learning contributes to improving U.S manufacturing decision making process. Moreover, the findings also show that the computer vision positively and significantly affects decision making. We concluded that hypothesis H3 (p-value < 0.05) is supported. These findings corroborate those of previous authors [39, 40] who indicate that examining data using computer vision tools can facilitate decision making. Our findings also indicate that decision making positively and significantly affects financial performance. These findings support hypothesis H4 (p-value < 0.05) and confirm those of [8–11]. However, hypothesis (H2) (p-value > 0.05) regarding the effect of natural neural network on financial performance was not supported. These findings contradict those of other studies [7, 34–38].

Examining mediating effect of the decision-making variable (DM) involves two steps, according to [51]. First, we examine the indirect effects using the bootstrap technique. Secondly, we conduct a bootstrap test of the confidence interval. Indeed, the zero must not be inside the low and high levels of this confidence interval. Once these two conditions have been verified, the DM variable will be a perfect mediator.

Table 4. Mediation test

Hypothesis	Path a	Path b	t-value	p-value	Confidence level		Result
					95%LL	95%UL	
H5: AI -> FP							
H5.1: ML -> FP	0.222	0.134	1.721	1.721	0.066	0.128	Mediation
H5.2: NNN -> FP	0.281	0.170	2.604	2.604	−0.071	0.081	No mediation
H5.3: CPV - >FP	0.328	0.199	1.808	1.808	0.057	0.233	Mediation

Table 4 shows that the confidence interval includes zero for the NNN variable. We therefore conclude that the mediating effect of the mediating variable exists for these two paths, meaning that the mediating links are established. However, the confidence interval does not include a zero for the DM and CPV variables. This means that hypotheses H5.1 and H5.3 are confirmed. These results corroborate those of [6, 32–40]. Hypothesis H5.2 is rejected, which contradicts the results of [7, 34–38].

5 Conclusion

This study is a proposal of a new theoretical model to be empirically studied in Moroccan industrial firms. In addition, this paper provides some important managerial contributions. Firstly, our findings will help executives to improve their financial performance by adopting the major AI components. Secondly, the findings can be used to mitigate risks relative to decision making. Managers need therefore to understand the benefits of introducing artificial intelligence in the decision-making process. As all papers, ours has some limits. First, our sample size of our study is limited, giver reluctance of respondents. Second, only financial performance is considered in the model so as not to bias the empirical analysis. Further research could be carried out on the basis of data collected from other Moroccan companies in different sectors. We also suggest a benchmark study between the Moroccan context and that of an emerging country.

References

1. Akhtar, P., Frynas, J.G., Mellahi, K., Ullah, S.: Big data-savvy teams' skills, big data-driven actions and business performance. Br. J. Manag. **30**(2), 252–271 (2019)

2. Balog, K.: The concept and competitiveness of agile organization in the fourth industrial revolution's drift. Strategic Manage. **25**(3), 14–27 (2020)
3. Russell, S., Norvig, P.: Artificial Intelligence: A Modern Approach, 4th edn. Pearson, London (2020)
4. Jirásek, M.: Daos: a lens on AI and bounded rationality. SSRN (2024)
5. Shrestha, Y.R., Ben-Menahem, S.M., Von Krogh, G.: Organizational decision-making structures in the age of artificial intelligence. Calif. Manage. Rev. **61**(4), 66–83 (2019)
6. Mullainathan, S., Spiess, J.: Machine learning: an applied econometric approach. e J. Econ. Perspect. **31**(2), 87–106 (2017)
7. Collins, E., Ghosh, C., Scofield, R.: An application of a multiple neural network learning system to emulation of mortgage underwriting judgements. In: IEEE 1988 International Conference on Neural Networks, vol. 2, no. 3, pp. 459–466 (1988)
8. Haveman, H.A.: Between a rock and a hard place: organizational change and performance under conditions of fundamental environmental transformation. Adm. Sci. Q. **37**(1), 48–75 (1992)
9. Fuller, C.W.F., Stopford, J.M.: Creating corporate entrepreneurship. Strategic Manage. J. **15**(7), 521–536 (1992)
10. Zajac, E.D., Kraatz, M.S.: A diametric forces model of strategic change: assessing the antecedents and consequences of restructuring in the higher education industry. Strateg. Manag. J. **14**(S1), 83–102 (1993)
11. Feitler, J.N., et al.: Strategic and performance changes among LTL Motor carriers: 1976–1993. Transp. J. **37**(4), 5–12 (1998)
12. Nilsson, N.J.: The Quest for Artificial Intelligence: A History of Ideas and Achievements. Cambridge University Press, Cambridge (2010)
13. Legg, S., Hutter, M.: A formal measure of machine intelligence. In: Proceedings of the 15th Annual Machine Learning Conference of Belgium and The Netherlands (Benelearn 2006), pp. 73–80 (2006)
14. Vedamuthu, T.: Artificial intelligence and human collaboration in project decision making (publication n° 28000743 [mémoire de maîtrise, The College of St. Scholastica]. ProQuest Dissertations and Theses Global (2020). https://www.proquest.com/docview/2423809554?pqorigsite=gscholar&fromopenview=true&sourcetype=Dissertations%20&%20Theses. Accessed 20 Sept 2024
15. LeCun, Y., Bengio, Y., Hinton, G.: Deep learning. Nature **521**(7553), 436–444 (2015)
16. Rumelhart, D.E., Hinton, G.E., Williams, R.J.: Learning representations by back-propagating errors. Nature **323**(6088), 533–536 (1986)
17. Wood, T.: Comprehensive review of backpropagation neural networks. Acad. J. Sci. Technol. **9**(1), 150–154 (2024)
18. Szeliski, R.: Computer Vision: Algorithms and Applications. Springer, London (2010)
19. Van Den Brink, A.M., Benyayer, L., Kupp, M.: Decision-making in organizations: should managers use AI. J. Bus. Strategy **45**(4) (2023)
20. March, J.G.: A Primer on Decision Making: How Decisions Happen. Free Press, New York (1994)
21. Dean, J.W., Jr., Sharfman, M.P.: Does decision process matter? A study of strategic decision-making effectiveness. Acad. Manag. J. **39**(2), 368–392 (1996)
22. Mintzberg, H., Raisinghani, D., Theoret, A.: The structure of" unstructured" decision processes. Adm. Sci. Q. **21**(2), 246–275 (1976)
23. Schwenk, C.R.: The cognitive perspective on strategic decision making. J. Manage. Stud. **25**(1), 41–55 (1988)
24. Wilson, D.C., Butler, R.J., Cray, D., Hickson, D.J., Mallory, G.R.: Breaking the bounds of organization in strategic decision making. Hum. Relat. **39**(4), 309–331 (1986)

25. Stahl, M.J., Grigsby, D.W.: Strategic Management for Decision Making. PWS-Kent Publishing Company, Indiana (1992)
26. Al-Matari, E.M., Kaid Al-Swidi, A., Bt Fadzil, F.H.: The measurements of firm performance's dimensions. Asian J. Financ. Account. **1**(6), 24–49 (2014)
27. Baba, S., Nasieku, T.: Effect of macroeconomic factors on financial banks in Nigeria. Int. J. Soc. Sci. Inf. Technol. **2**(6), 1278–1299 (2016)
28. Omondi, M.M., Muturi, W.: Factors affecting the financial performance of listed companies at the nairobi securities exchange in Kenya. Res. J. Financ. Account. **4**(15), 99–105 (2013)
29. Demertzis, M., Merler, S., Wolff, G.: Capital markets union and the fintech opportunity. Bruegel Policy Contribution (2017)
30. Malesev, S., Cherry, M.: Digital and social media marketing-growing market share for construction SMEs. Constr. Econ. Build. **21**(1), 65–82 (2021)
31. Pollák, F., Markovič, P.: Size of business unit as a factor influencing adoption of digital marketing: empirical analysis of SMEs operating in the central European market. Adm. Sci. **11**(3), 71 (2021)
32. Brynjolfsson, E., McElheran, K.: Data in action: data-driven decision making and predictive analytics in U.S. manufacturing. SSRN, pp. 1–49 (2019)
33. Worasucheep, C.: Forecasting currency exchange rates with an artificial bee colony-optimized neural network. In: Proceedings of the 2015 IEEE Congress on Evolutionary Computation (CEC), pp. 3319–3326. IEEE, Sendai (2015)
34. Dutta, S., Shekhar, S.: Bond rating: a nonconservative application of neural networks. In: IEEE 1988 International Conference on Neural Networks, vol. 2, no. 3, pp. 443–450 (1988)
35. Fletcher, D., Goss, E.: Forecasting with neural networks: an application using bankruptcy data. Inf. Manage. **24**(3), 159–167 (1993)
36. Krizhevsky, A., Sutskever, I., Hinton, G.E.: ImageNet classification with deep convolutional neural networks. Commun. ACM **60**(6), 84–90 (2017)
37. Kimoto, T., Asakawa, K., Yoda, M., Takeoda, M.: Stock market prediction system with modular neural networks. In: International Joint Conference on Neural Networks, vol. 1, no. 2, pp. 1–6 (1990)
38. Li, A.W., Bastos, G.S.: Stock market forecasting using deep learning and technical analysis: a systematic review. IEEE Access **8**(3), 185232–185242 (2020)
39. Brath, R., Peters, M.: Information of visualization for business past and future. DM Rev. **15**(1), 1–40 (2005)
40. Eick, S.G., Wills, G.: High interaction graphics. Eur. J. Oper. Res. **8**(3), 445–459 (2005)
41. Ma, L., Sun, B.: Machine learning and AI in marketing – connecting computing power to human insights .Int. J. Res. Mark. **37**(3), 481–504 (2020)
42. Stanwick, P.: The determinants of corporate social performance: an empirical examination. Am. Bus. Rev. **16**(41), 86–93 (1998)
43. Moore, G.: Corporate social and financial performance: an investigation in the U.K. supermarket industry. J. Bus. Ethics **34**(4), 299–315 (2001)
44. Barclay, D., Thompson, R., Higgins, C.: The partial least squares (PLS) approach to causal modeling: personal computer use as an illustration. Technol. Stud. **2**(2), 285–309 (1995)
45. Memon, M.A., Ting, H., Cheah, J.H., Thurasamy, R., Chuah, F., Cham, T.H.: Sample size for survey research: review and recommendations. J. Appl. Struct. Equ. Model. **4**(2), 1–20 (2020)
46. Hair, J.F., Anderson, R.E., Tatham, R.L., Black, W.C.: Multivariate Data Analysis, 5th edn. Prentice Hall International, Upper Saddle River (1998)
47. Hair, J.F., Black, W.C., Babin, B.J., Anderson, R.E., Tatham, R.: Multivariate Data Analysis, 7th edn. Pearson, New York (2010)
48. Hulland, J.: Use of partial least squares (PLS) in strategic management research: a review of four recent studies. Strateg. Manag. J. **20**(2), 195–204 (1999)

49. Chin, W.W.: Issues and opinion on structural equation modeling. MIS Q. **22**(1), 7–25 (1998)
50. Fornell, C., Larcker, D.F.: Evaluating structural equation models with unobservable variables and measurement error. J. Mark. Res. **18**(1), 39–50 (1981)
51. Preacher, K.J., Hayes, A.F.: Asymptotic and resampling strategies for assessing and comparing indirect effects in multiple mediator models. Behav. Res. Methods **40**(3), 879–891 (2008)

Intelligent Traffic Management Systems: A Literature Review on AI-Based Traffic Light Control

Chakib El Mokhi(✉), Hassane Erguig, Nabil Hmina, and Hanaa Hachimi

Ibn Tofail University, 14000 Kenitra, Morocco
chakib.elmokhi@uit.ac.ma

Abstract. The exponential increase in vehicles, rapid urbanization, and rising transportation demands have created unprecedented challenges for urban traffic management worldwide. Traditional fixed-time traffic signal systems have proven inadequate, with studies indicating they account for approximately 10% of global traffic delays and contribute to millions of hours lost in congestion annually. This comprehensive literature review examines the emerging field of intelligent traffic management systems that leverage artificial intelligence techniques to optimize traffic light control. The review categorizes and analyzes various AI approaches including Fuzzy Logic (FL), Metaheuristic algorithms (MH), Dynamic Programming (DP), Reinforcement Learning (RL), Deep Reinforcement Learning (DRL), and hybrid techniques. We explore both single-intersection and multiple-intersection implementations, examining their architectures, methodologies, and performance metrics. Simulation results from multiple studies demonstrate significant improvements in key performance indicators, with some implementations reducing queue lengths by 28%, average vehicle latency by 27%, and CO_2 emissions by 28%. The review also discusses microsimulation tools essential for evaluating these systems, including CORSIM, AIMSUN, VISSIM, and SUMO. Despite promising results, challenges remain in scalability, real-time implementation, and integration with existing infrastructure. Future research directions point toward integration with connected and autonomous vehicles, multi-modal traffic management, and smart city ecosystems.

Keywords: Smart City Traffic · AI in Urban Traffic Optimization · Intelligent Traffic Management Systems · Adaptive Traffic Signal Control · Reinforcement Learning for Traffic Control

1 Introduction

Road congestion is one of the most worrying challenges facing today's urban environments, a phenomenon amplified by growing urbanization worldwide, which is set to reach 68% by 2050 according to United Nations projections. This demographic trend is putting increasing pressure on existing transport infrastructures, particularly in metropolitan areas. Traditional traffic management systems, based mainly on fixed-time traffic light

C. El Mokhi et al. (Eds.): ICASET 2025, SUCI, pp. 154–171, 2025.
https://doi.org/10.1007/978-3-031-98334-4_15

cycles, are proving increasingly inadequate to cope with the dynamic complexity of today's traffic flows. Failure to adapt to real world traffic scenarios is resulting in under-utilized road capacity, increased travel times, heightened fuel and energy consumption, as well as greater arboreal emissions [1].

Economically, the repercussions are substantial. Studies approximate the cost estimation for developed countries to be an excess of 295 million annually adrift due to non-changing traffic light systems. As non-moving traffic creates further arteries of congestion, these also accumulate emissions of gasses, solidifying their contribution to environmental decline. Cities also face a crippling quality of life crisis due to growing travel duration and stress levels from raised traffic [2].

The most recent innovations in artificial intelligence (AI) have shown potential in mitigating these shortcomings, particularly with the creation of Intelligent Traffic Management Systems – ITMS. AI traffic light control is a departure from traditional, rigid methodologies since it permits real-time data assessment and adaptive alteration of light intervals. This work intends to define the state-of-the-art in algorithmic AI applications for traffic control using approaches such as fuzzy logic (FL), metaheuristic (MH) algorithms, dynamic programming (DP), reinforcement learning (RL), deep reinforcement learning (DRL), and hybrid techniques [3].

The study covers applications at the scale of isolated intersections as well as those in multi-intersection networks, analyzing architectures, simulation methodologies, performance criteria and implementation constraints. This work aims to provide an in-depth synthesis of theoretical frameworks, modeling tools, experimental results and future perspectives, thus constituting a baseline for transportation engineers, urban planners and AI researchers working towards a more efficient, sustainable and intelligent urban mobility system [1].

2 Theoretical Background

2.1 Traffic Signal Control Fundamentals

To manage traffic and the safety of pedestrians at road intersections (RIs) effectively, well-structured traffic signal control (TSC) systems are crucial for cities with permanently high traffic volume. Adaptive TSC systems are preferred to fixed-time systems as they allow for flexibility in servicing different traffic patterns. An equally important component of traffic control at RIs is the traffic signal timing (TST) for each approach direction, which directly impacts both the safety and capacity of the intersection [4].

For regulated intersections, traffic flow is managed using four-position signal lights with red, yellow, and green lights for each direction signaling right of way periods. A TSC system operates using several basic and guiding parameters which include [5]:

- **Signal Phase** is the interval during which a particular vehicle or pedestrian movement is permitted to enter and leave the intersection controlled by traffic signals. This permits partial independence of traffic on parallel approaches within a given signal period.
- **Signal Cycle** refers to the full rotation process that signals all traffic movements to ensure safe road intersection travel.

- **Signal Sequence** refers to the specific order in which signal phases are activated during an entire signal cycle.
- **Cycle Length** is the period of time required for complete movement through all the different traffic flow indications.
- **Green Time** represents the duration of seconds when an RI allows maximum flow of traffic through one specific traffic lane.
- **Timing of Phases** denotes the duration, measured in seconds, that a specific phase remains active for a given direction within one full cycle of traffic signals.
- **Change Interval** represents the yellow duration provided between the end of green movement signals.
- **Offset** refers to how synchronized phases should temporally relate to each other. When a specific movement extends its green phase duration it leads to less vehicles that stop and decreased delays. However, increasing the green period for one traffic flow typically increases delay and vehicle queuing for competing traffic lane flows.
- **Average Waiting Time** refers to how long it takes a vehicle to leave a road intersection after stopping at the intersection.

2.2 Performance Metrics in Traffic Management

Evaluating the effectiveness of traffic management systems requires specific performance metrics that quantify improvements in traffic flow and environmental impact. Common metrics include [1]:

- **Queue Length:** The number of vehicles waiting at an intersection [5]. Reduction in queue length is a primary objective of intelligent traffic systems.
- **Average Vehicle Delay:** The time vehicles spend waiting at intersections, typically measured in seconds or minutes [6].
- **Travel Time:** The total time required to travel between origin and destination points.
- **Throughput:** The number of vehicles that pass through an intersection during a specific time period [6].
- **Emissions:** The amount of pollutants (CO_2, NOx, etc.) produced by vehicles in the traffic network.
- Fuel Consumption: The amount of fuel consumed by vehicles, which correlates with both economic and environmental impacts [2].
- **Stop Rate**: The average number of stops a vehicle makes during its journey.

2.3 Microsimulation Tools for Traffic Analysis

General system performance evaluation relies on analytical models whereas simulation tools serve as necessary tools for dynamic traffic analysis of different use cases. Simulation allows the replication of real-world traffic scenarios within a controlled environment, thereby facilitating the evaluation and testing of several traffic management strategies [2, 5].

Microscopic Simulation. The microscopic simulation approach integrates drivers' behavior together with their responses to other traffic participants along with pedestrians. CORSIM, AIMSUN, PARAMICS, MATSim, VISSIM and SUMO constitute the list of most often used microsimulation tools for traffic modeling across urban environments [5].

The authors investigated multiple traffic simulation packages by identifying their core attributes while providing an evaluation of unique attributes.

- **AIMSUN** with **VISSIM** and **CORSIM** served as appropriate solutions for handling states routes along with arterial congestion as well as major street systems.
- **AIMSUN** along with **CORSIM** and **PARAMICS** proved the most suitable options for Intelligent Transportation Systems (ITS) applications.
- Through **MATSim** users obtain a logistical structure to conduct extensive agent-based transport simulation operations.

These microsimulation tools are classified based on [5]:

- Open Source availability
- System type (Discrete control system or Continuous control systems)
- Visualization capabilities (2D/3D)
- Scope of Application (Regional, City, Country)
- Output file format
- Capability of importing maps
- Programming language supported
- Level of programming proficiency required

Macroscopic Simulation. Macroscopic traffic flow frameworks establish mathematical ties between traffic flow dynamics with traffic density movements and traffic stream velocities and other elements using this model. The models normalize individual traffic properties to create system analogs from microscopic traffic flow simulations.

3 AI Techniques in Traffic Management

3.1 Overview of AI Approaches in Traffic Signal Control

The integration of artificial intelligence into traffic signal control represents a significant advancement over traditional fixed-time systems. AI-based Adaptive Traffic Signal Control (ATSC) systems can be broadly categorized based on the number of road intersections they manage (single-intersection vs. multiple-intersection) and the specific AI techniques employed. Real-time adjustments made by these systems to signal timings produce better results for traffic operations within urban roads in response to changing traffic demand patterns (Fig. 1).

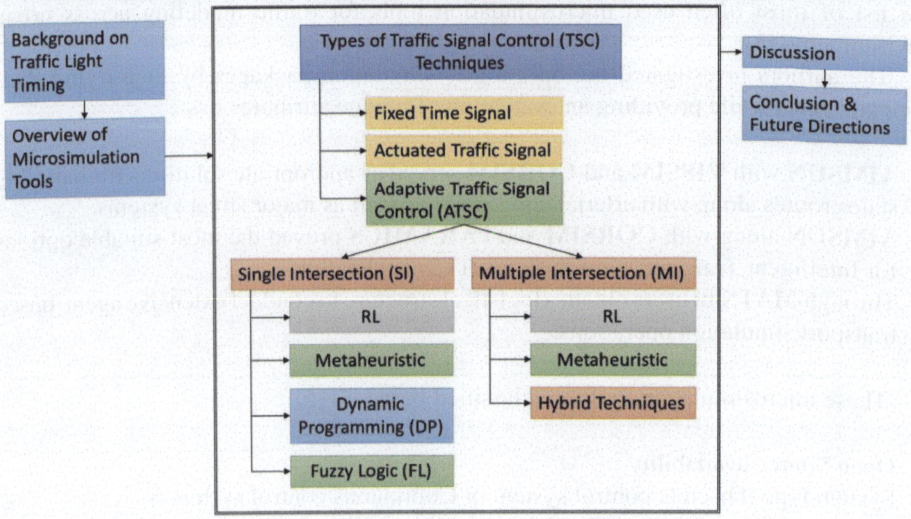

Fig. 1. Classification of Traffic Signal Control (TSC) techniques by intersection type and applied AI methods.

Several forms of artificial intelligence apply to traffic management through Fuzzy Logic (FL) and Metaheuristic algorithms (MH) and Dynamic Programming (DP) and Reinforcement Learning (RL) united with Deep Reinforcement Learning (DRL) and diverse combination methods. Traffic signal control problems benefit from specific advantages and drawbacks in each applied technique [3].

3.2 Single-Intersection vs. Multiple-Intersection Techniques

Single-Intersection Adaptive Traffic Signal Control (SI-ATSC). Single-intersection ATSC methods focus on improving traffic throughput at a single defined junction without factoring the neighboring intersections. The approach's strength lies in its ease of execution and quick computation time. However, this approach is suboptimal in most practical cases, especially in congested urban areas where disjointed control strategies tend to create boundless vehicle queues. Such queues are known to result in increased average waiting times and, therefore, reduce the benefits of localized optimization. Regardless of these limitations, SI-ATSC is widely addressed in foundational works and continues to serve as a point of reference for more complex multi intersection strategies [5].

Multiple-Intersection Adaptive Traffic Signal Control (MI-ATSC). On the other hand, MI-ATSC frameworks seek to exploit the relationships that exist among gaps from neighboring intersections to improve the overall performance of a road network [7]. These systems need more advanced coordination strategies and greater computational resources, but significantly outperform SI-ATSC in more complex urban areas. The advancement in performance is due to synchronized traffic signaling across multiple intersections. MI-ATSC systems are particularly effective in dense urban environments where intersections are closely spaced and traffic patterns at one intersection significantly impact neighboring intersections [3] (Fig. 2).

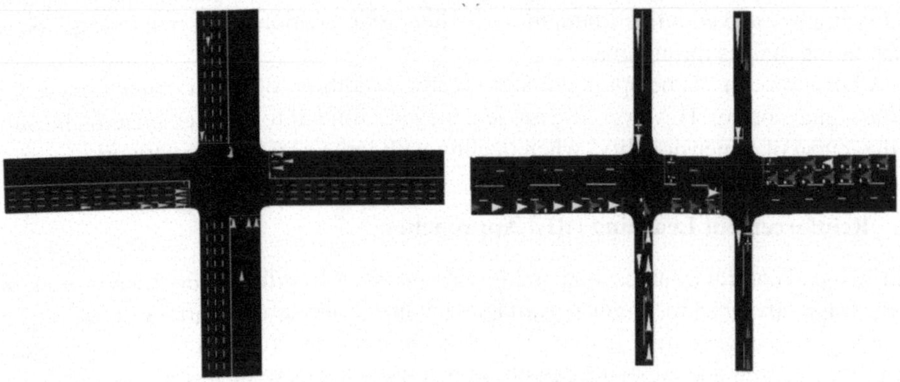

(a) SI (b) MI

Fig. 2. Comparison of traffic flow simulation for (a) Single Intersection (SI) and (b) Multiple Intersection (MI) using a microscopic simulator.

3.3 Fuzzy Logic (FL) Approaches

Fuzzy logic (FL) approaches capture the uncertainty and vagueness associated with traffic systems with the help of heuristics and linguistics. Usually, the input variables considered for these systems are traffic density, queue length, waiting time while green time extension is taken as the main output variable.

FL controllers are capable of managing the non-linear behavior of traffic systems without the need of accurate mathematical formulations. These controllers work well when expert knowledge is available and can be translated into fuzzy rules. The main problem is that fuzzy logic control strategies might struggle with optimality in more complicated scenarios and require extensive fine tuning of membership functions.

3.4 Metaheuristic (MH) Approaches

Traffic signals have been managed with the help of metaheuristic algorithms like Genetic algorithms (GA), Particle swarm optimization (PSO) and Ant colony optimization (ACO) in order to provide near optimal solutions in vast search spaces. These methods are best suited for longtime plans which need offline optimization and re-calibration of signal planning and timing.

Such MH techniques are helpful in finding appropriate solutions from a wide-ranging and complicated solution set without the need to use elaborate algorithmic frameworks. Their major drawback, however, is the lack of real-time applicability due to the need for extensive pre-computation modifications.

3.5 Dynamic Programming (DP) Methods

Dynamic Programming methods treat the problem of traffic signal control as a hierarchy of simpler tasks. This method resolves lower-level problems for each possible optimal timing strategy and combines them into a higher order solution – one that minimizes the delay or maximizes throughput.

A DP approach can be optimal under specific conditions of time control features for traffic signal control. However, in practice, they are limited to smaller systems because of the 'curse of dimensionality' when dealing with large-scale traffic networks.

3.6 Reinforcement Learning (RL) Approaches

Traffic signal control is increasingly taking advantage of Reinforcement Learning as one of the more advanced techniques available. Control policies are learned by RL agents through active engagement within the traffic environment, rewarding favorable flow-controlling actions and imposing penalties for actions that increase traffic congestion [3].

The Q-learning algorithm, a popular RL technique, updates the value function $Q(s, a)$ according to the equation [8]:

$$Q(s_t, a_t) \leftarrow Q(s_t, a_t) + \alpha\left[r_{t+1} + \gamma \cdot \max_a Q(s_{t+1}, a_t) - Q(s_t, a_t)\right] \tag{1}$$

where $Q(s_t, a_t)$ is the value of action a_t taken from state s_t, α is the learning rate, r_{t+1} is the reward received, and γ is the discount factor.

Unlike other approaches, RL techniques do not require explicit mathematical models of traffic dynamics because they can accommodate changing traffic conditions. Still, they can take a long time to train, and they can struggle with the exploration-exploitation balance in more complex environments [2].

3.7 Deep Reinforcement Learning (DRL) Approaches

Traffic signal control is a domain where DRL can be used as it enables learning from raw sensor information and allows for complex mapping between traffic states and control actions to be learned. This is achievable due to the use of deep neural networks which enable the handling of high-dimensional state spaces, something traditional neural networks could not do.

A common DRL approach is DQN or Deep Q-Network which uses a neural network to estimate the Q-value function for a given state-action pair.

The equation for DQN's update is as follows:

$$y_t^{DQN} = r_t + \gamma_{t+1} \cdot \max_a Q^T\left(s_{p+1}, a_{p+1}; \theta_t^-\right) \tag{2}$$

where Q^T refers to the target network.

Systems controlled by DRL approaches succeed at understanding complex states simultaneously with their ability to master advanced control practices. These systems need very powerful computational systems for their training process while their output interpretations and validation pose significant challenges.

3.8 Hybrid Approaches

Hybrid approaches unite various AI approaches to allow their individual advantages support each other while reducing their mutual limitations. The combination of FL-RL and DRL-MH represents two standard hybrid approaches that maintain fuzzy logic systems for uncertainty management and real-time learning through reinforcement learning respectively [3].

A hybrid system that includes multiple techniques proves more effective in terms of performance than using a single methodology because it resolves individual shortcomings. These combination methods add complexity that affects design stages as well as implementation and maintenance activities.

4 Simulation Results and Performance Analysis

4.1 Comparative Analysis of AI Techniques

The performance outcomes that AI methods display within traffic signal control systems differ substantially due to specific execution protocols and traffic conditions together with assessment criteria. Multiple studies underwent simulation tests which produced results that display both advantages and drawbacks associated with their techniques (Fig. 3).

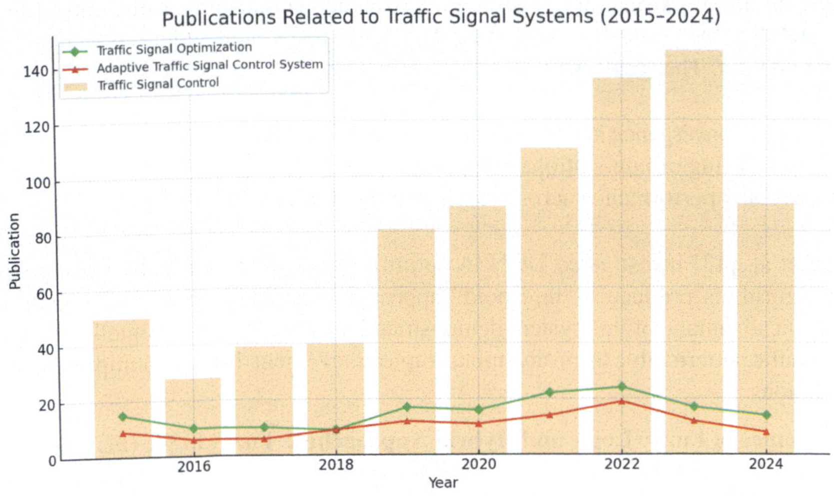

Fig. 3. Publication trends from 2015 to 2024 on traffic signal control and optimization techniques (source: dblp.org).

Performance of Reinforcement Learning Approaches. RL shows exceptional accomplishment when used for traffic signal control systems. El-Tantawy et al. [9] established a multi-phase traffic signal control system through stochastic performance evaluation in Downtown Toronto using RL closed-loop optimal control. The simulation outcome addressed through their study revealed substantial enhancements [3].

- 28% reduction in queue length
- 27% reduction in average latency
- 28% reduction in CO_2 emissions

The research outcomes show that RL establishes itself as a valuable method to achieve better traffic performance together with environmental sustainability benefits. The authors reported that transit vehicles were not included within this framework thus creating limitations for broad urban traffic control strategies [2].

Using Q-Learning at an isolated multi-phase intersection revealed the practical use of RL according to Touhbi et al. [10]. The authors developed a new state space model using distinct reward terms as they realized that reward system efficiency depends heavily on intersection traffic densities. Various traffic scenarios demonstrated high adaptability from their method yet the successful implementation depended on specific adjustments to reward systems.

Performance of Deep Reinforcement Learning Approaches. The utilization of Deep Reinforcement Learning (DRL) methods extends performance limits through its ability to handle sophisticated state data. DQN and IRL techniques implemented by Sahand et al. [11] produced driving results which averaged with human performance. Within their simulation framework the agent acquired lane-changing capabilities through autonomous learning after completing several rounds [3].

DRL processes by Juntao Gao et al. [12] established an ideal ATSC strategy through automatic extraction of information from actual real-time traffic data. The system's stability increased by using their target networks alongside experience replay implementation. The result was:

- Improved convergence speed compared to traditional RL
- Better handling of non-stationary traffic patterns
- More robust performance across varying traffic densities

Liu et al. [13] investigated DQN to optimize TSC policies and demonstrated that DQN algorithms produce a "threshold" approach in single-intersection scenarios. A significant advantage of this system demonstrated the development of intelligent "green wave" traffic patterns able to ensure uninterrupted movement through multiple connected intersections.

Performance of Fuzzy Logic and Hybrid Approaches. Fuzzy Logic (FL) approaches have shown particular strength in handling uncertainty in traffic patterns. Studies implementing FL controllers have reported:

- Improved adaptability to sudden changes in traffic volume
- Reduced average waiting time during peak hours
- Better handling of pedestrian crossing requirements

Hybrid approaches combining multiple AI techniques often outperform single-technique implementations. For example, systems combining FL with RL leverage FL's ability to handle uncertainty while benefiting from RL's adaptability. Similarly, integrations of DRL with metaheuristic optimization have shown promise in finding optimal hyperparameters for neural network architectures.

4.2 Performance Metrics Comparison

Table 1 summarizes the performance improvements reported across different studies implementing various AI techniques for traffic signal control.

Table 1. Performance comparison of AI techniques in traffic signal control.

Study	AI Technique	Queue Length Reduction	Delay Reduction	Emissions Reduction	Other Metrics
El-Tantawy et al. [9]	RL	28%	27%	28% (CO_2)	–
Sahand et al. [11]	DQN & IRL	22%	25%	Not reported	Human-like behavior
Juntao Gao et al. [12]	DRL	31%	29%	24% (CO_2)	Improved stability
Touhbi et al. [10]	Q-Learning	19%	23%	18% (CO_2)	Adaptable to traffic dynamics
Genders et al. [14]	Neural Network	26%	24%	20% (CO_2)	Three-state depictions
Liu et al. [13]	DQN	25%	27%	22% (CO_2)	"Green wave" formations

4.3 Case Studies and Real-World Implementations

While many AI-based traffic management systems remain in the simulation stage, several have progressed to real-world implementations with promising results [1].

Downtown Toronto Implementation. The implementation of RL-based traffic signal control in Downtown Toronto, as studied by El-Tantawy et al. [9], represents one of the most comprehensive real-world applications. The system was deployed at a remote intersection with multiple phases and demonstrated the practical viability of RL approaches in urban environments [3].

SCOOT and SCATS Enhancements. Traditional adaptive systems like SCOOT (Split Cycle Offset Optimization Technique) and SCATS (Sydney Coordinated Adaptive Traffic System) have been enhanced with AI capabilities in several cities. These enhancements have typically resulted in:

- 12–20% reduction in travel times
- 8–17% decrease in stops
- 5–15% reduction in fuel consumption

Smart City Integrations. AI-based traffic management functions as the essential foundation for different smart city implementation projects. Such implementations establish connections with various urban systems which support the delivery of complete solutions for urban mobility [1].

4.4 Simulation Tools and Environments

The assessment of AI-based traffic management systems heavily depends on selecting an appropriate simulation platform. Popular microsimulation tools include [1]:

- **SUMO (Simulation of Urban MObility)** functions as an open-source microscopic traffic simulation package that manages big road networks using trajectory control interface. The TraCI (Traffic Control Interface) available in SUMO is why this application holds universal popularity among researchers working with RL for traffic control [1].
- **VISSIM:** A commercial microscopic multi-modal traffic flow simulation software that offers detailed modeling of urban and highway traffic, including cars, trucks, buses, and pedestrians.
- **AIMSUN:** A traffic modeling software that can perform microscopic, mesoscopic, and macroscopic simulations, allowing for multi-level analysis of traffic systems.
- **CORSIM:** A microscopic traffic simulation software developed by the Federal Highway Administration (FHWA) that is widely used for analyzing traffic operations.
- **MATSim:** An open-source framework for implementing large-scale agent-based transport simulations, particularly suitable for modeling individual travel behavior.

The selection of appropriate simulation tools depends on the specific requirements of the study, including the scale of the traffic network, the level of detail required, and the computational resources available.

5 Challenges and Limitations

5.1 Scalability Issues

Social services face major obstacles in scaling up their use of artificial intelligence for traffic control because of limitations in system expansion. Controlled environments working with restricted intersections enable many approaches to achieve great performance but their effectiveness declines when implemented across bigger urban transportation systems. Multiple factors limit the practice of large-scale deployment of AI-based traffic management systems [1].

Computational Complexity. The computational needs scale exponentially when more intersections are present particularly when Dynamic Programming and certain Reinforcement Learning approaches are applied. The growing number of intersections creates a computational complexity that restricts the direct implementation of some AI methodologies when dealing with large traffic networks unless powerful computational resources are available for deployment [3].

Coordination Complexity. ATSC systems that operate across multiple intersections need to conduct time signal coordination among numerous intersections which possess distinct traffic patterns and operational restrictions. The coordination process becomes harder to manage because larger networks need complex algorithms which must optimize between network-wide efficiency and local needs.

Training Data Requirements. Deep Learning techniques need massive training datasets for attaining their maximum operational effectiveness. The assembly and handling of network data for major metropolitan areas proves challenging because cities equipped with inadequate sensing technologies must overcome both technical and logistical issues [15].

5.2 Real-Time Implementation Challenges

Real-world implementation following simulation brings forward the following practical issues:

Computational Latency. The process of real-time traffic management requires sensors to operate immediately followed by quick decision-making steps. Extended neural network applications used in AI can bring unacceptable time delays to decision making that hinders system performance during active traffic operations.

Reliability and Robustness. Operations of real-world traffic systems depend on continuous reliability under every condition alongside failures of sensors and interruptions in communication and extreme weather conditions. Public safety remains the primary challenge when developers strive to enhance the operational reliability of AI platforms in different conditions.

Safety Considerations. Public safety stands as the main target which traffic management systems influence directly. Traffic safety becomes a significant issue after implementing AI systems which function independently to manage traffic flow because such systems must interact with human-operated vehicles and established older infrastructure during the transition [1].

5.3 Data Requirements and Quality

Physically possible traffic management system performance strongly depends on both the quality and access to existing traffic data.

Sensor Infrastructure. To manage traffic efficiently one needs detailed sensor systems which gather current traffic status data. The lack of required infrastructure in numerous cities reduces the possibility of deploying modern AI-based systems.

Data Quality and Preprocessing. Traffic data includes several problems such as noise along with absent values and inconsistency which negatively affect the functionality of AI algorithms. System implementation becomes more complex because data preprocessing together with quality assurance measures prove essential to success.

Privacy Concerns. The monitoring of advanced traffic data may require collecting details which create privacy issues for vehicles and people while operating cameras. The implementation of detailed traffic data collection requires solutions for protecting privacy rights because these two needs create technical limitations and ethical dilemmas.

5.4 Integration with Existing Infrastructure

Cities possess major investments toward their current traffic infrastructure that require immediate replacement:

Legacy Systems Compatibility. AI traffic management programs need to connect with existing infrastructure systems that were built without consideration of integrating AI systems. The challenge with hardware and software incompatibility stops the implementation of modern features and demands expensive system upgrade projects.

Incremental Deployment. AI systems need to be deployed through incremental implementations rather than total replacements because operational needs and budget limitations exist in the present reality. Systems which need to operate properly with partial deployment requirements must face additional implementation difficulties.

Standardization Issues. Standardization problems within traffic management systems create difficulties for integration tasks especially during the process of combining products from various vendors and technology eras [1].

5.5 Validation and Verification

The evaluation of AI-based traffic management systems which operate correctly under all possible circumstances presents a difficult task [1].

Simulation-Reality Gap. Simulation environments tend to produce different performance results compared to real-world applications due to simplification choices found within simulation models. Veteran experts must verify the results of AI simulations through combination approaches which link simulation data with minimal testing on real-world conditions [2].

Long-term Performance Evaluation. The pattern of traffic changes through time because of regular changes in seasons and new urban growth as well as shifting travel habits among residents. The assessment of AI system longevity along with adaptability calls for continuous monitoring that becomes difficult to perform initially when systems begin deployment.

Explainability and Transparency. Advanced AI methods especially deep learning approaches maintain "black box" behavior because their decision-making operations maintain obscure interpretations. The inability to explain computer decisions makes it harder to validate systems and produces distrust in their operational performance particularly during unexpected system behavior [15].

6 Future Research Directions

6.1 Emerging Trends in AI for Traffic Management

AI-based traffic management continues to advance through multiple new tendencies which boost system performance and capabilities at a quick pace.

Advanced Deep Learning Architectures. Deep learning architectures specifically transformers and graph neural networks bring fresh opportunities to manage traffic signal control systems. Such modern architectures improve their ability to extract temporal and spatial dependencies within traffic systems thereby developing more adapted traffic controls. Scientists need to research the methods that advanced architectures could be modified to address traffic management needs [15].

Transfer Learning and Meta-learning. AI-based traffic management solutions demand major programs of data acquisition coupled with ample computing system resources. Transfer learning and meta-learning techniques can transfer learned knowledge across different intersections or cities so new deployments would need reduced training requirements. Research which explores effective methods of transferring traffic control knowledge between urban environments will enhance AI-based system scalability [2].

Explainable AI (XAI). The explainable nature of AI decisions is essential to gain public trust regarding traffic management and regulatory authorization because such systems directly influence commute safety. The development of explainable AI systems with clear traffic control decision explanations should become a future research priority because they must retain outstanding performance metrics.

6.2 Integration with Connected and Autonomous Vehicles

Connected and autonomous vehicles (CAVs) bring both opportunities and difficulties to traffic management yet they emerge in the market.

Vehicle-to-Infrastructure Communication. The communication system known as V2I provides vehicles and traffic infrastructure with a direct way to exchange information. New investigations must examine the dual flow of vehicle-infrastructure information to help traffic signal systems obtain better roadway state awareness and generate customized vehicle guidance recommendations.

Mixed Traffic Management. Autonomous vehicles will adopt the roads alongside human-driven vehicles so traffic systems require techniques to handle this dual-operational system during the development phase. Research must develop effective methods for AI systems to execute traffic management of heterogeneous traffic flows which may require separate operational strategies per vehicle type.

Cooperative Driving and Platooning. Self-governed vehicles have the ability to join together in platoons which leads to enhanced road capacity and decreases energy use. The deployment of future research needs to study combinations of platooning strategies with traffic signal control to maximize network efficiency.

6.3 Multi-modal Traffic Management

The existing AI-based traffic management systems primarily concentrate on motor vehicle control while citywide mobility needs attention to various transportation modes [1].

Public Transportation Priority. Public transportation priority integrated into AI-based traffic signal control systems will boost the operational quality of mass transit services. The development of AI technology requires serious investigation about its ability to simultaneously optimize public transportation flow and overall road traffic efficiency for sustainable city transportation [2].

Pedestrian and Cyclist Integration. Active walking and cycling form fundamental parts in achieving sustainable urban mobility. The exploration of solutions needs to combine improved road user protection with continuing traffic efficiency performance [1].

Mobility-as-a-Service Integration. The development of Mobility-as-a-Service (MaaS) platforms which unite different transportation choices provides excellent chances for improved traffic management systems. Research to find how AI traffic systems should work with MaaS platforms to optimize citywide transportation efficiency would be an important breakthrough in the field [1].

6.4 Edge Computing for Traffic Control

Edge computing resources that are becoming more accessible create fresh opportunities for spread-out traffic control systems [16].

Distributed AI Architectures. Edge computing as a data source next to AI processors contributes to both decreased latency and faster traffic control functions. Further development of distributed AI architectures must achieve proper balance between local traffic optimization and global coordination for the next-generation traffic management structures [16].

Real-time Learning and Adaptation. The ability to process data in real time at edges allows faster local traffic analysis which can lead to instant control method development and adjustment abilities. Academic investigators should study mechanisms which enable edge-based AI systems to update their traffic condition knowledge automatically while continuing to perform without centralized training sessions.

Resilient and Fault-Tolerant Systems. The deployment of distributed edge computing systems strengthens traffic management systems because it reduces their dependency on single centralized infrastructure elements. Research on how distributed systems maintain operational performance during partial failures and communication breakdowns must be studied to become practical [16].

6.5 Smart City Integration

A smart city initiative contains traffic management as its fundamental element.

Cross-Domain Optimization. Scientific studies should investigate ways to combine traffic management practices with city systems including energy systems and environmental units and emergency resources for maximizing shared optimization goals [2].

Data Fusion and Integration. Smart urban systems create various data flows which show potential to improve traffic administration systems through effective data combination. Research has to develop efficient data fusion strategies which merge traffic information and city-based data sources to create enhanced traffic control systems.

Sustainable Urban Mobility. Research should focus on merging environmental and sustainability orientation with traffic management practices. Future study needs to explore methods for which AI systems can maintain their focus on conventional traffic system efficiency measurements while achieving sustainable targets related to emission cuts and energy savings and urban living standards protection [1].

7 Conclusion

This comprehensive literature review has examined the rapidly evolving field of intelligent traffic management systems that leverage artificial intelligence for traffic light control. The review has traversed the theoretical foundations, implementation approaches, performance evaluations, challenges, and future directions of this critical domain at the intersection of transportation engineering and artificial intelligence [1].

Traditional fixed-time traffic signal systems have proven increasingly inadequate in addressing the dynamic nature of modern urban traffic flows. With studies indicating that fixed traffic signal latencies account for approximately 10% of global traffic delays and contribute to over 295 million hours of delay annually in developed countries alone, the need for more adaptive and intelligent approaches is evident. AI-based Adaptive Traffic Signal Control (ATSC) systems have emerged as a promising solution to this pressing urban challenge [5].

Our analysis of various AI techniques—including Fuzzy Logic, Metaheuristic algorithms, Dynamic Programming, Reinforcement Learning, Deep Reinforcement Learning, and hybrid approaches—reveals significant performance improvements across key metrics. Simulation results from multiple studies demonstrate reductions in queue length (19–31%), average vehicle delay (23–29%), and emissions (18–28%) compared to traditional fixed-time systems. These improvements translate to substantial economic, environmental, and quality-of-life benefits for urban populations [3].

ATSC system classifications regarding single or multiple intersection management provide key knowledge about system scalability alongside coordination abilities. The implementation of single-intersection approaches creates simple setups which assist researchers in algorithm testing yet multiple-intersection approaches deliver better results for optimizing interconnected traffic networks.

The extensive deployment of AI-based traffic management systems faces various substantial obstacles despite their promising starting outcomes. Continuous research and development need to resolve problems regarding scalability and real-time capabilities as well as data collection needs and existing infrastructure compatibility and system validity assessments. Management of complex urban network systems along with the need to maintain reliability especially under different circumstances and close simulation-reality gaps demands immediate action [1].

Installation of AI-based traffic management systems will proceed through an extended period instead of immediate implementation. AI-based traffic management systems will integrate through incremental development steps as new capabilities are combined with present systems then tested among actual traffic conditions. Proper integration of AI-based traffic management systems into everyday operations requires active alliance between transportation engineers, urban planners, policymakers and computer scientists to fulfill municipal needs of all neighborhood types [1].

AIT will establish itself as the base of traffic management through a lengthy process. The implementation will progress slowly through successive stages where AI functions are added to start and real-world confirmation processes follow. Specifically, this transition needs combined collaboration of transportation engineers and both computer scientists and urban planners with policymakers to satisfy the many needs of urban populations [17, 18].

The application of AI-based traffic management technologies exceeds traditional traffic control methods by creating a favorable path toward developing more effective and sustainable and liveable urban environments. Future urban mobility will continue to evolve through the quick development of AI alongside transportation technologies even though challenges persist. Further scientific advancement combined with new exploration of system possibilities will result in improved capabilities that lead to the widespread implementation of traffic management solutions for congested urban areas [1].

References

1. Amador, A., Dias, R., Dias, T., Canas, T.: Traffic management systems. In: Intelligent Transport Systems, pp. 247–271. Wiley (2015)
2. Miletić, M., Čakija, D., Vrbanić, F., Ivanjko, E.: Impact of connected vehicles on learning based adaptive traffic control systems. In: 2022 IEEE International Conference on Systems, Man, and Cybernetics (SMC), pp. 3311–3316 (2022)
3. Mei, X., et al.: Reinforcement learning-based intelligent traffic signal control considering sensing information of railway. IEEE Sens. J. **23**, 31125–31136 (2023). https://doi.org/10.1109/JSEN.2023.3327696
4. Datesh, J., Scherer, W.T., Smith, B.L.: Using k-means clustering to improve traffic signal efficacy in an IntelliDriveSM environment. In: 2011 IEEE Forum on Integrated and Sustainable Transportation Systems, pp. 122–127 (2011)
5. Agrahari, A., Dhabu, M.M., Deshpande, P.S., Tiwari, A., Baig, M.A., Sawarkar, A.D.: Artificial intelligence-based adaptive traffic signal control system: a comprehensive review. Electronics **13**, 3875 (2024). https://doi.org/10.3390/electronics13193875
6. Realistic Adaptive Signal Control Under Platooned Connected Vehicles Environment – ProQuest. https://www.proquest.com/docview/2869261164

7. Ge, H., Song, Y., Wu, C., Ren, J., Tan, G.: Cooperative deep Q-learning with Q-value transfer for multi-intersection signal control. IEEE Access **7**, 40797–40809 (2019). https://doi.org/10.1109/ACCESS.2019.2907618
8. Lehman, H., Valasek, J.: Application of computational intelligence for command & control of unmanned air systems. In: AIAA Scitech 2019 Forum. American Institute of Aeronautics and Astronautics (2019)
9. El-Tantawy, S., Abdulhai, B., Abdelgawad, H.: Design of reinforcement learning parameters for seamless application of adaptive traffic signal control. J. Intell. Transp. Syst. **18**, 227–245 (2014). https://doi.org/10.1080/15472450.2013.810991
10. Touhbi, S., et al.: Adaptive traffic signal control: exploring reward definition for reinforcement learning. Procedia Comput. Sci. **109**, 513–520 (2017). https://doi.org/10.1016/j.procs.2017.05.327
11. Sharifzadeh, S., Chiotellis, J., Triebel, R., Cremers, D.: Learning to drive using inverse reinforcement learning and deep Q-networks (2016)
12. Shen, Y., Liu, J., Ito, M., Shiratori, N.: Adaptive traffic signal control: deep reinforcement learning algorithm with experience replay and target network (2017). https://doi.org/10.48550/arXiv.1705.02755
13. Liu, C., Xu, X., Hu, D.: Multiobjective reinforcement learning: a comprehensive overview. IEEE Trans. Syst. Man Cybern.: Syst. **45**, 385–398 (2015). https://doi.org/10.1109/TSMC.2014.2358639
14. Optimizing Paths by Lowest CO 2 Emissions for Urban Transportation Networks – ProQuest. https://www.proquest.com/docview/3143190397
15. Hassan, M.A., Elhadef, M., Khan, M.U.G.: Collaborative traffic signal automation using deep Q-learning. IEEE Access **11**, 136015–136032 (2023). https://doi.org/10.1109/ACCESS.2023.3331317
16. Zhu, W., Zhu, C., Cao, Y., Ngai, E.C.H., Zhou, J.: An intelligent route guidance strategy based on congestion type for ITS. In: ICC 2022 - IEEE International Conference on Communications, pp. 5047–5052 (2022)
17. El Mokhi, C., Addaim, A.: Near-optimal cable layout design of a wind farm using genetic algorithm. In: Peng, S.-L., Hao, R.-X., Pal, S. (eds.) Proceedings of First International Conference on Mathematical Modeling and Computational Science. AISC, vol. 1292, pp. 47–56. Springer, Singapore (2021). https://doi.org/10.1007/978-981-33-4389-4_5
18. Satri, J., El Mokhi, C., Hachimi, H.: Artificial intelligence and machine learning for a better decision making in the public sector. Presented at the 8th International Conference on Optimization and Applications, ICOA 2022 - Proceedings (2022)

Advanced Technologies for Smart Cities Management

Data-Driven Approach to Seismic Classification of Moroccan Provinces: PCA, k-Means, and Hierarchical Clustering

Mouna El Mkhalet(✉) and Nouzha Lamdouar

Civil Engineering and Construction Laboratory GCC, Mohammadia School of Engineers, Mohammed V University, Avenue Ibn Sina B.P 765, 10090 Agdal, Rabat, Morocco
mouna.el.2019@gmail.com

Abstract. This study proposes a data-driven methodology for classifying Moroccan provinces based on seismic zoning criteria, specifically Peak Ground Velocity (PGV) and Peak Ground Acceleration (PGA). In the **Introduction**, an overview of earthquakes and seismic zoning is provided. The **results** section includes a detailed review of Principal Component Analysis (PCA), K-means clustering, and Hierarchical Clustering (AHC), discussing their theoretical foundations and implementation. In the **case study,** we evaluate the performance of individual clustering algorithms and hybrid clustering algorithms with PCA. In the **discussion,** we analyze performance indices obtained using cross-validation and quality metrics, showing that K-means clustering combined with PCA is the best approach. For cross-validation we found the average error intra-cluster (SSE), the standard deviation of intra-clusters distance, the average silhouette index, the calinski-harabasz index and davies-bouldin index are better compared to others. Additionally, we examine the sensitivity analysis and show that the model is not robust and we suggest future directions, including other parameters Geographic Information Systems (GIS) for enhanced seismic management and decision-making.

Keywords: PCA · K-means Algorithm · Hierarchical Clustering (AHC) · PGV · PGV seismic zoning · PGA seismic zoning

1 Introduction

According Earthquakes happen when there are movements along existing faults in the Earth's crust and can cause significant damage to infrastructure and loss of life [1–3]. Earthquakes are a manifestation of the sudden discharge of strain energy that has been built up over an extended period, providing direct insights into current tectonic activities, Areas where these phenomena are observed are often regarded as regions of significant seismic and tectonic activity [4].

Some measurements are essential for analyzing earthquakes. In the Moroccan context, in order to derive more accurate conclusions, it is essential to investigate how some seismic parameters vary across different regions of the country. An effective method for this analysis is clustering. In this context, how can we effectively apply algorithms such

C. El Mokhi et al. (Eds.): ICASET 2025, SUCI, pp. 175–182, 2025.
https://doi.org/10.1007/978-3-031-98334-4_16

as K-means and Hierarchical Clustering (AHC) to classify the provinces of Morocco based on PGV, PGV seismic zoning and PGA seismic zoning? Is it relevant to use PCA in conjunction with these two algorithms to ensure reliable results? What methods can we employ to determine the most appropriate model, and which performance evaluation techniques should we utilize to assess their effectiveness? These questions will guide our investigation to improve the understanding and classification of provinces in seismic zoning in Morocco.

2 Methodology

The Fig. 1 below presents the schematic representation detailing the steps of our study.

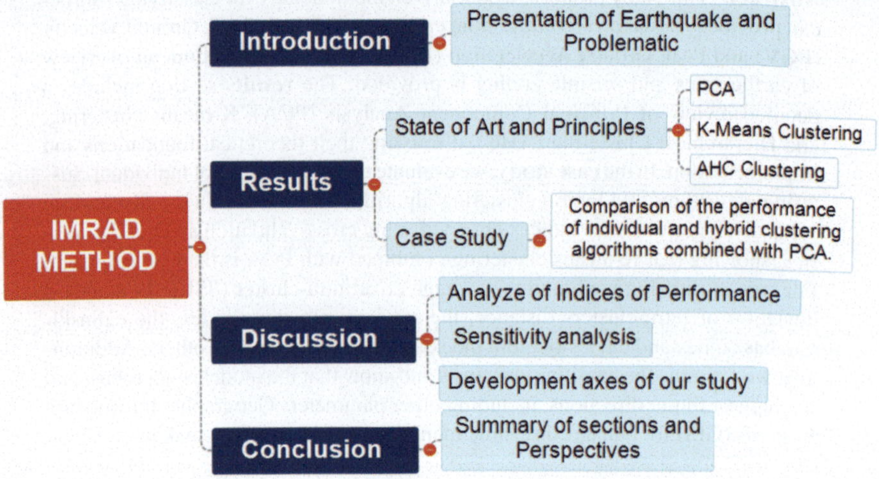

Fig. 1. The schematic representation detailing the steps of our study

3 Results

3.1 Literature Review

Dimensionality Reduction Approach PCA

It converts the initial data, consisting of various interconnected variables, into a new, independent space to extract main components [5]. PCA is a mathematical method used to reduce and analyze data, facilitating their analysis and visualization, in different areas such as engineering, physics and finance [6]. We present the different calculation steps using the PCA method:

According to Wall et al. [7] and Mustafa et al. [8].

Let A be a matrix,

$$A = \begin{pmatrix} a_{11} & \dots & a_{1m} \\ \vdots & \dots & \vdots \\ a_{n1} & \dots & a_{nm} \end{pmatrix}$$

PCA is characterized by performing an eigen decomposition on the covariance matrix $\left(A^T A\right)$

- Step1-Calculate the covariance matrix of $B = \left(A^T A\right)$
- Step2-Calculate the Eigenvalues μ, by solving the equation $\det(B - \mu I) = 0$ and Eigenvectors α, D is a Matrix of Eigenvectors, and organize the pairs according to their eigenvalues, starting with the highest and ending with the lowest.
- Step3- Select the elements and create the Eigenvectors, where the sequence of Eigenvectors aligns with the larger Eigenvalues.
- Step4- Generate the new dataset by multiplying A and V, where C represents the altered data matrix $C = AD$

K-Means Algorithm

The K-means clustering method is an algorithm that divides data into k clusters by optimizing a particular function [9, 10]. This approach typically segments datasets into k clusters by imitating a machine-like procedure; it begins by choosing k initial cluster centers and advances from that starting point [11]. According to Selim et Ismail [12] and Pérez-Ortega et al. [13] K-means involves iteratively dividing a group of n items into k ≥ 2 clusters, ensuring that the items within a cluster are alike and distinct from those in other clusters. The subsequent paragraphs formalize the clustering issue associated with K-means. Consider a set of M objects, $M = \{\alpha_1, \ldots, \alpha_m\}$, which need to be clustered based on their similarity using a similarity criterion. Each α_i is an element of the real number space $\Re^d$, where i ranges from 1 to m and d is greater than or equal to 1, representing the number of dimensions.

In addition, let's take an integer β that is greater than or equal to 2, with $\boldsymbol{\beta} = \{1, \ldots, \beta\}$. When it comes to a β-partition $\Omega = \{O(1), \ldots, O(\beta)\}$ of M, we can denote the centroid of cluster O(j) as V_j, where j is an element of $\boldsymbol{\beta}$. Furthermore, let's denote N as $\{V_1, \ldots, V_\beta\}$ and Z as $\{z_{11}, \ldots, z_{ij}\}$.

Hence, the issue of clustering can be expressed as a problem in optimization, which is illustrated by Equation:

Minimize

$$C(Z, N) = \sum_{i=1}^{m} \sum_{j=1}^{\beta} z_{ij} d\left(\alpha_{i,} V_j\right) \tag{1}$$

Subject to

$$\sum_{j=1}^{\beta} z_{ij} = 1 \, for \, i = 1 \ldots m \tag{2}$$

Subject to

$$z_{ij} = 0\, or\, 1, for\, i = 1 \ldots m\, and\, j = 1 \ldots \beta \quad (3)$$

where $z_{ij} = 1$ the statement indicates that the item α_i is a part of the cluster O(j), and $for i = 1 \ldots m$ and $j = 1 \ldots \beta$ (Table 1).

Table 1. Steps of K-Means Clustering

Step1	# Initialization:
	M = {α1, ..., αm}
	N = {v1, ..., vβ}

Step2	# Classification: For αi ∈ M and vj ∈ N $d(\alpha_i, \beta) = \sqrt{\sum_{j=1}^{n} (\alpha_{i,j} - \beta_j)^2}$ $\alpha_i \rightarrow v_j$ and $j = argmin_k d(\alpha_i, v_k)$
Step3	# Mean position calculation: Calculate centroid vj
Step4	# Convergence: If N = {v1, ..., vβ} remains unchanged in two consecutive iterations Stop the algorithm; else: Go to Classification End

AHC-Clustering

An agglomerative algorithm merges clusters by comparing their differences, starting with separate clusters and uniting them until all items are in one group, it employs methods such as single linkage, complete linkage, average linkage, and centroid linkage to calculate distances [14], Steps of AHC-Clustering: The two equations according to Lance and Williams [15]:

$$C_{ij} = C_i \cup C_j \quad (4)$$

The second equation:

$$D(C_{ij}, C_k) = \alpha_i D(C_i, C_k) + \alpha_j D(C_j, C_k) + \beta D(C_i, C_j) + |\mu D(C_i, C_k) - D(C_j, C_k)| \quad (5)$$

In the case of average Hierarchical agglomerative cluster, we will have the third equation:

$$\boldsymbol{\alpha_i} = \frac{|C_i|}{|C_i + C_j|} \boldsymbol{\alpha_j} = \frac{|C_j|}{|C_i + C_j|}, \beta = 0, \mu = 0 \quad (6)$$

Choice of the Average Method-Preferred hierarchical clustering techniques must be monotonic to ensure a consistent decrease in similarity, as non-monotonic methods like centroid and median can cause noticeable inversions in dendrograms [16]. Batagelj [17] proposed a theorem stating that hierarchical clustering methods derived from the Lance and Williams algorithm can guarantee monotonic dendrograms if certain conditions are respected. The only method that respects the three conditions in the following equation is the Average method [16]. The equation:

$$\begin{cases} \mu \geq -min(\alpha_1 + \alpha_2) \\ \alpha_1 + \alpha_2 \geq 0 \\ \alpha_1 + \alpha_2 + \beta \geq 1 \end{cases} \tag{7}$$

We will subsequently use the Euclidean distance in the following equation:

$$D_{ij} = \sqrt{\sum_{k=1}^{p} \left(x_{ik} - y_{jk}\right)^2} \tag{8}$$

3.2 Case Study -Application of Individuals Algorithms and Combined Algorithms with PCA

We will use the table provided in the Moroccan RPS 2011, which lists the provinces along with the PGV, PGV seismic zoning, and PGA seismic zoning. We use MATLAB to apply clustering methods and obtain the results as follows (Table 2):

Table 2. Comparison of Performance between individuals and Hybrid clustering Algorithms

	Number of clusters	Cross-validation	Clustering quality metrics
K-means clustering	3	Average Error intra-cluster (SSE): 31.1003 Standard Deviation of Intra-Clusters: 7.8231	Average Silhouette Index: 0.6254 Calinski-Harabasz Index: 30.8611 Davies-Bouldin Index: 0.7426
PCA + k means clustering	3	Average Error intra-cluster (SSE): 28.5645 Standard Deviation of Intra-Clusters Distance: 12.3647	Average Silhouette Index: 0.7057 Calinski-Harabasz Index: 33.2241 Davies-Bouldin Index: 0.5918
AHC with average linkage	3	Average Error intra-cluster (SSE):154.6667 Standard Deviation of Intra-Clusters: 1.5320	Average Silhouette Index: 0.4022 Calinski-Harabasz Index: 13.1108 Davies-Bouldin Index: 0.4684
PCA + AHC	3	Average Error intra-cluster (SSE): 41.4041 Standard Deviation of Intra-Clusters Distance: 1.0797	Average Silhouette Index: 0.3911 Calinski-Harabasz Index: 13.1205 Davies-Bouldin Index: 0.4683

4 Discussion

4.1 Analysis of the Indices

According to Table 4 PCA + K-means clustering outperforms K-means alone across all metrics: it achieves the lowest intra-cluster error (28.5645), the highest Silhouette Index (0.7057) indicating better cluster cohesion and separation, the highest Calinski-Harabasz Index (33.2241) demonstrating superior inter-cluster separation, and the lowest Davies-Bouldin Index (0.5918) reflecting better separation relative to dispersion.

4.2 Model Robustness by Sensitivity Analysis

To determine whether the model is robust or not, we proceed to do a sensitivity study by slightly modifying the initial parameters of the study cited at the start, the Table 3 shows that there are certain limitations, the model is not so robust since the number of clusters changes.

Table 3. Sensitivity Analysis

Number of initial Parameters	Number of Clusters
Initial study parameters	3
Increased initial parameter values by 1%	8
Increased initial parameter values by 2%	9
Increased initial parameter values by 3%	9
Increased initial parameter values by 5%	8

4.3 Perspectives

The Fig. 2 below presents the main steps to overcome the limitations found and develop our study with the chosen algorithm.

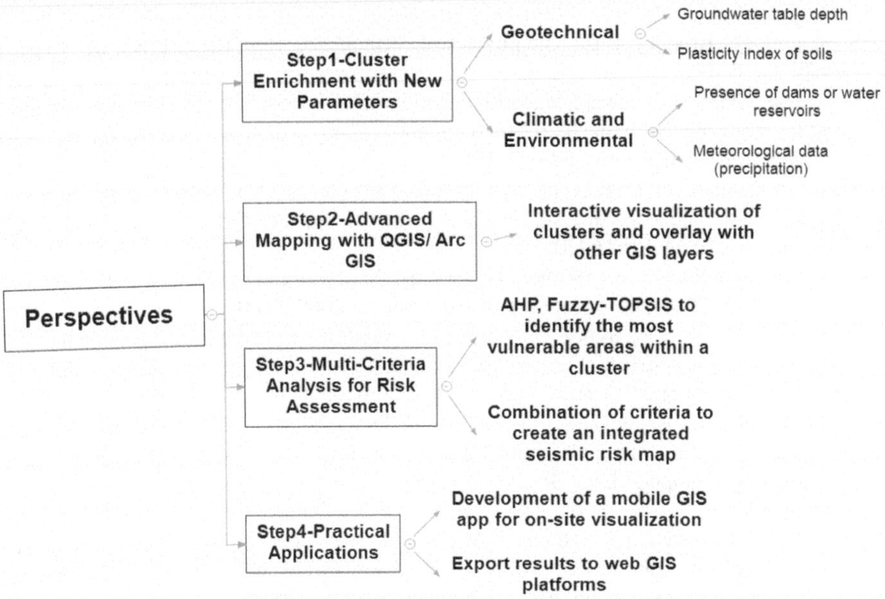

Fig. 2. Development axes of our study

5 Conclusion

In the introduction section, we will provide an overview of earthquakes, including PGV and PGA seismic zoning, while outlining the main research problem. The key questions focus on the application of K-means and AHC for classifying Moroccan provinces, the relevance of using PCA with these algorithms, and the methods for selecting the best model and evaluating performance.

The initial section of the Results will present an extensive literature review on Principal Component Analysis (PCA) along with two algorithms: K-means clustering and Hierarchical Clustering (AHC). This review will elaborate on their core principles and procedural methodologies, supported by pertinent mathematical formulations. The second section of the Results, which centers on a case study, will present we will first assess the performance of the four methods employed: K-means clustering, Hierarchical Clustering (AHC), and the two hybrid approaches—PCA combined with K-means clustering and PCA combined with AHC. This evaluation will utilize the average intra-cluster error (SSE) calculated through cross-validation, alongside three performance metrics: the average Silhouette Index, the Calinski-Harabasz Index, and the Davies-Bouldin Index.

In Discussion, we analyze the table of performance and showing the best performance of the hybrid method PCA combined with K-means clustering. The Table 3 shows that model is not robust. Finally, in perspective we plan to develop a GIS application by combining other different parameters with those that we have studied, where we can really have a map defining the seismic risk at different scales, which is really very important.

References

1. Kanamori, H., Brodsky, E.E.: The physics of earthquakes. Rep. Prog. Phys. **67**, 1429–1496 (2004)
2. Qasim, M., Shah, M., Shahzad, R., Jamjareegulgarn, P.: Atmospheric precursors from multiple satellites associated with the 2020 Mw 6.5 Idaho (USA) earthquake. Adv. Space Res. (2023). https://doi.org/10.1016/j.asr.2023.09.057
3. Haider, S.F., Shah, M., Li, B., et al.: Synchronized and co-located ionospheric and atmospheric anomalies associated with the 2023 Mw 7.8 Turkey earthquake. Remote Sens. **16**, 222 (2024). https://doi.org/10.3390/rs16020222
4. Caputo, R., Pavlides, S.B.: Earthquake geology: Methods and applications. Tectonophysics **453**(1–4), 1–6 (2008). https://doi.org/10.1016/j.tecto.2008.01.007
5. Zang, J.G., Huang, H.B., Sun, Z.G., Wang, D.S.: Subdomain principal component analysis for damage detection of structures subjected to changing environments. Eng. Struct. **288**, 116265 (2023). https://doi.org/10.1016/j.engstruct.2023.116265
6. Jin, J., Cauvin, L., Raghavan, B., Breitkopf, P., Dutta, S., Xiao, M.: Towards a data-driven paradigm for characterizing plastic anisotropy using principal components analysis and manifold learning. Comput. Mater. Sci. **235**, 112834 (2024). https://doi.org/10.1016/j.commatsci.2024.112834
7. Wall, M.E., Rechtsteiner, A., Rocha, L.M.: Singular value decomposition and principal component analysis. In: Berrar, D.P., Dubitzky, W., Granzow, M. (eds.) A Practical Approach to Microarray Data Analysis, pp. 91–109. Kluwer, Norwell (2003)
8. Mustafa, S.F.Z., et al.: Modelling of similarity characteristics of polycyclic aromatic hydrocarbons (PAHs) in Sungai Perak, Malaysia via rough set theory and principal component analysis (PCA). Chem. Phys. Lett. **828**, 140721 (2023). https://doi.org/10.1016/j.cplett.2023.140721
9. Lloyd, S.: Least squares quantization in PCM. IEEE Trans. Inf. Theory **28**(2), 129–137 (1982)
10. Forgy, E.W.: Cluster analysis of multivariate data: efficiency versus interpretability of classifications. Biometrics **21**, 768–769 (1965)
11. Zeebaree, D.Q., Haron, H., Abdulazeez, A.M., Zeebaree, S.R.M.: Combination of K-means clustering with genetic algorithm: a review. Int. J. Appl. Eng. Res. **12**(24), 14238–14245 (2017)
12. Selim, S.Z., Ismail, M.A.: K-means-type algorithms: a generalized convergence theorem and characterization of local optimality. IEEE Trans. Pattern Anal. Mach. Intell. **1**, 81–87 (1984)
13. Pérez-Ortega, J., Almanza-Ortega, N., Vega-Villalobos, A., Pazos-Rangel, R., Zavala-Díaz, C., Martínez-Rebollar, A.: The K-means algorithm evolution. In: Introduction to Data Science and Machine Learning. IntechOpen (2020). https://doi.org/10.5772/intechopen.85447
14. Müllner, D.: Fastcluster: fast hierarchical, agglomerative clustering routines for R and python. J. Stat. Softw. **53**, 1–18 (2013). https://doi.org/10.18637/jss.v053.i09
15. Lance, G.N., Williams, W.T.: A general theory of classificatory sorting strategies II. Clustering systems. Comput. J. **10**(3), 271–277 (1967)
16. Erman, N., Korosec, A., Suklan, J.: Performance of selected agglomerative hierarchical clustering methods. Innov. Issues Approaches Soc. Sci. **8**(1), 180–204 (2015)
17. Batagelj, V.: Note on ultrametric hierarchical clustering algorithms. Psychometrika **46**(3), 351–352 (1981)

Exploring Healthcare Professional's Engagement with E-Health in Oncology: Toward Smarter and Connected Healthcare System

Ouassima Erefai[1,2,3](✉), Hinde Hami[2], Sanae Elkafssaoui[4], Amina Boufars[5], Abdelmajid Soulaymani[2], Abdelrhani Mokhtari[2], Sara Sennani[2], and Rachid Fares[1,3]

[1] Higher Institute of Nursing Professions and Health Techniques of Rabat, Rabat, Morocco
rfares@um6ss.ma

[2] Laboratory of Biology and Health, Faculty of Science, Ibn Tofail University, Kenitra, Morocco
{ouassima.erefai,soulaymani,sara.sennani}@uit.ac.ma

[3] Mohammed VI Faculty of Nursing and Allied Professions, Mohammed VI University of Sciences and Health, Casablanca, Morocco

[4] Royal School of Military Health, Rabat, Morocco

[5] Department of Public Health, Faculty of Medicine and Pharmacy, Mohammed V University, Rabat, Morocco

Abstract. E-health has emerged as a transformative innovation in healthcare, with significant potential to enhance oncology care through improved accessibility, care coordination, and patient-centered services. However, its adoption remains limited, particularly in resource-constrained settings like Morocco, due to various barriers. This study assessed the perceptions, attitudes, and experiences of oncology healthcare professionals in Morocco regarding e-health and identify key barriers to its integration. This was a cross-sectional study conducted using a structured survey administered to oncology professionals. The survey explored sociodemographic characteristics, professional perceptions, training, and practical experiences with e-health technologies. A total of 130 professionals participated, predominantly women, with an average age of 29.9 years. Most participants (95.4%) stated never having received specific training in e-health, and 86.2% lacked practical experience with these tools. Despite this, most recognized the benefits of e-health, including improved accessibility, reduced waiting times and stronger relationships between patients and professionals. Major barriers identified included insufficient training, poor internet connectivity, and concerns about data security. However, over 60% expressed optimism about e-health's future role in oncology care. E-health holds great promise for transforming oncology care in Morocco. Addressing barriers through targeted training, improved infrastructure, and legal reforms is essential to realize its potential within smarter, connected healthcare systems.

Keywords: E-health · Oncology · Digital healthcare · Smart systems · Morocco

C. El Mokhi et al. (Eds.): ICASET 2025, SUCI, pp. 183–193, 2025.
https://doi.org/10.1007/978-3-031-98334-4_17

1 Introduction

Cancer remains among the world's most pressing health concerns, as its prevalence and impact continue to rise [1]. According to GLOBOCAN 2022, approximately 20 million new cancer cases have been registered globally, and almost 10 million succumbed to the disease, making cancer responsible for one in six deaths among non-communicable diseases [2]. This growing burden is driven by population growth, aging, and lifestyle changes [3]. The situation in Morocco reflects global trends, with 63,609 new cases of cancer recorded by 2022 [4]. The rising incidence of cancer places a growing burden on healthcare systems, particularly in resource-limited regions where access to rapid diagnosis and effective treatment remains a major challenge [5].

In Morocco, a lower-middle-income country, the healthcare system suffers from several organizational and logistical problems, with disparities in access to healthcare. These anomalies prevent optimal cancer management [6]. The latest national census conducted in 2024 revealed that around 37% of the population resides in rural areas, where healthcare resources are extremely inadequate [7]. These regions present a significant deficit in biomedical equipment, infrastructure, and healthcare professionals [8]. Such anomalies significantly hamper access to essential medical services, and hurt the timely diagnosis and treatment of cancer [9].

E-health, also known as digital health, broadly refers to the use of information and communication technologies in healthcare to optimize medical services. This includes many systems and services such as telemedicine, mobile health applications, connected "smart" devices, and digital platforms or networks for sharing information [10]. Building on these technologies, e-health provides innovative solutions for improving access to care, facilitating communication healthcare professionals and patients, and rationalizing the management of complex disease such as cancer [11]. In resource-limited settings such as Morocco, e-health can revolutionize oncology care by placing specialized services closer to patients and reducing the burden of travel for treatment [12].

Several countries have evaluated the impact of e-health in oncology, with promising results. Studies in high-income countries show that telemedicine improves patient satisfaction and reduces delays in cancer diagnosis and treatment [13]. In middle-income contexts, e-health has shown significant potential to transform care delivery, underscoring the urgency of digital health reforms to close gaps in access and quality of care [11]. Mobile health interventions in these settings have been particularly effective in improving patient adherence to treatment protocols and facilitating follow-up care. However, despite its benefits, adoption of e-health remains uneven, particularly in low- and middle-income countries, where infrastructural, technological, and cultural barriers persist [14].

The purpose of this study is to explore the knowledge, attitudes, and engagement of oncology healthcare professionals in Morocco with regard to e-health. By understanding their perspectives and identifying existing barriers, this investigation intends to help develop strategies for integrating e-health into oncology care, in order to align with the vision of smarter, more connected healthcare systems in the era of smart cities.

2 Materials and Methods

2.1 Study Design and Data Acquisition

This is a cross-sectional research based on a questionnaire distributed online and in person to healthcare professionals working in oncology throughout Morocco during October and November 2024. The questionnaire was developed in the light of a literature review [15–17] and structured into three thematic sections: socio-demographic characteristics; professionals' knowledge and familiarity with e-health; and perceptions and attitudes towards its use in oncology care. To assess the questionnaire's clarity, relevance and ease of use, a pre-test was carried out with oncology professionals. The questionnaire was then distributed to all tertiary oncology centers in Morocco, to ensure complete geographical representation. Reminders were also sent to participants to maximize the response rate. The first part of the questionnaire covered socio-demographic and professional data, including age, gender and years of experience in oncology. The second part involved questions relating to participants' knowledge and familiarity with e-health. These questions sought to determine whether participants had ever received formal training in e-health, the level of knowledge they claimed to have on the subject, and their familiarity with specific e-health technologies such as telemedicine, mobile health apps and electronic health records. We also asked participants if they had ever used e-health tools in their professional practice.

The last part explores participants' perceptions and attitudes regarding e-Health. We asked them whether they thought eHealth could improve the quality of oncology care, enhance collaboration between healthcare professionals, facilitate communication, strengthen the relationship between healthcare professionals and patients, and reduce the workload of healthcare providers. Participants were also asked to identify the main benefits of e-health, express their opinion on the possibility of it becoming an integral part of oncology care in the future, and say whether they would recommend e-health in their field. Finally, an open-ended question was proposed to gather their suggestions for improving the implementation and integration of eHealth in oncology care.

2.2 Statistical Analysis

The collected data were analyzed with descriptive analysis to synthetize the demographic and professional features of the participants, and their responses regarding knowledge, familiarity, and attitudes toward e-health. Categorical items have been expressed as frequencies and percentages.

2.3 Ethical Considerations

The research was conducted in line with the ethical principles outlined in the Declaration of Helsinki [18], which emphasizes the protection of participants' rights, safety, and well-being. All responding participated in the study voluntarily and were asked to give their informed consent before completing the questionnaire. Answers have been treated with confidentiality and anonymity.

3 Results

3.1 Sociodemographic and Professional Characteristics

Overall, 130 individuals were enrolled in the study. Table 1 summarizes the demographic and professional characteristics of the sample. Most participants were women (103/130, 79.2%), with an average age of 29.9 years. In work experience, 54/130 (41.5%) reported having between 1 and 5 years of experience. Educational qualifications showed that the majority held a bachelor's degree (101/130, 77.7%), followed by a smaller proportion with a master's degree (21/130, 16.2%).

Table 1. Socio-demographic and professional characteristics of the participants

Characteristics	No. Cases	%
Age		
<35 years	98	75.38
≥35 years	32	24.61
Gender		
Men	27	20.80
Women	103	79.20
Educational level		
Bachelor's degree	101	77.70%
Master's degree	21	16.20%
PhD	8	6.20%
Professional experience		
<10 years	93	71.54%
≥10 years	37	28.46%
Oncology experience		
<10 years	97	74.60%
≥10 years	33	25.40%

3.2 Training and Practical Experience

Most (95.4%) reported never receiving specific training in teleconsultation, tele-care, or e-health. Regarding self-assessed proficiency in these technologies, 98/130 (75.4%) considered themselves either "not informed" (49/130, 37.7%) or "not very informed" (49/130, 37.7%). Only a small proportion (10/130, 7.7%) rated their knowledge as "good" or "very good". Practical experience was also limited, with 112/130 (86.2%) indicating no prior use of this modality in their professional practice. Among the 18/130

(13.8%) who had experience, the majority (67.8%) provided a neutral evaluation. Positive experiences were reported by 30.5% of the participants, with 25.4% describing their experience as "good" and 5.1% as "very good". Table 2 presents a detailed breakdown of e-health training, self-assessed proficiency, and practical experience among participants.

Table 2. Training, knowledge, use, and evaluation of e-health among the study population

	n	%
Completion of e-health training (n = 130)		
No	124	95.38%
Yes	6	4.62%
Self-assessment of e-health knowledge (n = 130)		
Little informed	71	54.61%
Not informed	49	37.70%
Well informed	10	7.69%
Practical use of e-health (n = 130)		
No	112	86.2%
Yes	18	13.8%
Evaluation of e-health experience (n = 59)		
Good	18	30.50%
Neutral	40	67.80%
Poor	1	1.70%

3.3 Professional Perceptions and Attitudes

E-health was identified as a beneficial innovation by the majority of participants, with 77.7% recognizing its potential to improve oncology care through enhanced access to timely and specialized services. Many of the interviewed (79.2%) emphasized its ability to promote interprofessional collaboration, particularly within multidisciplinary teams where effective communication and coordination are essential. Furthermore, 73.8% indicated that e-health strengthens the patient-professional relationship by enabling continuous support and follow-up in a flexible, patient-centered manner. Despite these recognized benefits, a strong preference for face-to-face consultations was reported by 95.4% of participants, suggesting that teleconsultation is perceived as a complementary approach rather than a replacement for face-to-face care.

Participants identified five main advantages of e-health: improved accessibility to care, reduced waiting times, optimized care coordination, cost reduction, and enhanced continuity of care (Fig. 2). However, several barriers to implementation were noted. Technical challenges, such as unreliable Internet connectivity and suboptimal audio-visual quality, were frequently reported. Human factors, including communication difficulties with patients with limited technological literacy, were also highlighted. Additionally, 95.4% of respondents cited inadequate staff training as a critical barrier. Security concerns, particularly regarding data protection and patient confidentiality, were frequently mentioned as significant obstacles to broader adoption (Fig. 1).

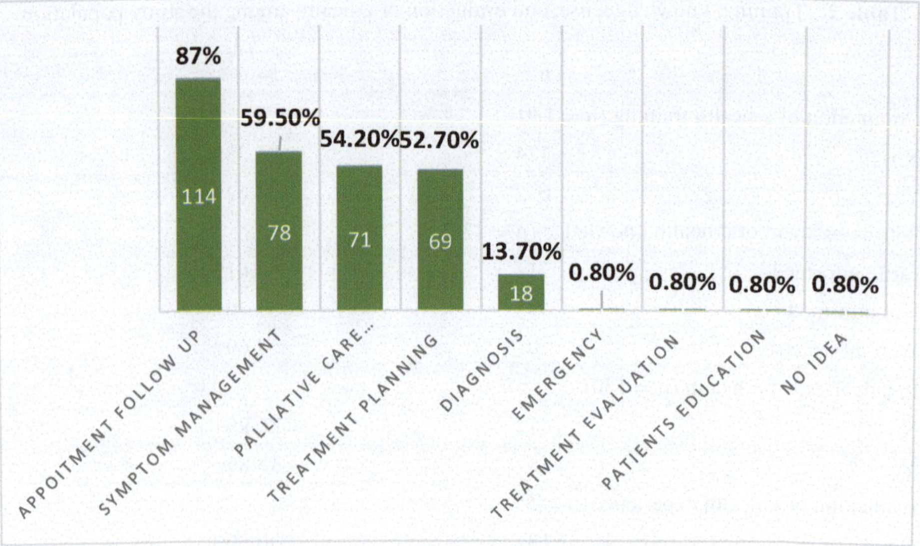

Fig. 1. e-health applications by study population

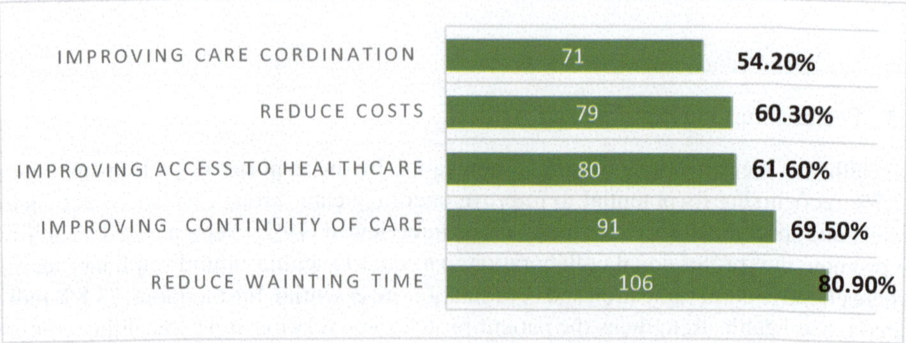

Fig. 2. The benefits of e-health as perceived by participants

3.4 Prospects for Improvement

The outlook for integrating e-health in oncology appears promising, with most participants (63.1%) believing this modality could become a critical component of oncology care. Furthermore, 66.2% of respondents indicated their willingness to recommend it as an option for medical follow-up, highlighting a significant opportunity for its adoption as part of routine clinical care. Nevertheless, a non-negligible proportion of participants (22,90%) demanded adequate training in e-health for healthcare professionals, and 12.27% thought that this training should include patients for greater acceptability. A minority of participants (n = 5) also mentioned closing the legal gap as a way of improving the use of e-health.

4 Discussion

In an era of digital transformation, where technology has integrated all domains, including healthcare systems, this study investigated the perceptions and engagement of oncology healthcare professionals in Morocco towards e-health. By examining their knowledge, attitudes, and practices, this research aimed to highlight potential catalysts and obstacles to the adoption of e-tools and to design appropriate strategies for promoting a connected oncology ecosystem, aligned with the vision of smart cities.

The findings revealed a significant gap in training and practical experience among oncology healthcare professionals regarding e-health technologies such as teleconsultation and telecare. Most participants reported never having received formal training, which aligns with their self-assessment of low proficiency. This lack of familiarity not only limits the adoption of e-health but also reflects broader systemic challenges, particularly in resource-constrained settings like Morocco.

This challenge is not specific to our country. Studies conducted in similar contexts have found the same obstacles, notably insufficient training and capacity building [19, 20]. In Latin America, the scarcity of training has been revealed as a major obstacle to exploiting the full potential of e-health to improve healthcare access [19]. Research in Africa points to a shortage of training opportunities for policymakers and healthcare professionals in e-health, which significantly hampers the adoption of these technologies [20].

Our results also showed a lack of practical experience with digital technology in healthcare among the participants, which further exacerbated the situation. Only a small proportion reported using e-health tools in their professional practice. Among those with prior experience, many expressed neutral feelings, suggesting discomfort or lack of confidence in using these tools. Nevertheless, it is encouraging that a notable fraction reported positive experiences, underlining the potential for success when adequate training and support systems are implemented. These outcomes are in line with existing literature, which demonstrates that healthcare professionals derive benefits from the use of digital tools, both on a professional and personal level, such as greater self-confidence, job satisfaction, and a sense of responsibility. To promote and grow smart healthcare systems, it is essential to share and encourage these positive experiences, as they can be catalysts for wider acceptance and engagement of these technologies by professionals [21, 22].

In this study, e-health emerges as a crucial innovation for enhancing specialized care access, promoting inter-professional collaboration, and strengthening patient-professional relationships. The results corroborate existing literature, underscoring e-health's pivotal role in fostering early disease detection, and enabling more effective health management by patients [23–25]. Moreover, a prospective qualitative study echoed these benefits, highlighting significant reductions in symptom burden, enhanced patient autonomy, and improved self-management, alongside stronger caregiver-patient relationships [26].

These benefits have been recognized across all areas of healthcare, but in oncology in particular, online health services offer a transformative potential for care delivery and patient management. Concrete examples include advanced imaging and predictive analysis, which have facilitated early detection and timely diagnosis of patients. Meanwhile, personalized treatment plans, now made possible through data-driven approaches, have significantly contributed to improving therapeutic precision and effectiveness [25, 27]. Additionally, smartphone applications have improved the quality of life and knowledge of patients with cancer, significantly reducing hospital readmissions and streamlining healthcare delivery [28, 29].

Despite the evident benefits, the uptake of e-health in oncology across Morocco remains limited, faced with substantial barriers. This study underscores the critical need for targeted training for both healthcare professionals and patients to enhance the acceptability and effective utilization of e-health technologies. A systematic review has revealed that e-health training for oncology professionals significantly enhances digital skills, optimizes the use of technology-assisted learning tools, and improves patient care and cancer treatment outcomes [30]. Another multicenter German study demonstrated the key role of patient education in maximizing the potential of e-health in oncology [31]. However, barriers such as workload, limited access to technology, socio-economic barriers including affordability and connectivity issues, which particularly affect disadvantaged populations, persist [32–35].

Furthermore, participants in this study identified urgent need to close legal and regulatory gaps to foster trust and ease of use of e-health applications. A scoping study highlights crucial issues such as data privacy, security, consent, and liability. Resolving these requires clear regulations and enhanced collaboration between stakeholders, to ensure the sustainable integration of e-health technologies in healthcare systems [36].

5 Limitation

Given the cross-sectional design of our study, we were only able to obtain a momentary overview of oncology professionals' knowledge, attitudes and perceptions of e-health. Future research is needed to address these limitations, including longitudinal studies, to evaluate the effectiveness of training programs and explore the perspectives of other stakeholders, in particular patients and policy-makers, to provide a global comprehension of the implementation of e-health in oncology care in our context.

6 Conclusion

Despite the transformative potential in oncology offered by e-health, our findings highlight major obstacles, including a lack of training, limited practical experience and regulatory gaps. A comprehensive and strategic approach is essential to fully capitalize on the potential of e-health in oncology, and align with the vision of smart systems. This includes investing in capacity-building initiatives for healthcare professionals and patients, improving the technology infrastructure, and ensuring robust legal and regulatory frameworks. By challenging these barriers, healthcare systems can evolve towards a connected, patient-centric oncology ecosystem, harnessing the power of digital transformation to improve outcomes and efficiency.

Funding. None.

Declaration of Competing Interest. None.

References

1. Schwartz, S.M.: Epidemiology of cancer. Clin. Chem. **70**, 140–149 (2024). https://doi.org/10.1093/clinchem/hvad202
2. Bray, F., et al.: Global cancer statistics 2022: GLOBOCAN estimates of incidence and mortality worldwide for 36 cancers in 185 countries. CA Cancer J. Clin. **74**, 229–263 (2024). https://doi.org/10.3322/caac.21834
3. Jeleff, M., et al.: Cancer risk factors and access to cancer prevention services for people experiencing homelessness. Lancet Public Health **9**, e128–e146 (2024). https://doi.org/10.1016/S2468-2667(23)00298-0
4. Global Cancer Observatory. https://gco.iarc.who.int/media/globocan/factsheets/populations/504-morocco-fact-sheet.pdf
5. Oyeniyi, J.: Telemedicine and its impact on breast cancer survival in Sub-Saharan Africa. IRJMETS **6**, 2582–5208 (2024). https://doi.org/10.56726/IRJMETS52066
6. Special Commission on the Development Model. The New Development Model: General Report (2021). https://www.hcp.ma/attachment/2327960/
7. High Commission for planning. https://www.hcp.ma/Population-legale-du-Royaume-du-Maroc-repartie-par-regions-provinces-et-prefectures-et-communes-selon-les-resultats-du_a3974.html
8. Mahdaoui, M., Kissani, N.: Morocco's healthcare system: achievements, challenges, and perspectives. Cureus **15**, e41143 (2023). https://doi.org/10.7759/cureus.41143
9. Sauvaget, C., et al.: The journey of cancer patients and the quest to equity: findings from Morocco. Public Health **223**, 33–41 (2023). https://doi.org/10.1016/j.puhe.2023.07.015
10. Stevenin, G., et al.: E-Health and environmental sustainability in vascular surgery. Semin vascul surg **37**, 333–341 (2024). https://doi.org/10.1053/j.semvascsurg.2024.08.005
11. Melhem, S.J., Nabhani-Gebara, S., Kayyali, R.: Leveraging e-health for enhanced cancer care service models in middle-income contexts: qualitative insights from oncology care providers. Digit. Health **10**, 1–13 (2024). https://doi.org/10.1177/20552076241237668
12. Akingbola, A., Adegbesan, A., Ojo, O., Otumara, J.U., Alao, U.H.: Artificial intelligence and cancer care in Africa. J Med Surg Public Health **3**, 100132 (2024). https://doi.org/10.1016/j.glmedi.2024.100132
13. Białczyk, A., Leśniak, G., Nadolny, F., Mrowiec, J., Otałęga, A.: Exploring digital health horizons: a narrative review of e-health innovations in Poland, Spain, Romania and Estonia. Prospects Pharm Sci. **22**, 32–37 (2024). https://doi.org/10.56782/pps.178
14. Dau, H., Kassam, F., Payne, B.A., Miller, H., Ogilvie, G.: Digital technology as a tool to provide social support to individuals with cancer in low-and middle-income countries: a scoping review. medRxiv **8** (2024).https://doi.org/10.1101/2024.08.09.24311634
15. Al-Harbi, A.: Healthcare providers' perceptions towards health information applications at King Abdul-Aziz medical city, Saudi Arabia. IJACSA **2**, 14–22 (2011)

16. Okyere Boadu, R., et al.: Healthcare providers' perception towards utilization of health information applications and its associated factors in healthcare delivery in health facilities in Cape Coast Metropolis. Ghana. PLoS One **19**, e0297388 (2024). https://doi.org/10.1371/journal.pone.0297388
17. Baradwan, S., Al-Hanawi, M.: Perceived knowledge, attitudes, and barriers toward the adoption of telemedicine services in the Kingdom of Saudi Arabia: cross-sectional study. JMIR Form Res **7**, e46446 (2023). https://doi.org/10.2196/46446
18. World Medical Association. World medical association declaration of Helsinki: ethical principles for medical research involving human subjects. JAMA **310**, 2191–2194 (2013). https://doi.org/10.1001/jama.2013.281053
19. Curioso, W.H.: Building capacity and training for digital health: challenges and opportunities in Latin America. J. Med. Internet Res. **21**, e16513 (2019). https://doi.org/10.2196/16513
20. Asah, F.N., Kaasbøll, J.J.: Challenges and strategies for enhancing eHealth capacity building programs in African nations. J. Pers. Med. **13**, 1463 (2023). https://doi.org/10.3390/jpm13101463
21. Wosny, M., Strasser, L.M., Hastings, J.: Experience of health care professionals using digital tools in the hospital: qualitative systematic review. JMIR Hum. Factors **10**, e50357 (2023). https://doi.org/10.2196/50357
22. Lampickienė, I., Davoody, N.: Healthcare professionals' experience of performing digital care visits-a scoping review. Life (Basel) **12**, 913 (2022). https://doi.org/10.3390/life12060913
23. De Louche, C.D., et al.: E-Health education for patients and health professionals in the field of vascular disease. Semin. Vasc. Surg. **37**, 350–356 (2024). https://doi.org/10.1053/j.semvascsurg.2024.08.001
24. Senbekov, M., et al.: The recent progress and applications of digital technologies in healthcare: a review. Int. J. Telemed. Appl. **3**, 8830200 (2020). https://doi.org/10.1155/2020/8830200
25. Borkar, S., Chakole, S., Prasad, R., Bansod, S.: Revolutionizing oncology: a comprehensive review of digital health applications. Cureus **16**, e59203 (2024). https://doi.org/10.7759/cureus.59203
26. Smith, C.S., et al.: Patient-perceived benefits and limitations of standard of care remote symptom monitoring during cancer treatment. Cancer Control **31**, 10732748241297368 (2024). https://doi.org/10.1177/10732748241297368
27. Ge, H., Li, L., Zhang, D., Ma, F.: Applications of digital medicine in oncology: prospects and challenges. Cancer Innov. **1**, 285–292 (2022). https://doi.org/10.1002/cai2.37
28. Seguin, L., Tassy, L.: E-santé, digitalisation ou transformation numérique: impact sur les soins de support en oncologie [E-health, digitalization, or digital transformation: impact on cancer supportive care]. Bull. Cancer **109**, 598–611 (2022). https://doi.org/10.1016/j.bulcan.2021.08.015
29. He, M., Chen, M., Ji, Y., Lu, G.: Effectiveness of smartphone app-based interventions after surgery on quality of recovery among cancer patients: a systematic review and meta-analysis. Ann. Med. **56**, 2390167 (2024). https://doi.org/10.1080/07853890
30. Kulaksız, T., Steinbacher, J., Kalz, M.: Technology-enhanced learning in the education of oncology medical professionals: a systematic literature review. J. Cancer Educ. **38**, 1743–1751 (2023). https://doi.org/10.1007/s13187-023-02329-1
31. Holderried, T.A.W., et al.: Unleashing the potential of eHealth in outpatient cancer care for patients undergoing immunotherapy-a quantitative study considering patients' needs and current healthcare challenges. Front. Digit. Health **6**, 1414442 (2024). https://doi.org/10.3389/fdgth.2024.1414442
32. Bimerew, M.: Barriers and enablers of nurses' adoption of digital health technology to facilitate healthcare delivery in resource-limited settings. Stud. Health Technol. Inform. **315**, 64–68 (2024). https://doi.org/10.3233/SHTI240107

33. Freeman, T., et al.: Barriers to digital health services among people living in areas of socioeconomic disadvantage: research from hospital diabetes and antenatal clinics. Health Promot J Austra **33**, 751–757 (2022). https://doi.org/10.1002/hpja.540
34. Linardon, J., Messer, M., Lee, S., Rosato, J.: Perspectives of e-health interventions for treating and preventing eating disorders: descriptive study of perceived advantages and barriers, help-seeking intentions, and preferred functionality. Eat. Weight Disord. **26**, 1097–1109 (2021). https://doi.org/10.1007/s40519-020-01005-3
35. Kautsch, M., Lichoń, M., Matuszak, N.: EHealth development in selected EU countries: barriers and opportunities. Int. J. Integr. Care **16**, A97 (2016). https://doi.org/10.5334/ijic.2645
36. Bente, B.E., Van Dongen, A., Verdaasdonk, R., van Gemert-Pijnen, L.: eHealth implementation in Europe: a scoping review on legal, ethical, financial, and technological aspects. Front. Digit. Health **6**, 1332707 (2024). https://doi.org/10.3389/fdgth.2024.1332707

Optimal Patient Distribution Enforced by Blockchain

Ahmed Khoumsi[1(✉)] and Mohammed Erradi[2]

[1] Université de Sherbrooke, Sherbrooke, Canada
Ahmed.Khoumsi@USherbrooke.ca
[2] ENSIAS, Université Mohammed V, Rabat, Morocco

Abstract. We consider patients in a city who need emergency hospital care and are *distributed* across city hospitals using an ambulance service $\mathcal{A}$. Patients come from all areas of the city and are categorized depending on the type of care they need to receive. A patient's hospitalization and transportation costs obviously depend on the health care he or she needs. We propose a method to *optimally* distribute patients so that the *total* cost to be paid to $\mathcal{A}$ and hospitals is minimized, while respecting their capacities. A pricing model is also developed to determine fair amounts of payments to hospitals and $\mathcal{A}$. The suggested patient distribution and payment protocol is imposed by a smart contract executed in the blockchain. The interest of using blockchain technology is that it guarantees that patient transport and hospitalization, as well as payments, follow established rules and are recorded in the blockchain in a secure, immutable and fully decentralized manner.

Keywords: Optimal patient distribution · payments · blockchain · smart contract (SC) · Ethereum

1 Introduction

We consider the case of patients in a city who need emergency hospital care and are distributed across city hospitals using an ambulance company $\mathcal{A}$. Here are our three main contributions: 1) we develop a fair pricing model of what is paid to $\mathcal{A}$ and each hospital, for transport and hospitalization costs, respectively; 2) we suggest a method to compute the *optimal patient distribution* (OPD), i.e. a hospital is selected for each patient such that the total cost to be paid to $\mathcal{A}$ and hospitals is minimized while respecting their capacities; and 3) we suggest a protocol executed by a smart contract in the blockchain which monitors patient distribution and enforces the corresponding payments. Although our work is in the field of healthcare, it is rather inspired by the use of blockchain in energy trading [1,2] and electric vehicle charging [3,4]. As this work is a *feasibility study* on the use of blockchain to efficiently implement patient distribution, we make several simplifications in modelling patient distribution (Sect. 2), in order to

C. El Mokhi et al. (Eds.): ICASET 2025, SUCI, pp. 194–203, 2025.
https://doi.org/10.1007/978-3-031-98334-4_18

give enough focus to the blockchain part (Sect. 3). In the conclusion, we suggest several relevant generalizations.

The remainder of the paper is organized as follows: contributions 1 and 2 are presented in Sect. 2, while Sect. 3 presents contribution 3. A conclusion is given in Sect. 4.

2 Optimal Patient Distribution

2.1 Definitions and Notations

$]a, b]$ is the set of real values x such that $a < x \leq b$; the city is partitioned into areas noted $z_1, z_2, \cdots$; the city's hospitals are noted $\mathrm{h}_1, \mathrm{h}_2, \cdots$; a patient of category k is noted patient$_k$; i, j, k are used as indices, respectively for city areas, hospitals and patient categories; I, J, K are the respective indexing sets of city areas, hospitals and patient categories; $\mathcal{A}$ is the ambulance service; and $\mathcal{I}$ is the insurance company that covers ambulance and hospital costs of patients.

In order to consider four quantities $(n_j^k, d_j^k, \pi_{i,j}^k, \pi_i^k)$ that continuously vary over time, we *sample* time at a constant period Δ, i.e. we consider time instants $\tau_0, \tau_1, \cdots$ such that $\tau_0 = 0$ and $\tau_{u+1} - \tau_u = \Delta$ for $u \geq 0$. Our study applies entirely to one τ_u at a time, and hence the considered time instant τ_u is absent from the notation of the following quantities $(n_j^k, d_j^k, \pi_{i,j}^k, \pi_i^k)$: n_j^k is the number of patients$_k$ in h_j at time τ_u; d_j^k is the number of patients$_k$ that leave h_j within $]\tau_u, \tau_{u+1}]$; $\pi_{i,j}^k$ is the number of patients$_k$ that are transported from z_i to h_j within $]\tau_u, \tau_{u+1}]$; and π_i^k is the number of patients$_k$ in z_i within $]\tau_u, \tau_{u+1}]$.

We also need to *partition* time into subsets called *seasons* in order to consider four quantities $(cap_j^k, cst_j^k, cap^k, cst_{i,j}^k)$ that remain constant within each season. Since our study applies entirely to *one* τ_u at a time, which is included in *one* season $\mathcal{S}_v$, τ_u and $\mathcal{S}_v$ are absent from the notation of the following quantities $(cap_j^k, cst_j^k, cap^k, cst_{i,j}^k)$: the capacity cap_j^k is the maximum number of patients$_k$ that can be at the same time in h_j during season $\mathcal{S}_v$; cst_j^k is the cost for h_j to hospitalize one patient$_k$ during season $\mathcal{S}_v$; the capacity cap^k is the maximum number of patients$_k$ that can be transported by $\mathcal{A}$ to hospitals within a period Δ during season $\mathcal{S}_v$; and $cst_{i,j}^k$ is the cost for $\mathcal{A}$ to transport one patient$_k$ from z_i to h_j during season $\mathcal{S}_v$.

2.2 Illustrative Example

As an illustrative example, we consider a city partitioned in two areas z_1 and z_2. Each of the two areas z_i contains one hospital h_i. Both hospitals accept two categories 1 and 2 of patients: non-critical cases (1) and critical cases (2). We consider given time instant τ_u and season $\mathcal{S}_v$, for which the quantities previously defined take values as specified below.

Non-critical Patient Category $(k = 1)$: $n_1^1 = n_2^1 = 10$, i.e. at time τ_u: each of h_1 and h_2 contains 10 patients$_1$. $d_1^1 = 2$, $d_2^1 = 1$, i.e. within $]\tau_u, \tau_{u+1}]$: 2 and 1 patients$_1$ leave h_1 and h_2 respectively. $\pi_1^1 = \pi_{1,1}^1 + \pi_{1,2}^1 = 5 + 2 = 7$,

$\pi_2^1 = \pi_{2,1}^1 + \pi_{2,2}^1 = 0 + 4 = 4$, i.e. within $]\tau_u, \tau_{u+1}]$: in areas z_1 and z_2, 7 and 4 patients$_1$ need hospitalizations respectively; of the 7 patients$_1$ from z_1, 5 are taken to h_1 and 2 to h_2; and the 4 patients$_1$ from z_2 are all taken to h_2. $cap_1^1 = cap_2^1 = 15$, $cst_1^1 = 5$, $cst_2^1 = 4$, $cap^1 = 7$, $cst_{1,2}^1 = cst_{2,1}^1 = 0.5$, $cst_{1,1}^1 = cst_{2,2}^1 = 0.3$, i.e. in season $\mathcal{S}_v$: each of h_1 and h_2 has a capacity of 15 patients$_1$, hospitalization costs per patient$_1$ are 5 and 4 in h_1 and h_2 respectively, $\mathcal{A}$ has the capacity to transport 7 patients$_1$ within $]\tau_u, \tau_{u+1}]$, the cost for inter-area transport of one patient$_1$ by $\mathcal{A}$ is 0.5, while the cost for intra-area transport of one patient$_1$ is 0.3.

Critical Patient Category *($k = 2$)*: $n_1^2 = 7$, $n_2^2 = 5$, i.e. at time τ_u: h_1 and h_2 contain 7 and 5 patients$_2$ respectively. $d_1^2 = 2$, $d_2^2 = 1$, i.e. within $]\tau_u, \tau_{u+1}]$: 2 and 1 patients$_2$ leave h_1 and h_2 respectively. $\pi_1^2 = \pi_{1,1}^2 + \pi_{1,2}^2 = 6 + 0 = 6$, $\pi_2^2 = \pi_{2,1}^2 + \pi_{2,2}^2 = 1 + 2 = 3$, i.e. within $]\tau_u, \tau_{u+1}]$: in areas z_1 and z_2, 6 and 3 patients$_2$ need hospitalizations respectively; the 6 patients$_2$ from z_1 are all taken to h_1; and of the 3 patients$_2$ from z_2, 1 is taken to h_1 and 2 to h_2. $cap_1^2 = 13$, $cap_2^2 = 10$, $cst_1^2 = 6$, $cst_2^2 = 7$, $cap^2 = 5$, $cst_{1,2}^2 = cst_{2,1}^2 = 0.7$, $cst_{1,1}^2 = cst_{2,2}^2 = 0.4$, i.e. in season $\mathcal{S}_v$: h_1 and h_2 have capacities of 13 and 10 patients$_2$ respectively, hospitalization costs per patient$_2$t are 6 and 7 in h_1 and h_2 respectively, $\mathcal{A}$ has the capacity to transport 5 patients$_2$ within $]\tau_u, \tau_{u+1}]$, the cost for inter-area transport of one patient$_2$ by $\mathcal{A}$ is 0.7, while the cost for intra-area transport of one patient$_2$ is 0.4.

2.3 Cost and Constraints Models

Since each h_j receives $\pi_{i,j}^k$ patients within $]\tau_u, \tau_{u+1}]$, all hospitals receive in total $\sum_{i \in I, j \in J, k \in K} \pi_{I,j}^k$ patients within $]\tau_u, \tau_{u+1}]$. The cost for all these hospitalizations is $\sum_{i \in I, j \in J, k \in K} cst_J^k \pi_{i,j}^k$, while the total cost for patient transportation is equal to $\sum_{i \in I, j \in J, k \in K} cst_{I,j}^k \pi_{i,j}^k$. Therefore, the total cost for patient transportation and hospitalization is specified by (1):

$$\text{cost} = \sum_{i \in I, j \in J, k \in K} (cst_{I,j}^k + cst_j^k) \pi_{i,j}^k \tag{1}$$

Equation (2) specifies that *all* patients of any category k in any area i are transported by $\mathcal{A}$ to the city's hospitals.

$$\forall i \in I, \forall k \in K : \sum_{j \in J} \pi_{I,j}^k = \pi_i^k \tag{2}$$

Equation (3) specifies that for every z_i, h_j and patient category k, the number of patients$_k$ that are transported from z_i to h_j within $]\tau_u, \tau_{u+1}]$ is ≥ 0:

$$\forall i \in I, \forall j \in J, \forall k \in K : \pi_{i,j}^k \geq 0 \tag{3}$$

Equation (4) specifies that, for each h_j and patient category k, the number of patients$_k$ in h_j after departure and arrival of patients$_k$ within $]\tau_u, \tau_{u+1}]$ (hence the number of patients$_k$ in h$_j$ at time τ_{u+1}) is ≥ 0 and does not exceed the capacity of h$_j$:

$$\forall j \in J, \forall k \in K : 0 \leq n_j^k - d_j^k + \sum_{i \in I} \pi_{I,j}^k \leq cap_j^k \quad (4)$$

Equation (5) specifies that for each patient category k, the total number of patients$_k$ in the city within $]\tau_u, \tau_{u+1}]$ is ≥ 0 and does not exceed the capacity of $\mathcal{A}$:

$$\forall k \in K : 0 \leq \sum_{i \in I} \pi_I^k \leq cap^k \quad (5)$$

2.4 Minimization of the Total Cost

Linear programming is used to compute the patient distribution that minimizes the total cost, while respecting capacities of hospitalization and transportation. The *expression to minimize* is specified by (1). The *unknown vector* to be solved (*output* of the minimization problem) is specified by all patient transports $(\pi_{i,j}^k)_{i \in I, j \in J, k \in K}$. The *known vector* (*input* of the minimization problem) is specified by the other quantities $(n_j^k, d_j^k, \pi_{i,j}^k, \pi_i^k, cap_j^k, cst_j^k, cap^k, cst_{i,j}^k)_{i \in I, j \in J, k \in K}$. Among the four constraints (2, 3, 4, 5), the first three depend on the *unknown* vector while (5) only depends on *known* quantities. Hence only (2, 3, 4) are used to solve the minimization problem, while (5) is verified initially, as explained below:

- If (5) is not satisfied, the minimization problem has *no* solution. Intuitively, $\mathcal{A}$ does not have enough capacity to transport all patients to hospitals.
- If (5) is satisfied, then the minimization problem is solved in order to find the unknown vectors $(\pi_{i,j}^k)_{i \in I, j \in J, k \in K}$, if any, that minimize (1) while respecting (2, 3, 4). Any of such resulting vectors is called *optimal patient distribution* (OPD). There may result none, one or more OPD.

For simplicity of the presentation, we assume that a single OPD results from the minimization problem.

For the example of Sect. 2.2, there exists indeed a single OPD that minimizes the overall cost, which consists precisely of the distribution given as example, that is, $\pi_{1,1}^1 = 5$, $\pi_{1,2}^1 = 2$, $\pi_{2,1}^1 = 0$, $\pi_{2,2}^1 = 4$, $\pi_{1,1}^2 = 6$, $\pi_{1,2}^2 = 0$, $\pi_{2,1}^2 = 1$, $\pi_{2,2}^2 = 2$.

2.5 Pricing

We assume that all patients in the city are insured by the same insurance company that covers all their ambulance and hospital costs. Let then $\mathcal{I}$ denote our unique insurance company. Note that as our model is developed for a given τ_u, the payments specified below cover only the patients transported by $\mathcal{A}$ and received by hospitals within $]\tau_u, \tau_{u+1}]$. Equation (6) specifies the amount paidTo($\mathcal{A}$) to be paid by $\mathcal{I}$ to $\mathcal{A}$, which covers all transport costs. Equation (7) specifies the amount paidTo(h_j) to be paid by $\mathcal{I}$ to h_j, which covers all hospital costs of h_j. Equation (8) specifies the total amount paidBy($\mathcal{I}$) to be paid by $\mathcal{I}$, which is the total cost specified in (1).

$$\text{paidTo}(\mathcal{A}) = \sum_{i \in I, j \in J, k \in K} cst^k_{I,j} \pi^k_{i,j} \tag{6}$$

$$\forall j \in J : \text{paidTo}(\text{h}_j) = \sum_{i \in I, k \in K} cst^k_{J} \pi^k_{i,j} \tag{7}$$

$$\text{paidBy}(\mathcal{I}) = \text{paidTo}(\mathcal{A}) + \sum_{j \in J} \text{paidTo}(\text{h}_j) = \text{cost} \tag{8}$$

3 Patients Distribution Management Using Blockchain

This section shows how blockchain [5] is used to manage the OPD and corresponding payments computed as shown in Sect. 2. We consider more specifically Ethereum [6], a blockchain equipped with the programming language Solidity [7] which is used to program any logic. A program that implements a given logic is executed in Ethereum and called *smart contract* (SC). Henceforth, with the term SC we refer to the *single* SC used to realize the patient distribution logic.

$\mathcal{I}$, $\mathcal{A}$ and hospitals register as *external* entities of Ethereum, while SC is an *internal* entity of Ethereum from the time it is deployed. Every entity X (whether internal or external) has an account in Ethereum identified by an address noted $@_X$.

3.1 HAC-coin Cryptocurrency

Payments are made in Ethereum by transferring values of a cryptocurrency. Instead of using Ether, Ethereum's default cryptocurrency, we can consider using a dedicated cryptocurrency HAC-coin, to make payments related to Hospitalization and Ambulance Costs (HAC) applications. In Ethereum, a cryptocurrency is usually created as a smart contract that implements an interface named ERC20 [6].

3.2 Deployment of SCand initial operations

After SC deployment in Ethereum, its deployer receives an event from Ethereum containing information such as the address $@_{SC}$ of SC. The SC constructor is automatically executed when SC is deployed. After some initializations, SC constructor publishes the *specification* and *audit* of SC and other information, including information useful for $\mathcal{I}$, $\mathcal{A}$ and hospitals to apply for registration. SC's *specification* describes the logic and rules SC is supposed to impose, while SC's *audit* certifies that SC's implementation conforms to SC's specification. There are several specialized companies for auditing SC [8–10].

3.3 Registrations of $\mathcal{I}$, $\mathcal{A}$ and Hospitals

Figure 1 illustrates the registration of X (i.e. $\mathcal{I}$, $\mathcal{A}$ or any hospital) by representing the interactions between SC and X and internal operations of SC. In the diagrams of Figs. 1, 2 and 3: time evolves from top to bottom; a plain arrow means a *transaction* sent by X for SC, which implies the execution of a function of SC; a dashed arrow means an *event* emitted by SC and detected by X; and a looped arrow means a processing of SC without communication.

Let us first explain the registration of $\mathcal{I}$. After consulting the information published by SC, $\mathcal{I}$ sends a specific transaction for SC to apply for registration. The transaction infers the execution of a function of SC that: sets $\mathcal{I}$ as a listener of SC (so that SC can communicate with $\mathcal{I}$ using events), and then emits an event to require from $\mathcal{I}$ the payment of a deposit $\mathcal{D}_{\mathcal{I}}$. Then, $\mathcal{I}$ sends a specific transaction for SC to transfer the amount $\mathcal{D}_{\mathcal{I}}$ of HAC-coin from $@_{\mathcal{I}}$ to $@_{\text{SC}}$. Then SC creates a balance $\$_{\mathcal{I}}$ which is initialized with the value of $\mathcal{D}_{\mathcal{I}}$. Finally, SC registers $\mathcal{I}$ before emitting an event to inform $\mathcal{I}$ that its registration request is accepted.

Registration of $\mathcal{A}$ and any hospital h_j is similar to that of $\mathcal{I}$, except that deposits $\mathcal{D}_{\mathcal{A}}$ and $\mathcal{D}_{\text{h}_j}$ are null, and balances $\$_{\mathcal{A}}$ and $\$_{\text{h}_j}$ are initialized to zero.

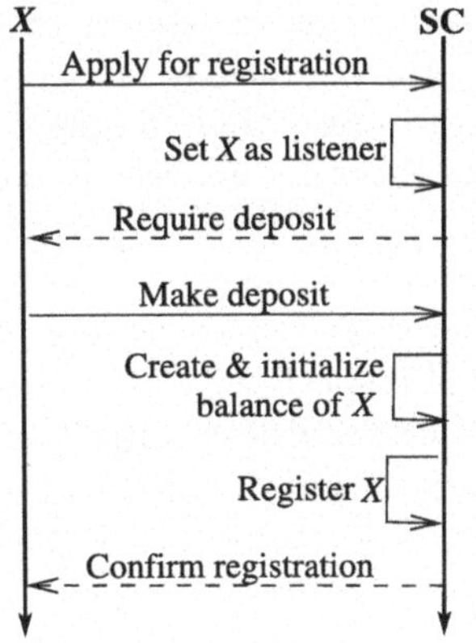

Fig. 1. Interactions with SC for registration of X (X is $\mathcal{I}$, $\mathcal{A}$ or any hospital).

3.4 Payment and Balance Management

Since $\mathcal{I}$ has to pay for transportation and hospitalization costs, the amount $\mathcal{D}_{\mathcal{I}}$ deposited by $\mathcal{I}$ is indeed used by SC to pay $\mathcal{A}$ and hospitals for future transportation and hospitalization costs. As for $\mathcal{A}$ and hospitals, since a priori they are paid and have no payments to make to $\mathcal{I}$, they should not make a deposit. That is why $\mathcal{D}_{\mathcal{A}}$ and $\mathcal{D}_{\text{h}_j}$ are null. However, a deposit is *explicitly* specified for $\mathcal{A}$ and each h_j (although with a null value) to facilitate a future transition to a version that will support *faulty* behaviors of $\mathcal{A}$ and hospitals. In such a case, the deposit is used to apply financial penalties to any perpetrator of faulty behavior.

Balance $\$_X$ ($X = \mathcal{I}, \mathcal{A}$ or some h_j) is a variable that represents the current debt that SC temporarily owes to X. Balances are used by SC to manage payments of $\mathcal{I}$ to Y ($Y = \mathcal{A}$ or some h_j) as follows: paying an amount x by $\mathcal{I}$ to Y consists in decreasing the value of $\$_\mathcal{I}$ by x and increasing the value of $\$_Y$ by x.

We have therefore seen that a *deposit* is realized by a transfer of cryptocurrency to the account of SC, while a *payment* consists in updating balances managed by SC. However, to have an *actual* payment to Y ($Y = \mathcal{A}$ or some h_j), a cryptocurrency transfer to $@_Y$ should be made at some point, instead of only updating $\$_Y$. Such a transfer typically occurs when the amount $\$_Y$ reaches a given threshold or after a given number of periods Δ have elapsed. The transfer is implemented by: transferring the amount $\$_Y$ from $@_{\mathrm{SC}}$ to $@_Y$, and then setting $\$_Y$ to zero. Intuitively, such a cryptocurrency transfer corresponds to the payment by SC of an amount x that SC owes to Y.

3.5 Patient Distribution Protocol

In this section, we present the protocol used in Ethereum to realize the OPD and corresponding payments. We assume that SC has been deployed (Sect. 3.2) and each of $\mathcal{I}$, $\mathcal{A}$ and every hospital has registered (Sect. 3.3). We will therefore only consider in the following the operations carried out after deployment of SC and all registrations. Instead of presenting the protocol in the general case, we will instead directly illustrate it with a generic scenario involving two hospitals represented in Fig. 2.

As we have explained in Sect. 2, time is partitioned into seasons and sampled into times $\tau_0, \tau_1, \cdots$. The first three transactions in Fig. 2 ("SC receives stable info") are sent by h_1, h_2 and $\mathcal{A}$ respectively, to inform SC of the following values *for each season*: each h_j provides $(cap_j^k, cst_j^k)_{k\in K}$, and $\mathcal{A}$ provides $(cap^k)_{k\in K}$ and $(cst_{i,j}^k)_{i\in I, j\in\{1,2\}, k\in K}$. The information is qualified as *stable*, because it is provided once for a long period.

Note that: *computing* the OPD and corresponding payments is *repeated* at instants $(\tau_u)_{u=0,1,\cdots}$; and after each computation of the OPD and corresponding payments at time τ_u, the computed OPD is *executed* by $\mathcal{A}$ and $(\mathrm{h}_j)_{j=1,2}$ within $]\tau_u, \tau_{u+1}]$. Execution of the OPD means that: $\mathcal{A}$ transports the patients which are received by $(\mathrm{h}_j)_{j=1,2}$ respecting the OPD, and then $\mathcal{A}$ and $(\mathrm{h}_j)_{j=1,2}$ are paid for the transport and hospitalization of the patients, respectively. In Fig. 2, the pattern in the grey box represents an instance of the *repetitive* process of *computing and executing* the OPD and corresponding payments, which starts at τ_u and terminates at τ_{u+1}, and hence lasts $\Delta = \tau_{u+1} - \tau_u$. As for the blue box "Execution of OPD" it represents the part of the process which corresponds to the *execution* of the OPD. Let us explain the operations performed inside the grey box.

1. SC emits an event for both h_j ($j = 1, 2$) inviting them to send the current values of $(n_j^k, d_j^k)_{k\in K}$ for the current time τ_u. SC also emits an event for $\mathcal{I}$

inviting them to send information $(\pi_i^k)_{i\in I,k\in K}$ for the current time τ_u. The 2 events are represented by the 2 dashed arrows "SC" asks for dynamic info. The information is qualified as *dynamic*, because it is provided at every $\tau_0, \tau_1, \cdots$.
2. Each of h_1, h_2 and $\mathcal{I}$ provides the requested information by sending a transaction for SC. The 3 transactions are represented by the 3 plain arrows "SC receives dynamic info".
3. SC computes the OPD $(\pi_{i,j}^k)_{i\in I,j\in\{1,2\},k\in K}$ and the corresponding payments paidBy($\mathcal{I}$), paidTo($\mathcal{A}$), paidTo(h_1) and paidTo(h_2), as shown in Sects. 2.4 and 2.5. Such computing is represented by the looped arrow "Compute OPD & payments".
4. SC emits an event for both h_j ($j = 1, 2$) and $\mathcal{A}$, containing the computed value of the OPD, that is, $(\pi_{i,j}^k)_{i\in I,j\in\{1,2\},k\in K}$. The event is represented by the dashed arrow "SC sends OPD info".
5. For each $\mathcal{I}$, $\mathcal{A}$, h_1 and h_2, SC emits an event containing the payment to be given or received. More precisely the 4 transactions contain the values of paidBy($\mathcal{I}$), paidTo($\mathcal{A}$), paidTo(h_1) and paidTo(h_2), respectively. The 4 events are represented by the 4 dashed arrows "SC sends payment info".
6. Then, the OPD is executed until the end of $]\tau_u, \tau_{u+1}]$. This corresponds to the blue box "Execution of OPD" in Fig. 2.

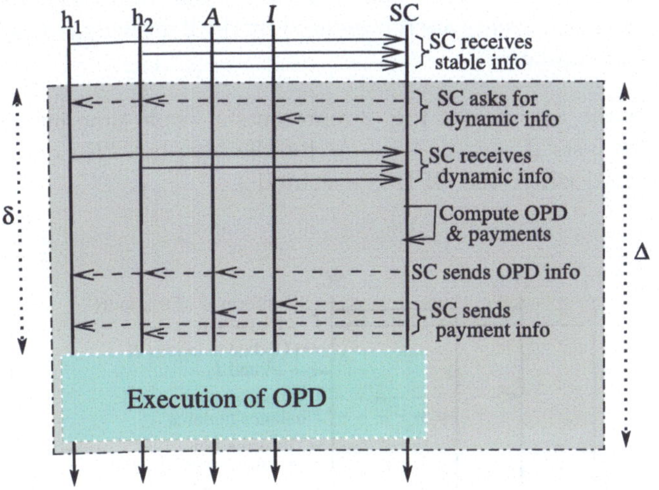

Fig. 2. Interactions with SC for OPD computation & execution.

During the "Execution of OPD" of Fig. 2, the interactions represented in Fig. 3 occur each time a patient is transported by $\mathcal{A}$ and hospitalized by some h_j, as explained in the following:
When a patient is transported by $\mathcal{A}$ to h_j, the interactions in the yellow part of Fig. 3 occur as follows:

- Each of $\mathcal{A}$ and h_j informs SC by generating a transaction for SC which specifies the patient category k, the hospital h_j that receives the patient, and the area z_i from which the patient is transported. The 2 transactions are represented by the 2 plain arrows "SC informed of transport of 1 patient".
- SC reacts by updating balances as follows: $\$_{\mathcal{A}}$ is incremented by $cst^k_{i,j}$ and $\$_{\mathcal{I}}$ is decremented by $cst^k_{i,j}$. This balance update is represented by the looped arrow "Update balances of $\mathcal{A}$ and $\mathcal{I}$".
- Then SC emits an event to each of $\mathcal{A}$ and $\mathcal{I}$ informing them of the balance updates. The 2 events are represented by the 2 dashed arrows "SC informs of balance updates".

When h_j confirms acceptation of the received patient, the interactions in the green part of Fig. 3 occur as follows:

- h_j informs SC by generating a transaction for SC which specifies the patient category k, the hospital h_j that accepts the patient, and the area z_i from which the patient is transported. The transaction is represented by the plain arrow "SC informed of hospitalization of 1 patient".
- SC reacts by updating balances as follows: $\$_{h_j}$ is incremented by cst^k_j and $\$_{\mathcal{I}}$ is decremented by cst^k_j. This balance update is represented by the looped arrow "Update balances of h_j and $\mathcal{I}$".
- Then SC emits an event to each of h_j and $\mathcal{I}$ informing them of the balance updates. The 2 events are represented by the 2 dashed arrows "SC informs of balance updates".

Recall that the grey box of Fig. 2 represents a whole time interval $]\tau_u, \tau_{u+1}]$ which therefore lasts $\Delta = \tau_{u+1} - \tau_u$. As for the blue box "Execution of OPD", it lasts the period where the OPD is executed.

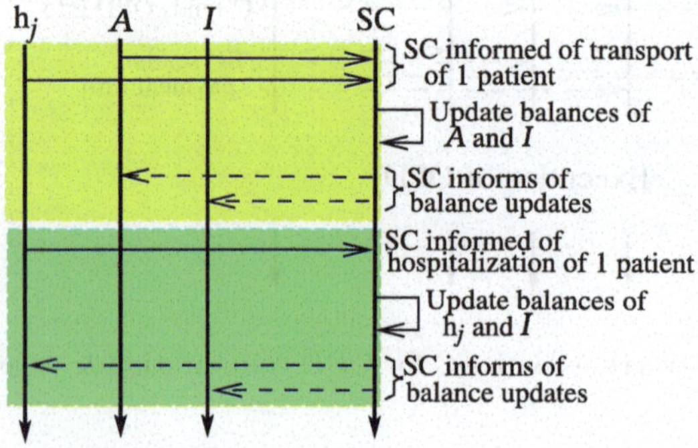

Fig. 3. Interactions with SC for transport (yellow) & hospitalization (green) of a patient.

4 Conclusion

We have considered the case of patients in a city needing emergency hospitalization. The objective of this study has been to efficiently distribute patients across city hospitals. Our main contributions are: 1) we proposed a rigorous method to optimally distribute patients to hospitals in a city, and suggested a pricing model to determine fair amounts of payments to hospitals and the ambulance service; and 2) we suggested a blockchain-based approach to implement our optimal patient distribution and the corresponding payments.

As already mentioned, the main objective of this work was to carry out a feasibility study on the use of blockchain to efficiently implement patient distribution. For that purpose, we used simplified versions of several aspects that are involved in our efficient patient distribution problem, such as insurance coverage and cost models. Since the results obtained in this simplified version of the problem are very satisfactory, we plan to go further with our study by considering the problem in gradually more general versions. Here is a non-exhaustive list of 5 ideas of future work: 1) We assumed that: a) the behaviors of the ambulance service and hospitals are *non-faulty* (i.e. they behave as indicated in the OPD), and b) patient predictions are *perfect* (i.e. the number of each patient category in each area of the city is perfectly predicted); we are now working on the development of a solution that removes assumptions a and b. 2) We plan to *formally* describe and validate all the steps involving blockchain. 3) We plan to implement our proposal on Ethereum, and evaluate its *scalability* in real life scenarios. 4) We assumed that only one ambulance service is used for patient transport; we plan to consider other ambulance service models, for example each hospital having its own ambulance service. 5) We assumed that all patients are covered by the same insurance company that covers all their transportation and hospitalization costs; we plan to consider other insurance models.

References

1. Petrovic, N., Kocic, D.: Framework for efficient energy trading in smart grids. In: 27th Telecommunications Forum (TELFOR). Belgrade (2019)
2. Khoumsi, A.: Efficient collaborative energy trading between smart grids using blockchain and smart contracts. In: International Conference on Advanced Intelligent Systems for Sustainable Development (AI2D). Tangier (2020)
3. Kim, M., et al.: A Secure charging system for electric vehicles based on blockchain. Sensors **33**(3028), 1–22 (2019)
4. Khoumsi, A.: An efficient blockchain-based electric vehicle charging management system. In: 2nd Workshop on Blockchain theoRy and ApplicatIoNs (BRAIN). Athens (2021)
5. Ushmani, A.: Blockchain insight. Int. J. Comput. Sci. Trends Technol. (IJCST) **7**(2), 1–3 (2019)
6. Ethereum whitepaper. https://ethereum.org/en/whitepaper/
7. Solidity. https://docs.soliditylang.org/
8. Blockchain security - security audits and services. https://chainsecurity.com/
9. Deepening the science of security. https://www.trailofbits.com/
10. Certik foundation. https://certik.org/

Toward A Generic Congestion Forecasting Framework

Zakaria Charouh[1](✉), Amal Ezzouhri[2], Mounir Ghogho[3], and Zouhair Guennoun[4]

[1] Giantlink, Fez, Morocco
z.charouh@gmail.com
[2] L3IA Laboratory, Sidi Mohamed Ben Abdellah University, Fez, Morocco
[3] TICLab, International University of Rabat, Rabat, Morocco
[4] ERSC Team, EMI, Mohammed V University, Rabat, Morocco

Abstract. Accurate traffic forecasting is crucial for efficient travel planning, congestion management, and urban mobility optimization. Traditional forecasting methods rely heavily on extensive historical datasets, requiring continuous data collection and processing, which can be costly and impractical in certain scenarios. This study introduces a generic congestion forecasting framework that operates without the need for historical traffic data. Instead of relying on continuous monitoring, the proposed approach leverages a machine learning-based model that extracts relevant information from road characteristics to predict congestion levels for specific time slots. Extensive experiments were conducted to evaluate the effectiveness of our approach. With an accuracy of 90%, our framework provides a cost-effective and scalable alternative to conventional solutions by reducing dependence on large-scale data collection. The proposed method can be seamlessly integrated into smart city applications, enabling real-time traffic management and improved decision-making for urban planning.

Keywords: driving range · congestion · urban planning · electric vehicles · traffic information

1 Introduction

The integration of Intelligent Transportation Systems (ITS) with data analytics presents a significant opportunity to address various transportation challenges, including travel monitoring, urban planning, advanced behavioral studies, sophisticated road-violation detection, and, most notably, traffic congestion forecasting—the central focus of this study [1].

Traffic Information Estimation (TIE) refers to the process of obtaining traffic states (e.g., mean speed, density, and volume) from historical data collected through trained observers or machine-based sensors [2]. While a driver's profile can be inferred from speed and acceleration patterns, traffic information provides

C. El Mokhi et al. (Eds.): ICASET 2025, SUCI, pp. 204–212, 2025.
https://doi.org/10.1007/978-3-031-98334-4_19

a broader understanding of driving maneuvers. Additionally, TIE plays a crucial role in urban planning by facilitating road expansion through the accurate identification of frequently congested roadways [3,4].

Traffic Information Prediction (TIP) is an online retrieval process that leverages pre-installed sensors and real-time data processing techniques. Many mobile applications, such as Google Maps, Waze, and TomTom, provide TIP services, with some extending their capabilities to future-state Traffic Information Forecasting (TIF).

One key application of TIP and TIF is alleviating range anxiety among electric vehicle (EV) users, particularly those with limited battery capacity. By offering accurate travel time and congestion forecasts, these systems help drivers optimize their routes and departure times, thereby reducing uncertainty and improving trip efficiency. According to the literature, TIP and TIF rely on various sensing technologies, including cameras, inductive loops, smartphones, GPS, and Bluetooth [5].

Existing TIF frameworks predominantly depend on historical data. Traditional approaches employ machine learning-based techniques and spatiotemporal datasets to construct predictive models. Moreover, advanced forecasting techniques necessitate continuous data collection for model updating (i.e., online learning) or rely on historical data windows for interactive forecasting [6]. As a result, these methods require permanent sensor installations with sufficient spatiotemporal resolution or a high density of GPS users traversing the studied road segments.

In contrast, this work proposes a novel TIF framework that eliminates the need for historical data, making it adaptable to any road segment or region without requiring prior traffic records.

The rest of the paper is organized as follows. In Sect. 2, we discuss the effective methods from the literature used for TIP and TIF. In Sect. 3, we describe the data collection strategy and the proposed datasets, and we present the proposed framework for TIF. The experimental results and benchmarks are presented in Sect. 4. Finally, a conclusion is drawn in Sect. 5.

2 Related Work

Traffic congestion forecasting has been extensively studied using various approaches. These methods primarily rely on structured datasets collected from different sources, such as fixed and mobile sensors.

Clustering-based methods have been widely applied to identify congestion patterns. Several studies employ Principal Component Analysis (PCA) combined with Gaussian Mixture Models (GMMs) and k-means clustering to classify traffic days based on historical congestion data [7]. Similarly, dynamic clustering techniques applied to vehicular trajectory data have proven effective in detecting congestion hotspots, integrating adaptive forgetting factors to enhance real-time traffic state estimation [8]. Another approach utilizes self-organizing maps (SOMs) to cluster congestion patterns based on time-varying traffic performance indices (TPI), providing deeper insights into traffic trends [9].

In addition to clustering techniques, deep learning models have gained popularity for traffic forecasting. Recent studies leverage Convolutional Neural Networks (CNNs) and Vision Transformers (VTs) to extract spatial and temporal congestion features from traffic images, significantly improving anomaly detection and forecasting accuracy [10]. Furthermore, Long Short-Term Memory (LSTM) models have been proposed to predict congestion propagation directions, demonstrating strong performance in both short- and long-term forecasting horizons [11]. Other studies introduce hybrid architectures, such as Graph Convolutional Networks (GCNs) combined with LSTM models, to capture spatial and temporal dependencies in traffic data, achieving robust performance across different forecasting horizons [12]. Machine learning methods, including Support Vector Machines (SVMs), Decision Trees (DTs), and K-Nearest Neighbors (KNN), have also been extensively explored for congestion prediction. Research on Vehicular Ad hoc Networks (VANETs) has shown that these models effectively detect and forecast congestion patterns, with ensemble learning approaches yielding high accuracy and robustness in intelligent traffic management systems [13]. Additionally, some studies have integrated Hidden Markov Models (HMMs) with contrast-based measures to improve forecasting reliability for prediction horizons ranging from 5 to 90 min [14]. Beyond traffic flow analysis, socio-demographic and economic factors have been shown to influence congestion levels. Research indicates that variables such as employment rates, fuel prices, rental vacancy rates, and economic conditions significantly impact congestion metrics, including travel time indices and congested hours [15]. These findings highlight the importance of incorporating external factors into traffic forecasting models. Regarding data sources, existing studies rely on two main categories: fixed and mobile sensors. Fixed sensors include inductive loop detectors [16] and vision-based sensors [17], which provide continuous traffic flow measurements at specific locations. Mobile sensors, such as GPS-equipped vehicles and cellular data, offer a more flexible and scalable solution for large-scale congestion monitoring [18]. Some approaches apply data preprocessing modules, such as converting raw speed and volume data into 2D matrices before applying CNN-based models for traffic information forecasting (TIF) [19]. Others employ clustering-based preprocessing techniques, such as k-means algorithms, combined with short-term forecasting using Random Forest models to enhance predictive performance in urban areas [20].

Despite the effectiveness of existing approaches, many require large-scale structured datasets and significant computational resources. While machine learning models such as SVMs and Decision Trees offer scalability and robustness, they often require extensive feature engineering to generalize across different traffic conditions. Our proposed framework addresses these limitations by introducing a generic congestion forecasting model that minimizes the need for continuous data collection. Instead, it leverages abstracted road features, making it adaptable to various regions. Furthermore, we present a novel dataset covering Moroccan roads, serving as a valuable benchmark for future research in congestion forecasting.

3 Methodology

Our approach to developing a generic framework for Traffic Information Forecasting (TIF) involved defining key road segment features and establishing data collection mechanisms for each attribute. Subsequently, we explored multiple machine learning models and fine-tuned their hyperparameters to identify the most suitable architecture.

This section details the methodology used to construct the proposed system, encompassing feature selection, data collection strategies, and framework design.

3.1 Feature Selection

A **road segment** represents a subdivision of a road characterized by homogeneous attributes. The trajectory is therefore divided into multiple road segments.

Road segments are described by two categories of features:

- **Static features**: Invariant properties such as starting and ending coordinates, length, slope angle, road type, number of lanes, street name, and city.
- **Dynamic features**: Time-dependent factors affecting congestion levels, including the day of the week, weather conditions, and time of day.

3.2 Data Collection

The data collection process was structured around these two feature categories. While static features were recorded once per road segment, dynamic features were monitored over three months to capture temporal variations in traffic conditions.

Static Features. The first step in dataset generation was selecting representative road segments from diverse Moroccan regions. A total of 25 cities were chosen: Rabat, Casablanca, Kenitra, Khemisset, Meknes, Fes, Beni Mellal, Marrakech, Benguerir, Imintanoute, Agadir, Ouarzazate, Essaouira, Khenifra, Taza, Errachidia, Oujda, Nador, Al Hoceima, Chefchaouen, Sidi Kacem, Asilah, Guelmim, Tan-Tan, and Laayoune.

Using the Overpass API [21], we randomly extracted road points within selected areas, as illustrated in Fig. 1). Segments were then generated by applying a length constraint—ranging from a minimum of 200 m to a maximum of 1km—computed using the Haversine formula. Next, the TomTom API [22] was utilized to filter road segments by type, selecting those classified as FRC0 to FRC4:

- **FR-C0**: Highways
- **FR-C1**: Major roads
- **FR-C2**: Other major roads
- **FR-C3**: Secondary roads
- **FR-C4**: Local connecting roads

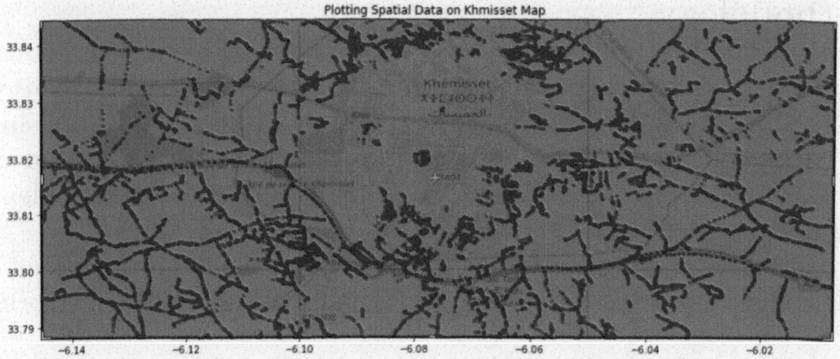

Fig. 1. A demonstration of the proposed points from the selected region

Slope angle computations were performed using NASA's Shuttle Radar Topography Mission (SRTM) [23]. Elevation data for the segment's endpoints were extracted, and the slope was determined accordingly.

Since congestion levels are also influenced by city-wide traffic flow, we introduced a **City Importance (CI) metric**. Traffic conditions in highly populated cities, such as Casablanca, are inherently different from those in smaller cities like Tan-Tan. Inspired by [24], where urban mobility patterns were analyzed using cellular signaling and license plate recognition data, we formulated CI as:

$$CI = \frac{\text{City Population}}{\text{Country Population}} \tag{1}$$

This metric accounts for relative urban activity and its potential impact on congestion.

Dynamic Features. Dynamic features include:

- **Day of the week**
- **Observation time**
- **Weather conditions** (temperature, precipitation, wind speed, surface shortwave radiation)

Weather data were collected via the ClimaCell API [25] at a 5-min resolution, with a forecast window of up to six hours. To ensure accurate labeling, observation timestamps were synchronized with congestion levels and corresponding weather conditions.

Congestion Level Extraction. To acquire the congestion level, we developed a method that accurately extracts traffic conditions from Google Maps. Google determines congestion levels using its users' GPS data, projects the results onto a map, and assigns a color to each road segment (e.g., green for free-flowing traffic and brown for highly congested areas).

In our approach, we define road segments rather than relying on individual points to ensure that the acquired congestion data corresponds to the correct direction. Each road segment is represented by a starting point, an ending point, and a middle point, with the rendered map image always centered at the middle point, as illustrated in Fig. 2).

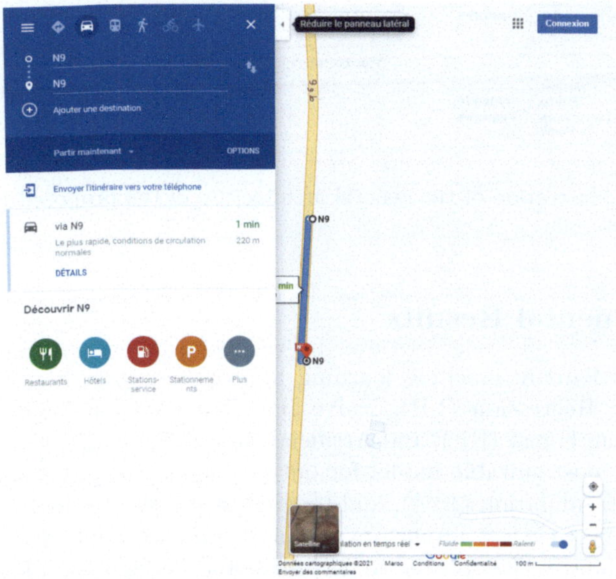

Fig. 2. An illustration of the recorded images

Since congestion information within a single road segment may vary, a direct extraction could lead to inconsistencies. To address this, we applied a color clustering technique to determine the dominant congestion level for each segment, ensuring a more reliable representation of traffic conditions, as depicted in Fig. 3).

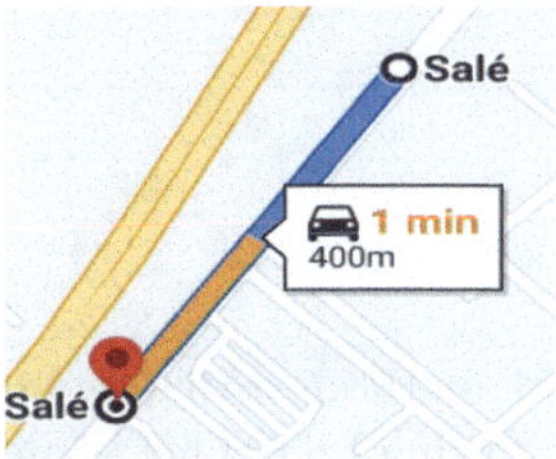

Fig. 3. An illustration of multiple congestion levels for one segment

The observation period was divided into eight time slots: early morning (5–7 AM), rush morning (7–9 AM), morning (9 AM–12 PM), rush afternoon (12–2

PM), afternoon (2–5 PM), rush evening (5–7 PM), evening (7–11 PM), and night (11 PM–5 AM). The overall architecture is summarized in Fig. 4.

Details on machine learning model selection and hyperparameter tuning are provided in the following section.

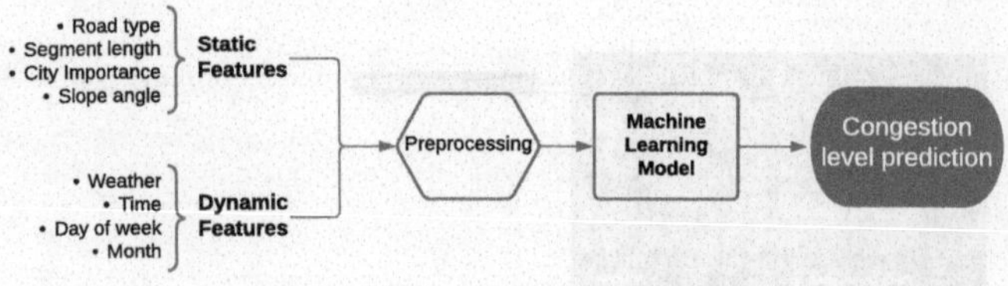

Fig. 4. An illustration of the general architecture of the proposed framework

4 Experimental Results

We investigated many machine learning approaches, including Decision Tree (DT), Logistic Regression (LR), Naive Bayesian (NB), K-Nearest Neighbors (KNN), Random Forest (RF), and Artificial Neural Network (ANN), in order to determine the most suitable model for our problem. We split the data set into training (70%), validation (10%), and testing (20%). We applied the exhaustive search technique to select important features. And we applied the grid search technique to identify the optimal hyperparameter configuration for each model, as summarized in Table 1).

Table 1. A summary of the feature selection results and the obtained accuracy scores

ML	time	day	length	road type	slope	wind	precip.	radiation	Flow	Accuracy
RF	x	x	x	x		x		x	x	0.89
DT	x			x	x			x	x	0.86
LR	x	x		x				x		0.82
NB	x				x	x				0.79
KNN	x	x	x	x		x		x	x	0.88
ANN	x	x		x	x			x	x	0.90

To compare the proposed method with the literature, we implemented a short-term forecasting method using LSTM for a 5 min time horizon; the training set was composed of road segments from just one city (as conventional methods). The training was done on 20 previous congestion levels. The obtained results were 92%. And when compared to the proposed method, we achieved 90% of accuracy on a testing set composed of different cities and different time horizons (from 5 min to 6 h).

5 Conclusion

In this work, we proposed a generic framework for congestion forecasting. Unlike conventional methods, the proposed framework can be applied to any region and does not require continuous data collection to construct window data for traffic impact forecasting (TIF). The framework leverages abstracted features describing a given road segment, including road topology, weather conditions, and city population. The core idea is to employ a generic classification method for forecasting congestion levels. We evaluated various machine learning models, achieving an overall accuracy of approximately 90% across different time horizons ranging from 5 min to 6 h. Notably, when LSTM-based time series forecasting was used, accuracy improved to about 92%. Although the proposed framework is slightly less accurate and reliable than conventional solutions, its results, when compared to the cost of conventional methods, remain highly promising.

References

1. Huang, L., Qin, J., Wu, T.: Multisource data fusion with graph convolutional neural networks for node-level traffic flow prediction. J. Adv. Transp. **2024** (2024). https://doi.org/10.1155/atr/7109780
2. Wang, P., Zhang, Y., Hu, T., Zhang, T.: Urban traffic flow prediction: a dynamic temporal graph network considering missing values. Int. J. Geograph. Inf. Sci. **37**, 885–912 (2022). https://doi.org/10.1080/13658816.2022.2146120
3. Abideen, Z.U., Sun, X., Sun, C.: Traffic flow prediction: a 3D adaptive multi-module joint modeling approach integrating spatial-temporal patterns to capture global features. J. Forecast. (2024). https://doi.org/10.1002/for.3147
4. Modi, Y., Teli, R., Mehta, A., et al.: A comprehensive review on intelligent traffic management using machine learning algorithms. Innov. Infrastruct. Solut. **7**, 128 (2022). https://doi.org/10.1007/s41062-021-00718-3
5. Tang, W.M., Yiu, K., Chen, K.: Conjoining congestion speed-cycle patterns and deep learning neural network for short-term traffic speed forecasting. Appl. Soft Comput. **138**, 110154 (2023). https://doi.org/10.1016/j.asoc.2023.110154
6. Zhu, Z., Xu, M., Ke, J., Yang, H., Chen, X.: A Bayesian clustering ensemble Gaussian process model for network-wide traffic flow clustering and prediction. Transp. Res. Part C-Emerg. Technol. **148**, 104032 (2023). https://doi.org/10.1016/j.trc.2023.104032
7. Chiabaut, N., Faitout, R.: Traffic congestion and travel time prediction based on historical congestion maps and identification of consensual days. Transp. Res. Part C: Emerg. Technol. **124**, 102920 (2021)
8. Reyes, G., Tolozano-Benites, R., Lanzarini, L., Estrebou, C., Bariviera, A.F., Barzola-Monteses, J.: Methodology for the identification of vehicle congestion based on dynamic clustering. Sustainability **15**(24), 16575 (2023)
9. Zang, J., Jiao, P., Liu, S., Zhang, X., Song, G., Yu, L.: Identifying traffic congestion patterns of urban road network based on traffic performance index. Sustainability **15**(2), 948 (2023)
10. Ramana, K., et al.: A vision transformer approach for traffic congestion prediction in urban areas. IEEE Trans. Intell. Transp. Syst. **24**(4), 3922–3934 (2023)

11. Lee, C., Kim, Y., Jin, S.: A visual analytics system for exploring, monitoring, and forecasting road traffic Congestion. IEEE Trans. Visual Comput. Graphics **26**(11), 3133–3146 (2020)
12. Bogaerts, T., Masegosa, A.D., Angarita-Zapata, J.S., Onieva, E., Hellinckx, P.: A graph CNN-LSTM neural network for short and long-term traffic forecasting based on trajectory data. Transp. Res. Part C: Emerg. Technol. **112**, 62–77 (2020)
13. Jasim, M.S., Zaghden, N., Salim Bouhlel, M.: Improving detection and prediction of traffic congestion in VANETs: an examination of machine learning. Int. J. Comput. Digital Syst. **15**(1), 947–960 (2024)
14. Zaki, J.F., Ali-Eldin, A., Hussein, S.E., Saraya, S.F., Areed, F.F.: Traffic congestion prediction based on Hidden Markov models and contrast measure. Ain Shams Eng. J. **11**(3), 535 (2020)
15. Dadashova, B., Li, X., Turner, S., Koeneman, P.: Multivariate time series analysis of traffic congestion measures in urban areas as they relate to socioeconomic indicators. Socioecon. Plann. Sci. **75**, 100877 (2021)
16. Zargari, S.A., Khorshidi, N., Mirzahossein, H., Shakoori, S., Jin, X.: Estimating travel time in the Helsinki region utilising sequential Bayesian inference. In Proceedings of the Institution of Civil Engineers-Municipal Engineer, vol. 177, no. 3, pp. 144–156. Emerald Publishing Limited (2024)
17. Chen, L., Zheng, L., Xia, L., Liu, W., Sun, D.: Detecting and analyzing unlicensed taxis: a case study of Chongqing city. XXPhys. A **584**, 126324 (2021)
18. Akhtar, M., Moridpour, S.: A review of traffic congestion prediction using artificial intelligence. J. Adv. Transp. (2021)
19. Ke, R., Li, W., Cui, Z., Wang, Y.: Two-stream multichannel convolutional neural network for multi-lane traffic speed prediction considering traffic volume impact. Transp. Res. Rec. J. Transp. Res. Board **2674**(4), 459–470 (2020)
20. Chen, Z., Jiang, Y., Sun, D., Liu, X.: Discrimination and prediction of traffic congestion states of urban road network based on spatio-temporal correlation. IEEE Access **8**, 3330–3342 (2020)
21. Overpass API User's Manual. https://dev.overpass-api.de/overpass-doc/en/. Accessed 30 Jan 2025
22. Tomtom Flow Segment Data. https://developer.tomtom.com/traffic-api/. Accessed 30 Jan 2025
23. Shuttle Radar Topography Mission. www.earthdata.nasa.gov/data/instruments/srtm. Accessed 30 Jan 2025
24. Chen, H., Cai, M., Xiong, C.: Research on human travel correlation for urban transport planning based on multisource data. Sensors **21**(1), 195 (2021)
25. ClimaCell API Documentation. https://www.postman.com/api-evangelist/weather/documentation/q11253f/climacell-api. Accessed 30 Jan 2025

Mathematics and Optimization for Sustainability

Modeling of Schedulers Switching Dynamics in a Multimode System Using Switched Linear Differential Systems

Zineb Elqabli(✉), Oulaid Kamach, and Youness Chater

Laboratory of Innovative Technology, Tangier, Morocco
zineb.elqabli@etu.uae.ac.ma

Abstract. Multimode systems are an efficient tool in organizations' endeavor of efficiency, productivity and competitiveness, as they have a crucial role in ensuring an uninterrupted production under normal or degraded conditions. This article presents an important research domain for industries seeking to thrive in an increasingly complex industrial world. Despite the existing contributions on multimode systems found in the literature, the researchers have identified a significant gap in addressing the transition between the different schedulers of a multimode system when a failure occurs. The primary objective is therefore to address the previously identified gap by proposing a formal modeling framework using the theory SLDS (Switched Linear Differential System). The effectiveness of the proposed framework has been validated through an example of a multimode production system where considerable emphasis was placed on gluing conditions, as they establish the necessary parameters for maintaining continuity before and after the switch. These results present significant implications in studying multimode approach since each mode is associated to a specific scheduler, making it crucial to handle the transitions between these schedulers effectively.

Keywords: Multimode systems · Transitions · Scheduling · Gluing conditions · Switched Linear Differential System

1 Introduction

Multimode production systems are designed to handle the increasing complexity and variability in contemporary production environments, since they can adapt to different manufacturing requirements. They are conceived to operate flexibly and reliably in different modes, under both normal and degraded conditions. Under normal conditions, the system operates in nominal mode. On the other hand, if an unforeseen event occurs, the system is impacted and switches directly to one of the degraded modes. The primary aim of these systems is to maintain production regardless of conditions.

Hence, multimode systems are characterized by their ability to manage several production modes simultaneously, which necessitates advanced control and planning strategies.

C. El Mokhi et al. (Eds.): ICASET 2025, SUCI, pp. 215–224, 2025.
https://doi.org/10.1007/978-3-031-98334-4_20

Although multimode systems are a powerful tool for managing complexity in modern industrial facilities. Managing their different modes is a challenge that needs to be faced to avoid the unexpected side-effects of these tools. It mainly involves mastering the mechanisms governing the transition from one mode to another, and enabling it to be coherent and compatible. The elaboration of appropriate transition management strategies which consider both current system states, as well as potential future scenarios and transition trajectories, facilitates understanding of the complex dynamics often present in real systems, and consequently ensures optimal exploitation of multimode systems.

This article is structured as follows: The next section presents an overview of the literature review. Section 3 elucidates the theory and the concepts of the SLDS approach. Section 4 outlines the application of the proposed approach to production scheduling field and discusses the results obtained. Section 5 is a conclusion of all the findings presented in this paper and a discussion of future perspectives and research directions.

2 Literature Review

Numerous articles exist on multimode systems; however, a limited number of these address transitions between the different modes. This transition has been addressed in [1] in a static manner. The researchers focused for the compatible state when the behavior of the physical system changes from one operating mode to another. They therefore developed an algorithm that applies the projection activation sequence to calculate the occurrence vector, then it elaborates the state function on the basis of the calculation of the compatibility matrix of the active mode. And finally, it determines the compatible state thanks to its activity vector and its matrix description. Transition rules between modes are designed in [2] and a new projection function identifying the compatible state is developed. The authors of [3] have developed a new approach called multi-model scheduling to address the challenge of modeling the mode switching. Researchers in [4] managed transitions between modes using RL techniques. Moreover, they introduced the concept of a digital twin into the studied system to avoid wrong decisions. The combination of Reinforcement learning techniques and Digital Twins was beneficial in providing a risk-free real-time management of mode transitions. The researchers [5] controlled the utilization of each mode and the transition between all modes in a multimode load reservoir through a hybrid Petri net that efficiently describes the system's states and transitions, integrating resource constraints. Optimal management of the various operational modes was ensured in [6] by developing an intelligent system based on multi-agent systems (MAS). The MAS proposed leverages the expertise of specialists and operators to quickly identify and detect faults, facilitating the management of the operational modes within an automated manufacturing system. A multimode energy system was addressed in [7]. This article focused also on multi-agent systems (MAS) to develop a hierarchical three-level decentralized hybrid control strategy to ensure intelligent switching of operating modes. In a similar context, the researchers in [8] continued the previous research and introduced the three level- MAS strategy while focusing on the middle-level agents, which utilize event-triggered hybrid controllers conceived with Differential Hybrid Petri-Net models in order to enable flexible and coordinated transitions of the different modes. The researchers in [9] addressed mode switching control strategies of a distributed generation system composed of four operating modes, based-on Petri net (PN) models. A tool based on hybrid automata was proposed in [10] to study the management of operating modes in industrial applications. Researchers in [11] consider different operation modes by using multimode approach, which represents each mode through a model. Discrete-event systems was elucidated in [12], where the researchers used a formal framework based on supervision control theory in order to monitor multiple operating modes.

To the best of the authors' knowledge, hardly any research work has addressed managing of the transition between the different schedulers in a multimode production system. Each operating mode is associated with a specific scheduler designed to optimize its particular processes and requirements, and to maintain a continuous flow of operations, while minimizing downtime and maximizing throughput. The oversight in managing the transition between these schedulers is a deficiency gap that was identified by the researchers. The aim of this article is to study and formally design how switching between schedulers can be defined when a change of mode is required.

This paper applies Switched Linear Differential System (SLDS) theory as a formal modeling framework to ensure smooth transitions between schedulers in each mode. SLDS theory has already been adopted in the literature in other fields, notably Energy Distribution Networks and electrical systems. The researchers in [13] used the switched linear differential systems framework to model electrical devices with switching dynamics. Their research was founded on modularity as a significant feature for modeling the dynamics of complex systems. The results revealed that their new proposed approach for modeling switched electrical systems greatly outperforms traditional representation-oriented theory. Switched Linear Differential System has been studied in [14] from a behavioral perspective. They have studied a bank of behaviors in order to define the switched behavioral system; and provided sufficient conditions for the stability of those systems. They have proved that the existence of a common behavioral Lyapunov function ensures the stability under switching. The authors of [15] have employed SLDS to study the dynamics of energy distribution networks. They have showed that quadratic differential forms and linear differential behaviors are two primary elements of the behavioral system theory.

3 SLDS Approach: Theory and Concepts

SLDS can be explored using a variety of approaches, providing distinct advantages: The abstract approach and the matrix approach.

The abstract approach, often utilized in systems theory, facilitates understanding of the interactions between the elements of a system, avoiding distraction by technical specificities or complex details. Furthermore, it offers the possibility of generalizing results for application in different contexts.

The matrix approach relies exclusively on matrix manipulations to characterize specific numerical representations. It is an efficient method for quantitative analysis, equation solving and simulation.

In this paper, the researchers have opted to apply only the abstract approach in their analysis of SLDS for the following reasons: avoiding the constraints of limited and rigid modeling, which could limit the scope of the results.

The matrix approach, although powerful for the quantitative analysis of linear systems, can sometimes restrict the vision as it can neglect qualitative aspects and underlying dynamics unsuitable for matrix representation. Consequently, the researchers' choice is justified since the abstract approach rhymes with the objectives pursued through this research.

The researchers refer to the classic definition of SLDS and the basic assumption that the number of switching instants in any finite interval of $\mathbb{R}$ is finite (see [13, 14] and [15]).

Definition: A switched linear differential system (SLDS) is a quadruple $\Sigma = \{P, \mathcal{F}, S, G\}$. Where:

$P = \{1, 2, \ldots, N\} \subset \mathbb{N}$ is the set of indices.

$\mathcal{F} = \{\mathcal{B}_1, \mathcal{B}_2, \ldots, \mathcal{B}_N\}$ is the bank of behaviors, with $\mathcal{B}_j \in \mathcal{L}^W$ for $j \in P$ and $\mathcal{L}^W$ is the set of linear differential behaviors.

$S = \{s : \mathbb{R} \to P\}$ is the set of admissible switching signals, with s is a piecewise constant and right-continuous.

$G = \{G^-_{k\to\ell}(\xi), G^+_{k\to\ell}(\xi) \in \mathbb{R}^{\cdot\times W}[\xi] \times \mathbb{R}^{\cdot\times W}[\xi] \,|\, 1 \le k, \ell \le N, k \ne \ell\}$ is the set of gluing conditions.

$\mathbb{T}_s := \{t \in \mathbb{R} | \lim_{\tau \nearrow t} s(\tau) \ne s(t)\} = \{t_1, t_2, \ldots\}$ is the set of switching instants associated with $s \in S$; where $t_i < t_{i+1}$.

4 Production System Application: Analysis and Modeling

In order to illustrate the application of SLDS approach in production scheduling field, the following production system will be treated: It's a production system consisting of four machines: M_1, M_2, M_3 and M_4. The system studied is a multimode system initially configured to operate in nominal mode. The term "mode" refers to a specific manner of utilization of the system's resources. Some resources can be shared between several modes, while others can only be utilized after a mode transition. Following the occurrence of an unexpected event, such as a failure, the system exits its default mode, the nominal mode, and switches to one of the degraded modes. A degraded mode is a mode activated only in response to the unexpected event occurred, in order to guarantee continuity of production in all circumstances, and it refers to an altered functioning condition of the system. Machines in degraded mode function at a reduced capacity and rate compared to their equivalents in nominal mode. M_i represents the basic machine in the production chain with $i \in \{1, \ldots, 4\}$. M_i' represents the replacement machine that replaces the faulty basic machine (Fig. 1). b_i denote that machine M_i takes the part from the previous stock to begin its processing and e_i means that machine M_i has completed its task and delivered the part to the next stock. Between each machine pair, there is a stock S.

In this paper, the researchers are interested in finding compatibility between schedulers at the moment of switching between modes. The primary objective in this research is to propose a formal approach that ensures the new scheduler starts its mission from the same terminal state as the previous scheduler, therefore providing scheduling continuity under degraded conditions as well.

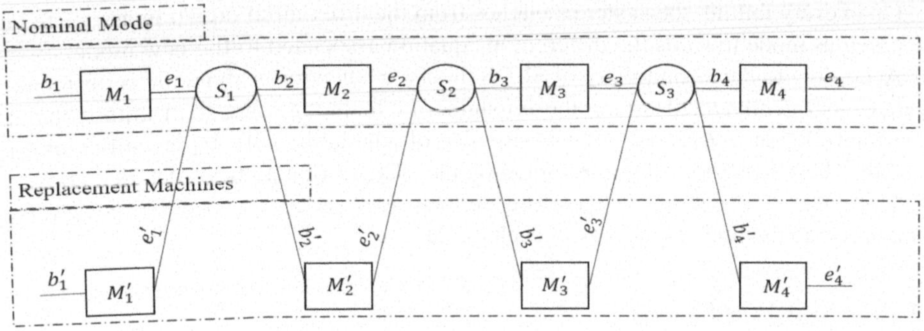

Fig. 1. Multimode production system

4.1 The Set of Indices

Let $\rangle = \{0, 1, 2, 3, 4\} and \rangle \in \mathcal{P}$ the set of indices that represents the system's modes.

$i = 0$ represents mode N_o: The nominal mode.
$i = 1$ represents mode D_1: The system mode when M_1 is replaced by M_1' upon the occurrence of a failure in machine M_1.
$i = 2$ represents mode D_2: The system mode when M_2 is replaced by M_2' upon the occurrence of a failure in machine M_2.
$i = 3$ represents mode D_3: The system mode when M_3 is replaced by M_3' upon the occurrence of a failure in machine M_3.
$i = 4$ represents mode D_4: The system mode when M_4 is replaced by M_4' upon the occurrence of a failure in machine M_4.

That's means:
D_i: is the mode with only one machine M_i defective.
$D_{i,j}$: is the mode where both machines M_i and M_j are failed at the same time.

4.2 The Bank of Behaviors

In a multimode system, every mode is assigned a specific scheduler S_i. For every mode $i \in \mathcal{P}$; the dynamic behavior (scheduler S_i) is modelized by a differential equation. These equations define the way the system evolves over time in each mode. They illustrate the functioning of a scheduler by representing the scheduling of job J_i in a given mode:

$$\frac{dx_i(t)}{dt} = A_i x_i(t) + B_i u_i(t) \tag{1}$$

where:

$x_i(t)$: a state vector representing the state of job J_i at time t in mode i.
$u_i(t)$: Control signal representing scheduler activity. According to the current mode, a particular scheduler can be activated or deactivated by this signal.
A_i is a coefficient representing the degraded performance of M_i. It reflects the impact of degradation on system dynamics.

Upon every failure, the system switches from the differential equation associated to the previous mode towards the differential equation associated to the new mode.

According to the terminology of SLDS theory, a behavior or dynamic is designated by B_i. Consequently, B_0 illustrates the dynamics of scheduling when all machines operate at optimal efficiency. B_1 models the progression of scheduling with M_1' less efficient, and so forth. These behaviors are represented by the systems of equations detailed above.

The bank of behaviors is therefore defined as a set $\mathcal{F} = \{\mathcal{B}_0, \mathcal{B}_1, \mathcal{B}_2, \mathcal{B}_3, \mathcal{B}_4\}$ containing the dynamic behaviors of each mode.

4.3 The Switching Signals

Switching signals represent the instants when a transition between two modes (and therefore between two schedulers) occurs, triggered by a failure and replacement of a defective machine by its alternate.

$S = \{s : \mathbb{R} \rightarrow$ P$|s(t) = i$ signifies that mode i is active at time t}.

For example, if $s(t) = 0$ for $t < t_1$ and $s(t) = 1$ for $t \geq t_1$, this signifies that M_1 has failed att_1, and has subsequently been supplanted byM_1', leading to a switch from nominal mode to mode 1.

4.4 The Gluing Conditions

Each switch between two modes means a transition from the old scheduler to the new one. Gluing conditions ensure consistency during this switchover. Which means, upon switching, the work already done in the previous mode must be considered by the new scheduler:

$G_{i \rightarrow j} = \boldsymbol{G}^{-}_{i \rightarrow j}(\boldsymbol{\xi}), \boldsymbol{G}^{+}_{i \rightarrow j}(\boldsymbol{\xi})$, where: $\boldsymbol{G}^{-}_{i \rightarrow j}(\boldsymbol{\xi})$ is the schedule before the transition to mode j, i.e., the state of progress at the moment of exit from mode i, $\boldsymbol{G}^{+}_{i \rightarrow j}(\boldsymbol{\xi})$ is the schedule after the transition to mode j, or the state of initial progress in mode j, and $\boldsymbol{\xi}$ denotes the amount of work either completed or currently in progress.

For a multimode production system, three cases are identified: transition between nominal and degraded modes, transition between degraded modes, and reverting to nominal mode after repair. In the following, the researchers will investigate Gluing conditions for each case.

Transition Between Nominal and Degraded Modes. Suppose the M_i machine has failed. The system will therefore switch from N_o mode to D_i mode. Consequently, for $i \in \mathcal{P}$, $s(t) =$ i, the gluing conditions are formalized as follows:

$$x_N\left(t_f^+\right) = x_N\left(t_f^-\right) \tag{2}$$

$$x_i\left(t_f^+\right) = \frac{1}{2} x_N\left(t_f^-\right) \tag{3}$$

Gluing conditions ensure that the progress of the job J_i remains continuous across the switching point (Eq. 2). Moreover, they illustrate that the process of the job J_i on the degraded mode D_i is initialized at half the progress of J_i before switching, since the alternate machines are characterized by reduced efficiency (Eq. 3).

To conclude this first case, gluing conditions ensure that planning resumes at the same point, there is no loss of progress, although efficiency is reduced.

Transition Between Degraded Modes. Assume that the system is operating in a degraded D_i mode. Another failure occurs allowing the system to switch to $D_{i,j}$ mode.

The objective is to ensure that the work in progress under mode D_i is preserved:

$$\frac{dx_i(t)}{dt} = A_i x_i(t) + B_i u_i(t) \tag{4}$$

$$x_i\left(t_f^+\right) = x_i\left(t_f^-\right) \tag{5}$$

$$x_i\left(t_f^+\right) = x_{ij}\left(t_f^-\right) \tag{6}$$

$$\frac{dx_{ij}(t)}{dt} = A_{ij} x_{ij}(t) + B_{ij} u_{ij}(t) \tag{7}$$

These gluing conditions ensure that, regardless of the time or combination of failures, production continues without interruption (Eqs. (4) and (5)) and jobs running in D_i mode are transferred to $D_{i,j}$ mode (Eqs. (6) and (7)).

Reverting to Nominal Mode Following Repair. In this section, three cases are identified:

Case 1: Transition from a degraded mode D_i to nominal mode N_o.

Case 2: Transition from a mode $D_{i,j}$ to nominal mode N_o.

Case 3: Progressive transition to N_o.

In the entire section, we denote by t_r^-: just before the repair and t_r^+: just after the repair.

Case 1: $D_i \rightarrow N_o$. A M_i machine is defective (D_i mode), once repaired the system has to return to nominal mode N_o. Gluing conditions are therefore expressed as follows:

$$x_i\left(t_r^+\right) = x_i\left(t_r^-\right) \tag{8}$$

$$\frac{1}{2} x_i\left(t_r^-\right) = x_N\left(t_r^+\right) \tag{9}$$

Equation (8) ensures that the progress of the job should be continued after the repair. Equation (9) indicates that the D_i mode is less efficient than the nominal mode.

Case 2: $D_{i,j} \rightarrow N_o$. If M_i and M_j had failed ($D_{i,j}$ mode), and both machines are repaired, the system has to return to nominal mode N_o. Equations (10) and (11) shows the progress of the job J_i remains continuous across the switching point, and Eqs. (12) and (13) represents the reduced efficiency of $D_{i,j}$ mode.

$$\frac{dx_{i,j}(t)}{dt} = A_{ij} x_{ij}(t) + B_{ij} u_{ij}(t) \tag{10}$$

$$x_{ij}\left(t_r^+\right) = x_{ij}\left(t_r^-\right) \tag{11}$$

$$\frac{1}{2}x_{ij}\left(t_r^-\right) = x_N\left(t_r^+\right) \tag{12}$$

$$\frac{dx_N(t)}{dt} = A_N x_N(t) + B_N u_N(t) \tag{13}$$

Case 3: Progressive transition to N_o. The system operates in $D_{i,j}$ mode (M_i and M_j are both defective). If M_j is repaired before M_i, the system switches first to degraded mode D_i, then to nominal mode N_o. Equations (14), (15), (16) and (17) shows the continuous of the progress of the job J_i across the switching point. Equation (18) illustrates the reduced efficiency of D_i mode.

$\mathrm{D_{i,j}} \rightarrow \mathrm{D_i}$:

$$\frac{dx_{i,j}(t)}{dt} = A_{ij}x_{ij}(t) + B_{ij}u_{ij}(t) \tag{14}$$

$$\frac{1}{2}x_{ij}\left(t_r^+\right) = \frac{1}{2}x_{ij}\left(t_r^-\right) \tag{15}$$

$\mathrm{D_i} \rightarrow \mathrm{N_o}$:

$$\frac{dx_i(t)}{dt} = A_i x_i(t) + B_i u_i(t) \tag{16}$$

$$x_i\left(t_r^+\right) = x_i\left(t_r^-\right) \tag{17}$$

$$\frac{1}{2}x_i\left(t_r^-\right) = x_N\left(t_r^+\right) \tag{18}$$

5 Conclusion and Perspectives

In this research, the researchers introduced a novel approach for modelling the switching within multimode systems. They presented the application of SLDS theory on multimode production system in order to formally investigate the transition between schedulers after the occurrence of an unexpected event.

The researchers focused on detailing glowing conditions and studied their relevance in ensuring the concatenation at the switching instants, since they relate the system variables before and after switching.

The proposed approach demonstrates that quadratic differential form has presented an effective tool to study the transition discussed above. Future research should explore the performance offered by s-switched linear differential systems applied on multimode production systems in terms of stability, robustness, and control and their comparison to the traditional switched linear differential systems. Additionally, this study will be extended by conducting practical experiments using real industrial datasets and benchmarking the proposed approach against other advanced techniques, such as Petri nets, multi-agent systems, and machine learning methods. These comparisons would provide a more comprehensive assessment of the effectiveness of the framework in handling complex multimode systems. Then, evaluating the robustness and stability of the model in the presence of significant disruptions would further validate its applicability in dynamic industrial environments. However, it is important to emphasize that this study primarily focuses on establishing a formal framework for modeling multimode systems, which has been validated through a detailed illustrative example. This formalization provides a solid theoretical foundation that can serve as a basis for future empirical validations, ensuring a structured and rigorous approach before moving to large-scale implementations.

References

1. El Ghadouali, A., Kamach, O., Amami, B.: Static approach for switching between different operating modes. In: 2nd International Conference on Communications, Computing and Control Applications (CCCA), pp 1–6. IEEE (2012). https://doi.org/10.1109/CCCA.2012.6417878
2. Kamach, O., Pietrac, L., Niel, E.: Multi-model approach to discrete events systems: Application to operating mode management. Math. Comput. Simul **70**, 394–407 (2006). https://doi.org/10.1016/j.matcom.2005.11.008
3. Elqabli, Z., Chater, Y., Kamach, O.: Operation modes scheduling: a formal framework for identification of the compatible state. In: 14th International Colloquium of Logistics and Supply Chain Management (LOGISTIQUA2022), pp. 1–6. IEEE (2022). https://doi.org/10.1109/LOGISTIQUA55056.2022.9938021
4. Elqabli, Z., Kamach, O., Chater, Y.: Multimode approach using reinforcement learning and digital twin for operating mode management. Prod. Eng. Arch. **31**(1), 116–128 (2025). https://doi.org/10.30657/pea.2025.31.11
5. An, Y., Wu, N.: Scheduling of crude oil operations for minimizing the usage of simultaneously-charging-and-feeding mode. In: 15th International Conference on Networking, Sensing And Control (ICNSC), pp. 1–6. IEEE (2018). https://doi.org/10.1109/ICNSC.2018.8361347
6. Mrabet, W., Ladhari, T.: Towards a multi_agent system for manufacturing reconfiguration. In: 5th International Conference on Modeling, Simulation and Applied Optimization (ICMSAO), pp. 1–6. IEEE (2013). https://doi.org/10.1109/ICMSAO.2013.6552724
7. Dou, C., Liu, B.: Hierarchical management and control based on MAS for distribution grid via intelligent mode switching. Int. J. Electr. Power Energy Syst. **54**, 352–366 (2014). https://doi.org/10.1016/j.ijepes.2013.07.029
8. Yu, J., Dou, C., Li, X.: MAS-based energy management strategies for a hybrid energy generation system. IEEE Trans. Industr. Electron. **63**, 3756–3764 (2016). https://doi.org/10.1109/TIE.2016.2524411
9. Dou, C., Wang, W.Q., Hao, D.W., Li, X.B.: MAS-based solution to energy management strategy of distributed generation system. Int. J. Electr. Power Energy Syst. **69**, 354–366 (2015). https://doi.org/10.1016/j.ijepes.2015.01.026
10. Azzabi, O., Ben Njima, C., Messaoud, H.: Diagnosis of a dynamic hybrid system by hybrid timed automata. In: International Conference on Control, Decision and Information Technologies (CODIT), pp 1–6. IEEE (2016). https://doi.org/10.1109/CoDIT.2016.7593633
11. Kamach, O., Chafik, S., Pietrac, L., Niel, E.: Representation of a reactive system with different models. In: International Conference on Systems, Man and Cybernetics, vol. 4, p. 5. IEEE (2002). https://doi.org/10.1109/ICSMC.2002.1173293
12. Kamach, O., Chafik, S., Pietrac, L., Niel, E.: Supervisory uniqueness for operating mode systems. IFAC Proc. Vol. **38**(1), 110–115 (2005). https://doi.org/10.3182/20050703-6-CZ-1902.01443

13. Mayo-Maldonado, J.C., Rapisarda, P.: Modelling of switching dynamics in electrical systems. In: Proceedings of the Mathematical Theory of Networks and Systems Symposium (MTNS), Groningen, NL (2014)
14. Rocha, P., Willems, J.C., Rapisarda, P., Napp, D.: On the stability of switched behavioral systems. In: 50th IEEE Conference on Decision and Control and European Control Conference, pp. 1534–1538. IEEE (2011). https://doi.org/10.1109/CDC.2011.6160860
15. Mayo-Maldonado, J.C., Rapisarda, P.: Modeling and analysis of energy distribution networks using switched differential systems. In: Camlibel, M., Julius, A., Pasumarthy, R., Scherpen, J. (eds.) Mathematical Control Theory I. Lecture Notes in Control and Information Sciences, vol. 461. Springer, Cham (2015). https://doi.org/10.1007/978-3-319-20988-3_11

Estimating Moroccan Yield Curve Parameters with the Nelson-Siegel-Svensson Model on Historical Data

Hind Bourass[1](✉), Outmane Soussi Noufail[2], Hanaa Hachimi[3], and Ithri Fettahi[4]

[1] Multidisciplinary Research Laboratory in Economics and Management, Faculty of Economics and Management Beni Mellal, Sultan Moulay Slimane University of Beni Mellal, Beni-Mellal, Morocco
bourasshind1@gmail.com

[2] Finance, Entrepreneurship, and Development Laboratory, Faculty of Legal, Economic, and Social Sciences of Salé, Mohammed V University in Rabat, Rabat, Morocco
n.soussi@um5r.ac.ma

[3] Advanced Systems Engineering Laboratory, National School of Applied Sciences, Ibn Tofail University, Kenitra, Morocco
hanaa.hachimi@uit.ac.ma

[4] Faculty of Legal, Economic, and Social Sciences of Mohammedia, Hassan II University, Casablanca, Morocco

Abstract. We apply the Nelson-Siegel-Svensson (NSS) model to historical Moroccan yield curves across different maturities. The model parameters, expressed as a linear combination of four latent factors, capture yield curve dynamics. This approach enables a flexible and efficient modeling of interest rates, facilitating the analysis of market trends and expectations. Based on 70 estimated curves, our results show that interest rate variations align with the Moroccan central bank's published rate (TMPjj). The analysis highlights the stability of long-term rates, a consistently high slope factor indicating a higher term premium for long-term bonds, and moderate long-term rate movements, suggesting minimal yield curve convexity. To assess the model's robustness, we conducted a sensitivity analysis using the Augmented Dickey-Fuller (ADF) methodology. These findings provide the Moroccan central bank with a valuable tool to refine its monetary policy and better anticipate interest rate developments. Furthermore, a deeper understanding of yield curve dynamics contributes to strengthening financial stability and optimizing investment decisions in the Moroccan economy.

Keywords: Moroccan yield curves · Nelson-Siegel-Svensson · Latent factors · Econometric estimation · Interest rates

1 Introduction

The interest rate curve is a crucial tool for central banks because it provides a visual representation of interest rates over different maturities. It not only helps assess current economic conditions but also plays a central role in developing monetary policies.

C. El Mokhi et al. (Eds.): ICASET 2025, SUCI, pp. 225–233, 2025.
https://doi.org/10.1007/978-3-031-98334-4_21

However, the yield curve is not always directly observable due to various factors. Its construction involves approximating empirical data by collecting Treasury and corporate bond rates. Choosing the appropriate mathematical functions is crucial at this stage.

Parametric models, such as the one introduced by Nelson Siegel and Svensson (NSS) in 1994, offer advantages over existing models. The NSS model provides estimates for different maturities, even those not directly covered by observed data, and its estimated parameters are economically interpretable, facilitating their use in the analysis of monetary policies and financial markets. Despite the challenges associated with calibrating the NSS model, it remains widely used by government institutions, central banks, and financial institutions worldwide.

This study aims to deepen the understanding of the behavior and dynamics of the yield curve using Treasury bond rates from the Moroccan secondary market across various maturities. The NSS model will be employed to estimate yield curves for different maturities in the sample, thereby allowing for the decomposition of the yield curve dynamics into four key elements representing the parameters of the NSS model.

2 Methods for Constructing the Yield Curve and Their Limitations

There are different methods to construct the yield curve, which can generally be classified into two main categories:

- *Classical Methods*: These methods primarily involve directly calculating central tendencies [1]. The most common method is weighted averaging, where interest rates are weighted based on transaction sizes in the financial market for each maturity. However, these methods often face several limitations. They assume a normal distribution of interest rates, which may not hold true in real market conditions. Additionally, they frequently overlook the volatility and non-linearity of interest rates, potentially leading to biased estimates. Moreover, traditional methods may struggle with missing or irregular data, resulting in information loss and reduced accuracy in yield curve construction.
- *Mathematical Methods*: These methods are primarily employed to extrapolate interest rate data across various maturities when collecting data for all maturities is challenging [1]. Several mathematical models are utilized [2], including parametric models, cubic splines, matrix factorizations, and kernel techniques. Additionally, deterministic and/or stochastic mathematical models [2], such as the Vasicek model, are used to simulate short- and long-term interest rates. However, these models often rely on simplifying assumptions that can oversimplify the complexity of the real interest rate market. They can be sensitive to the chosen parameters, potentially resulting in unstable and unreliable estimates, especially in dynamic market environments.

3 Estimating the Yield Curve and Decomposing Its Dynamics

The dataset comprises Treasury bond rates from the Moroccan secondary bond market across various maturities, including rates for 13 weeks, 26 weeks, 52 weeks, 2 years, 5 years, 10 years, 15 years, and 20 years.

3.1 Nelson-Siegel-Svensson Model

The Nelson-Siegel model, developed by Nelson and Siegel [7–9] in 1987, is widely used for modeling yield curves due to its effective representation of their shape, including slope and curvature [2]. Through specific coefficients, it allows for flexible estimation and extrapolation of interest rates over different periods. The Nelson-Siegel-Svensson (NSS) model extends the original model to enhance its flexibility and ability to represent long-term interest rate variations [7].

The mathematical function used to describe the yield curve is given by:

$$r(\tau) = \beta_0 + \beta_1 \left[\frac{1 - e^{\frac{-\tau}{\lambda_1}}}{\frac{\tau}{\lambda_1}} \right] + \beta_2 \left[\frac{1 - e^{\frac{-\tau}{\lambda_1}}}{\frac{\tau}{\lambda_1}} - e^{\frac{-\tau}{\lambda_1}} \right] + \beta_3 \left[\frac{1 - e^{\frac{-\tau}{\lambda_2}}}{\frac{\tau}{\lambda_2}} - e^{\frac{-\tau}{\lambda_2}} \right] \quad (1)$$

In the equation, $r(\tau)$ represents the interest rate at maturity T, and the parameters are defined as follows: β_0 is the long-term scale factor. β_1 is the rotation factor representing the cyclical component. β_2 is the slope factor representing the change in trajectory. β_3 is the additional curvature factor.

λ_1 is the first scale parameter. λ_2 is the second scale parameter.

The NSS model introduces two additional parameters: a second scale parameter, λ_2, and a new parameter, β_3, which accounts for the curvature shapes.

Note that: $\lim_{\tau \to \infty} = \beta_0$ and $\lim_{\tau \to 0} = \beta_0 + \beta_1$.

The algorithm for the Nelson-Siegel-Svensson (NSS) model starts by initializing its parameters and then optimizes them to minimize the loss function, often the sum of squared residuals, using methods like gradient descent or the Levenberg-Marquardt algorithm. Typically, ordinary least squares (OLS) is used to minimize the sum of squared differences between observed interest rate values and those predicted by the NSS model.

$$Min \sum \epsilon_t^2 = Min \sum_{t=1}^{T} \left(r(\tau)_t^o - r(\tau)_t^{NSS} \right)^2 \quad (2)$$

With: $r(\tau)_t^o$ is the rate observed for each maturity and $r(\tau)_t^{NSS}$ is the rate calculated by the NSS method.

3.2 Results

This iterative process continues until convergence, where parameter changes become minimal and the loss function no longer significantly decreases. The final estimated parameters describe the yield curve and can be used for further analyses, such as spread estimation or evaluating interest rate convergence speed.

The model was implemented using R-Studio software to estimate the parameters (β_0, β_1, β_2, β_3, λ_1, λ_2). Estimates were computed for each quarter from 2004 to 2022, with each row in the database corresponding to a quarter [9, 10].

We estimated 70 yield curves based on the studied sample. The results of the NSS model can be compared by calculating the geometric mean of the interest rate estimates for different dates relative to the rates published by the Moroccan central bank, known as the Weighted Average Rate of Operations, day-to-day, for liquidity lending and borrowing on the Moroccan interbank market (TMPjj) (Fig. 1).

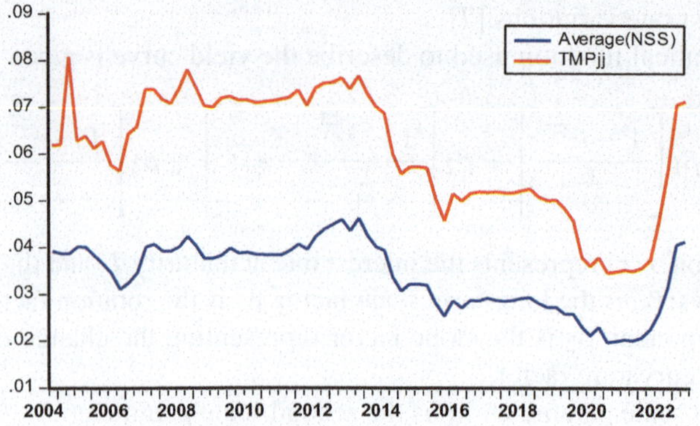

Fig. 1. Comparing TMPjj and Average Estimated Rates Curve

Both series exhibit the same long-term trend, suggesting that the NSS model provides a more accurate representation of yield curve dynamics. Discrepancies between NSS model-estimated interest rates and those published by the central bank offer insights into unaccounted factors, potentially including risk premiums or market-specific influences. Additionally, exploring parameter distributions over time provides valuable insights into their evolution, responses to economic events, and market volatility. Comparing these distributions with other economic indicators enhances understanding of their impact on interest rates and the broader economy [11–18].

The table below displays the descriptive statistics for the distribution of each model parameter: (see Table 1).

Table 1. Descriptive Statistics for NSS Parameters

	β_0	β_2	β_2	β_3	λ_1	λ_2
Mean	0,043471	−0,017235	0,007388	−0,017505	1,366208	4,688136
Skewness	−1.748.800	1,838723	2,000291	2,284689	−0,394614	1,308586
Kurtosis	8,769137	8,532102	7,911014	9,365690	1,479967	3,475687
Jarque-Bera	132,7554	128,7060	117,0246	179,0869	8,555691	20,63795

The skewness statistic indicates asymmetry in the distribution of the level and slope factors, with β_0 and λ_1 skewed leftward, and the other parameters skewed rightward. High kurtosis values suggest these distributions are more concentrated around the mean with thicker tails compared to a normal distribution. The Jarque-Bera test confirms the non-normality of the parameter distributions, with all cases yielding low probability values (0.00). Additionally, there is a disparity in the standard deviations of these variables.

Sensitivity analysis evaluates the robustness of the NSS model estimates by analyzing how the parameters fluctuate across different market conditions and time periods. This assessment relies on stationarity analysis and unit root testing. The summarized results are presented in the following table [19–24].

The Table 2 analysis reveals that the parameters β_1, β_2, and β_3 are stationary at the level without trend or intercept, while β_0 is stationary at the level according to the M2 model. This indicates that these parameters fluctuate around a fixed mean without systematic drift or structural trend. Consequently, the level, slope, and curvature factors of the NSS model exhibit a mean-reverting behavior, meaning they tend to return to a stable average value aftermarket shock. This suggests structural stability in the yield curve characteristics, where observed variations are solely due to temporary market fluctuations, without an underlying long-term evolution.

Table 2. Unit root test

	Test for Unit root in	Model	Parameter	P-Value	Results
β_0	Level	M2: Intercept	−3.61	0.0078	I(0)
β_1	Level	M1: None	−3.35	0.0011	I(0)
β_2	Level	M1: None	−6.06	0.0000	I(0)
β_3	Level	M1: None	−3.75	0.0003	I(0)
λ_1	1st difference	M1: None	−13.74	0.000	I(1)
λ_2	1st difference	M1: None	−16.75	0.0000	I(1)

In contrast, the speed of convergence parameters is more sensitive to transient market shocks, which explains why they become stationary only after differencing. We can now proceed with the interpretation of the dynamics of each parameter.

Global Level Factor: The parameter β_0 means the level of long-term interest rates. An increase in β_0 indicates an increase in long-term interest rates. The Fig. 2 demonstrates the stationary nature of the distribution of the rate curve level factor over time. However, these effects seem temporary, since we see a return to the mean after 2016. In addition, the values are mostly positive, except during the shocks of 2012–2013.

Global Slope Factor: The yield curve slope factor is a key indicator of interest rate movements over time, reflecting changes in the slope of the yield curve. Analyzing its dynamics across dates reveals shifts in direction and magnitude, aiding in understanding market trends. The overall slope factor is estimated using the following formula:

$$(\beta_1)_t = c_1 + c_2(\beta_1)_{t-1} + \epsilon_t \tag{3}$$

The results using the OLS method give: $(\beta_1)_t = -0.29 - 0.004(\beta_1)_{t-1}$.

The slope factor of our model follows the downward-sloping direction of the yield curve, as demonstrated by Diebold and Li (2006), meaning that high values correspond to a flat or inverted yield curve.

Spread Analysis: To calculate the spread from the parameters of the Nelson-Siegel-Svensson (NSS) model, you can utilize the following formula:

$$Spread = 2 \times (\beta_2 - \beta_1) \tag{4}$$

In this formula, β_1 represents the slope parameter, and β_2 represents the curvature parameter. The spread, which signifies the difference between long-term and short-term interest rates according to the NSS model, is obtained by subtracting the value of β_1 from that of β_2 and then multiplying the result by 2. (see Fig. 2).

Spread values mostly exceed 0, indicating a normal or sloping yield curve where long-term rates surpass short-term rates. However, from 2014, a flattening curve is observed, indicating accommodative monetary policy with central banks aiming to stimulate the economy by keeping short-term rates low, thus reducing the spread with longer-term rates.

During 2012–2014, a negative spread indicated investors anticipated an economic slowdown, accepting lower long-term returns to safeguard capital. Peaks in the spread, such as in 2007, show divergence between short- and long-term expectations. A peak in early 2012 indicates anticipation of increased short-term interest rates.

Convergence Speed: The Nelson-Siegel-Svensson (NSS) parameters also provide insight into the rate at which interest rates converge towards their long-term equilibrium. In the NSS model, λ_1 and λ_2 represent the time factors that dictate this convergence rate.

To be more specific, λ_1 determines how quickly short-term interest rates converge towards their long-term equilibrium, while λ_2 dictates the rate at which long-term interest rates converge towards their long-term equilibrium.

Smaller values of λ_1 and λ_2 indicate a faster convergence of interest rates toward their long-term equilibrium, whereas larger values suggest a slower convergence. These parameters offer insights into how quickly the interest rate market responds to changes and reverts to its long-term equilibrium.

In general, interest rates in the bond market tend to exhibit lower values of these parameters. This suggests a relatively moderate volatility of yield curves throughout the study period.

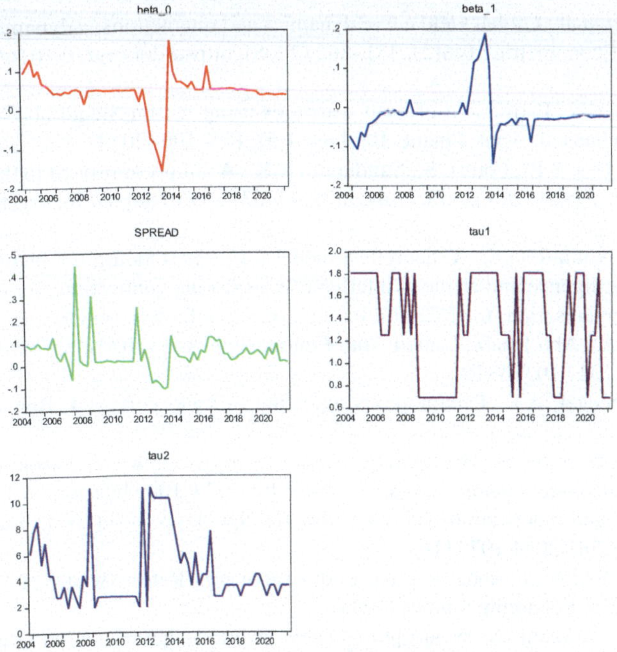

Fig. 2. Distribution of NSS model parameters

4 Conclusion

This study employs the Nelson-Siegel-Svensson model to decompose the yield curve dynamics into key elements (level, slope, convexity, trend), providing crucial insights for financial market analysis and modeling, aiding in investment decision-making.

Quarterly data from Morocco's secondary bond market from 2004:1 to 2022:1 is utilized. The estimated rates (NSS) align with the trend of the aggregate rate published by the Moroccan central bank. The results underscore the significant influence of interest rate level and slope factors on the yield curve structure across different horizons. Moreover, slope factors appear to be correlated with macroeconomic fundamentals, such as observed shocks to inflation and real activity. These results align with the findings of Carter and Kohn (1994) [3].

Future research will aim to expand these results by incorporating data from other countries, exploring interactions with additional macroeconomic and financial indicators.

References

1. Cox, J., Ingersoll, J., Ross, S.: A theory of term structure of interest rates. Econometrica **53**(2), 385–408 (1985)
2. Dai, Q., Singleton, K.J.: Specification analysis of affine term structure models. J. Financ. **55**(5), 1943–1978 (2002). https://doi.org/10.1111/0022-1082.00278

3. Diebold, F.X., et al.: Global yield curve dynamics and interactions: a dynamic Nelson–Siegel approach. J. Econometrics **146**(2), 351–363 (2008). https://doi.org/10.1016/j.jeconom.2008.08.017
4. Muvingi, J., et al.: Estimation of term structures using nelson Siegel and NSS: a case of a Zimbabwean bank. J. Appl. Financ. Banking **4**(6), 155–190 (2014)
5. Balduzzi, P., Das, S.R., Foresi, S., Sundaram, R.K.: A simple approach to three-factor affine term structure models. J. Fixed Income **6**(3), 43–53 (1996). https://doi.org/10.3905/jfi.1996.408181
6. Bourass, H., Noufail, O.S.: A theoretical model for understanding the relationship between economic intelligence and artificial intelligence: exploring connections for decision-making in financial organisation. In: 2023 9th International Conference on Optimization and Applications (ICOA), Abu Dhabi, United Arab Emirates, pp. 1–6 (2023). https://doi.org/10.1109/ICOA58279.2023.10376326
7. Nelson, C., Siegel, A.F.: Parsimonious modeling of yield curves. J. Bus. **60**(4), 473–489 (1987)
8. Noufail, O.S., Bourass, H., Hachimi, H.: Analyzing the dynamics of Nelson-Siegel-Svensson term structure models using time series data. In: 2024 10th International Conference on Optimization and Applications (ICOA), Almeria, Spain, pp. 1–6 (2024). https://doi.org/10.1109/ICOA62581.2024.10754443
9. Svensson, L.: Estimating and Interpreting Forward Interest Rates: Sweden 1992 1994. Institute for International Economic Studies (1994)
10. El Mokhi, C., Addaim, A.: Near-optimal cable layout design of a wind farm using genetic algorithm. In: Peng, S.-L., Hao, R.-X., Pal, S. (eds.) Proceedings of First International Conference on Mathematical Modeling and Computational Science. AISC, vol. 1292, pp. 47–56. Springer, Singapore (2021). https://doi.org/10.1007/978-981-33-4389-4_5
11. Satri, J., El Mokhi, C., Hachimi, H.: Artificial intelligence and machine learning for a better decision making in the public sector. Presented at the 8th International Conference on Optimization and Applications, ICOA 2022 - Proceedings (2022)
12. Haddouch, S., Hachimi, H., Hmina, N.: Modeling the flow of road traffic with the SUMO simulator. Presented at the Proceedings of the 2018 International Conference on Optimization and Applications, ICOA 2018 (2018)
13. Eddaoui, I., Itmi, M., Hami, A.E., Hmina, N., Mazri, T.: A deterministic approach for systems-of-systems resilience quantification. Int. J. Crit. Infrastruct. **14**, 80–99 (2018). https://doi.org/10.1504/IJCIS.2018.090654
14. Akhrif, O., El Idrissi, Y.E., Hmina, N.: Smart university: SOC-based study. Presented at the ACM International Conference Proceeding Series (2018)
15. El Mahjoubi, A., Hmina, N., Mazri, T.: M2M and eMTC communications via NB-IoT, Morocco first testbed experimental results and RF deployment scenario: new approach to improve main 5G KPIs and performances. Presented at the Proceedings - 2017 International Conference on Wireless Networks and Mobile Communications, WINCOM 2017 (2017)
16. Amrani, L.E., Mazri, T., Hmina, N.: Specific absorption rate (SAR) in human body exposed to wireless base station fields. Presented at the Proceedings of the 9th International Conference on Electronics, Computers and Artificial Intelligence, ECAI 2017 (2017)
17. Ed-Daoui, I., El Hami, A., Itmi, M., Hmina, N., Mazri, T.: Unstructured peer-to-peer systems: towards swift routing. Int. J. Eng. Technol. (UAE) **7**, 33–36 (2018). https://doi.org/10.14419/ijet.v7i2.3.9963
18. Abouelmajd, M., et al.: Experimental analysis and optimization of mechanical properties of FDM-processed polylactic acid using Taguchi design of experiment. Int. J. Simul. Multi. Des. Optim. **12**, 30 (2021). https://doi.org/10.1051/smdo/2021031

19. Elkhechafi, M., Benmamoun, Z., Hachimi, H., Amine, A., Elkettani, Y.: Firefly algorithm for supply chain optimization. Lobachevskii J. Math. **39**, 355–367 (2018). https://doi.org/10.1134/S1995080218030125
20. Ngadi, M., Amine, A., Hachimi, H., El-Attar, A.: A new optimal approach using NSVC for breast cancer diagnosis classification. Int. J. Imaging Robot. **16**, 24–36 (2016)
21. Chand, M., Hachimi, H., Rani, R.: New extension of beta function and its applications. Int. J. Math. Math. Sci. **2018**, 1–25 (2018). https://doi.org/10.1155/2018/6451592
22. Hachimi, H., Ellaia, R., Elhami, A.: A new hybrid genetic algorithm and particle swarm optimization. Key Eng. Mater. **498**, 115–125 (2012). https://doi.org/10.4028/www.scientific.net/KEM.498.115
23. Ngadi, M., Amine, A., Nassih, B., Hachimi, H., El-Attar, A.: Uniformed two local binary pattern combined with neighboring support vector classifier for classification. Int. J. Artif. Intell. **15**, 102–115 (2017)
24. Elkhechafi, M., Hachimi, H., Elkettani, Y.: A new hybrid cuckoo search and firefly optimization. Monte Carlo Methods Appl. **24**, 71–77 (2018). https://doi.org/10.1515/mcma-2018-0003

Energy Efficiency and Cost Reduction Through Parameters Cut Optimization for Finishing Phase of Multi-pass Turning Operation

Badreddine Lahlali(✉), Fatima Zohra El Abdelaoui, Nabil Moujibi, and Abdelouahhab Jabri

Laboratory of Mechanical Engineering, Faculty of Sciences and Techniques, University of Sidi Mohamed Ben Abdellah, Fez, Morocco
lahlali.badreddine@usmba.ac.ma

Abstract. In the machining processes, parameters like depth of cut d, cutting speed v and feed rate f are crucial in determining the optimal balance between productivity, quality, and sustainability. The aim of the study is to examine and minimize the impact of machining parameters on total processing timing, total energy consumption, and total cost production. The primary focus of this study will be on analyzing how energy consumption varies with changes in depth of cut during finishing phase. This approach enables a deeper understanding of energy efficiency within the machining process, aligning with the main goal of this study: reducing environmental impact while maintaining optimal performance and cost-effectiveness. To demonstrate in practical applicability the efficiency of proposed approach, it is applied to an illustrative example from the literature. This study has demonstrated positive results during finishing phase. Reducing the depth of cut and increasing the cutting speed led to lower energy consumption, reduced production costs, and shorter processing times, while maintaining high surface quality.

Keywords: Turning · Optimization · Cutting parameters · Energy efficiency · cost · machining time

1 Introduction

Global warming is a critical environmental issue caused by high carbon emissions. In 2007, the International Energy Agency stated that industrial activities contribute to around 36% of CO_2 emissions while utilizing about one-third of the global energy resources [1]. Therefore, in today's competitive manufacturing industry, delivering high-quality productions at lower costs and in less time is crucial. Optimization cutting parameters is the key for balancing these factors, achieving low-carbon manufacturing, and saving the use of Energy.

Principally, machining operations like turning are affected by a range of input factors, such as cutting velocity, feed rate, depth of cut, and the number of passes. These factors impact directly output data like production costs and time, energy consumption, tool life,

C. El Mokhi et al. (Eds.): ICASET 2025, SUCI, pp. 234–244, 2025.
https://doi.org/10.1007/978-3-031-98334-4_22

the quality of surface and cutting loads. Choosing the right parameters is mandatory to ensure producing high-quality, reducing cost, saving time, and minimizing energy usage. Many studies have been focused to optimize these parameters using different methods, and a review summarizing the studies conducted over the past decade was establish by [2]. Additionally, [3] noted that Shin and Joo's study was the first to address the parameter selection complications in multi-pass machining, leading to further research. Researchers from diverse fields have explored the optimization of this issue using various optimization methods [3]. The objective of this study is not to introduce a new optimization model or approach, but rather to analyze and discuss results achieved in real-world applications. The focus is on understanding how these outcomes can improve manufacturing processes. The key objective is to ensure that the findings provide tangible benefits and can lead to manufacturing efficiency, reducing costs, and minimizing energy consumption. The following sections are arranged as follows: Sect. 3 and 4 describes mathematical formulation and constraints adopted, in Sect. 5 an illustrative example is solved and discussed. Finaly, this paper ends with a conclusion.

2 Nomenclature

D, L	Diameter and length of work pieces (mm)
d_{rU}, d_{rL}	Upper and lower bound of depth of cut in rough machining (mm)
d_{sU}, d_{sL}	Upper and lower bound of depth of cut in finish machining (mm)
f_{rU}, f_{rL}	Upper and lower bound of feed rate in rough machining (mm/rev)
f_{sU}, f_{sL}	Upper and lower bound of feed rate in finish machining (mm/rev)
v_{rU}, v_{rL}	Upper and lower bound of cutting speed in rough machining (m/min)
v_{sU}, v_{sL}	Upper and lower bound of cutting speed in finish machining (m/min)
R_a	Roughness (μm)
t_a, t_r, t_e, t_c	Parameters of processing time
$\mu, \vartheta, \theta, k_f, h_1, h_2$	Coefficients and constants of machining condition
$k_e k_c k_t k_o$	Parameters of production cost
C_0, P, q, r	Coefficients associated with tool life
$A_1, A_2, P_O, P_w, v_0, \Delta v, x_e, \delta_f, n_i, \rho$	Parameters of Energy consumption

3 Turning Process Model

To evaluate the effectiveness of the proposed framework, an illustration from [4] is adopted. The elements of the presented approach have been implemented on Python 3.13.

3.1 Total Machining Duration

The total machining duration is defined by [3] and [5]:

$$T_P = t_a + t_r + \frac{t_e.t_m}{T_t} + t_i \tag{1}$$

- t_a startup duration.
- t_r tool setting duration.
- t_i machine idle phase.
- t_m cutting time.
- t_e tool change time.
- T_t tool life.

According to [4], the inactive machine duration (t_i) is calculated by adding fixed duration for loading and unloading operation (t_c), to the time variation spent on inactive tool movement (t_v), as shown in Eqs. (2) and (3):

$$t_i = t_c + t_v \tag{2}$$

$$t_v = (h_1 L + h_2)(n + 1) \tag{3}$$

Cutting time (t_m) is determined by factors such as cutting parameters, workpiece dimension [6]. The total cutting time, as defined by [3], is represented in Eq. (4), with the cutting times for the roughing, and finishing process given by Eqs. (5) and (6).

$$t_m = t_{mr} + t_{ms} \tag{4}$$

$$t_{mr} = \frac{\pi DL}{1000 v_r f_r} n = \frac{\pi DL}{1000 v_r f_r} \frac{d_t - d_s}{d_r} \tag{5}$$

$$t_{ms} = \frac{\pi DL}{1000 v_s f_s} \tag{6}$$

"According to [6], tool life (T_t) is calculated using Taylor's formula, which consists of two parts: roughing (T_r) and finishing (T_s) as described in the following Eqs. (7), (8), and (9)"

$$T_t = \theta T_r + (1 - \theta) T_s \tag{7}$$

$$T_r = \frac{C_0}{v_r^p f_r^q d_r^r} \tag{8}$$

$$T_s = \frac{C_0}{v_s^p f_s^q d_s^r} \tag{9}$$

3.2 Energy Consumption

According to [4], Eq. (10) establishes the formula of total power consumption:

$$E = E_m + E_i + E_r + E_t + E_c \tag{10}$$

where:

- E_m: cutting energy consumption

$$E_{mr} = E_{mr} + E_{ms} \tag{11}$$

$$E_{mr} = \frac{k_f f_r^{\mu} d_r^{\nu} v_r}{6120\eta} \times \frac{\pi DL}{1000 v_r f_r} (\frac{d_t - d_s}{d_r}) \tag{12}$$

$$E_{ms} = \frac{k_f f_s^{\mu} d_s^{\nu} v_s}{6120\eta} \times \frac{\pi DL}{1000 v_s f_s} \tag{13}$$

- E_i: energy consumption during machine idle time

$$E_i = P_u(h_1 L + h_2)(n + 1) + P_o t_c \tag{14}$$

$$P_u = 10^{-3} \times \left(40.6 + A_1 n_i + A_2 n_i^2\right) \tag{15}$$

- E_r: energy required during tool replacement

$$E_r = P_o t_e \left(\frac{t_m}{T_t}\right) \tag{16}$$

- E_t: cutting tool's energy usage

$$E_t = P_w \left(\frac{t_m}{T_t}\right) \tag{17}$$

- E_c: cutting fluid energy consumption

$$E_c = \frac{T_p}{T_c} \times e\rho\left((v_0 + \Delta v) + \frac{(v_0 + \Delta v)}{\delta_f}\right) \tag{18}$$

3.3 Production Cost

Unit production cost refers to [1–6, 7], is divided into six costs:

$$Uc = C_m + C_i + C_r + C_t + C_e + C_c \tag{19}$$

where:

- C_m: machining cost

$$C_m = k_o t_m \tag{20}$$

$$C_m = k_o \left[\frac{\pi DL}{1000 v_r f_r}\left(\frac{d_t - d_s}{d_r}\right) + \frac{\pi DL}{1000 v_s f_s}\right] \tag{21}$$

- C_i: tool change cost

$$C_i = k_o t_i \tag{22}$$

$$C_i = k_o t_c + (h_1 L + h_2)(n + 1) \tag{23}$$

- C_r: tool change cost

$$C_r = k_o t_e \left(\frac{t_m}{T_t} \right) \tag{24}$$

- C_t: tool usage cost

$$C_t = k_t \left(\frac{t_m}{T_t} \right)^{T_t} \tag{25}$$

- C_e: energy cost

$$C_e = k_e E_{total} \tag{26}$$

- C_c: cutting fluid cost

$$C_c = k_c \frac{T_p}{T_c} (v_0 + \Delta v) \tag{27}$$

3.4 Surface Quality Function

Quality of surface function is presented by [8]. The primary factor used to evaluate the quality of surface is the level of roughness, which is defined as:

$$R_a = k v^{x_1} {f_s}^{x_2} {d_s}^{x_3} \tag{28}$$

"where x_1, x_2, x_3 and k are constants associated with a particular tool-workpiece combination [8]".

4 Cutting Condition Constraints

The proposed limitation constraints that must be considered during roughing and finishing are presented from (29) to (36) according to [6] and [3]:

$$v_{rL} \leq v_r \leq v_{rU} \tag{29}$$

$$f_{rL} \leq f_r \leq f_{rU} \tag{30}$$

$$d_{rL} \leq d_r \leq d_{rU} \tag{31}$$

$$v_{sL} \leq v_s \leq v_{sU} \tag{32}$$

$$f_{sL} \leq f_s \leq f_{sU} \tag{33}$$

$$d_{sL} \leq d_s \leq d_{sU} \tag{34}$$

"Furthermore, the combined depths of cut in the rough passes and the finish cut should be the same as the total material depth that needs to be eliminated [3]":

$$d_t = nd_r + d_s \tag{35}$$

Consequently, the total count of rough passes can be represented as:

$$n = \frac{d_t - d_s}{d_r} \tag{36}$$

5 Results and Discussion

5.1 Illustrative Application Example

This section presents an example adopted from [4] to demonstrate the applicability of the proposed approach. "As noted in [4], the turning operation involves a C45 carbon steel workpiece, with a diameter (D) of 50 mm and a length (L) of 300 mm. The cutting speeds vary between 50 to 500 m/min."As specified by the machine's specifications, the maximum power of the engine (PU) is 5 kW, with a power efficiency (η) of 85%. The corresponding data is presented in Table 1 based on [4]".

Table 1. Cutting model data of [4]

Parameter/constraint	Value	Parameter /constraint	Value	Parameter /constraint	Value	Parameter /constraint	Value
D	50 mm	v_{rU}	500 m/min	k_t	2.5 \$/edge	A_1	0.227
L	300 mm	v_{rL}	50 m/min	k_e	6.91 \$/kWh	A_2	$-0.667 \cdot 10^{-6}$
d_t	6 mm	v_{sU}	500 m/min	k_c	8.75 \$/L	v_0	30 L
d_{rL}	1 mm	v_{sL}	50 m/min	k_f	108	Δv	6 L
d_{rU}	3 mm	t_a	0.17 min	n_i	1200	δ_f	0.05
d_{sL}	1 mm	t_r	3 min	η	0.85	P_O	3.6 kW
d_{sU}	3 mm	t_e	1.5 min/edge	C_0	$6 \cdot 10^{11}$	P_w	5.3 MJ/kg
R_a	6.3 µm	t_c	0.75 min/unit	P	5	x_e	72.885 kJ/kg
f_{rL}	0.1 mm/rev	k_o	123.61 \$/min	q	1.75	ρ	7.8 g/cm^3
f_{rU}	0.9 mm/rev	μ	0.75	r	0.75	x_1	−0.5
f_{sL}	0.1 mm/rev	ϑ	0.95	h_1	$7 \cdot 10^{-4}$ min/mm	x_2	1
f_{sU}	0.9 mm/rev	θ	0.8	h_2	0.3 min	x_3	0.2
K	1.5						

5.2 Discussion

Taking in consideration input data in Table 1, the same result was obtained as in the benchmarked research as it is shown in Table 2 using Python 3.13:

Table 2. Result from previous research

	Tp	E	Uc
$v_r = 84.43$ m/min	4.953 min	5.434 MJ	395.21 \$
$v_s = 182.48$ m/min			
$f_r = 0.9$ mm/rev			
$f_s = 0.36$ mm/rev			
$d_r = 3$ mm			
$d_s = 3$ mm			

Impact of Depth of Cut on Energy Consumption, Machining Time, and Cost. The most remarkable aspect of this result study is the high cutting depth in finition phase. In practice, roughing is performed first to shape the part and remove as much material as quickly. Consequently, according to the current study, a high depth of cut ($d_r = 3$ mm) is used to maximize material removal (MRR). On the contrary, the goal during finishing is to achieve the desired shape with a high surface quality. Thus, the depth in finishing should be lower than roughing, although in the current study found both to be equal ($d_s = 3$ mm), which, in practice, is not correct.

According to the above-mentioned study [4] the comparison with similar previous results indicates that, although different optimization methods such as MOORA, GA and

Goal programming were achieved a finishing depth of cut of 3 mm. However, in one study using MOGA method, the depth of cut in finishing was lower, at 0.93 mm. In [6], the optimal parameters were determined using Genetic Algorithms (GA), which resulted in a roughing cutting speed of 109.663 m/min, a finishing cutting speed of 169.986 m/min, and a cutting depth of 3 mm for both roughing and finishing. The Model presented by [7], solved through Goal programming, determined the optimal roughing cutting speed to be 50m/min, with a corresponding value of 374.3870 m/min for finishing cutting speed, 3 mm for roughing depth of cut, 2.9999 mm for finishing depth of cut. "As discussed in [9], the Multi-Objective Genetic Algorithm (MOGA) method in MATLAB R2016b was used to optimize the solution proposed." Results of the optimal parameters are shown in the Table 3 below:

Table 3. MOGA Optimization results

Parameters	Value
v_r	78.08 mm/min
v_s	308.09 mm/min
f_r	0.84 mm/rev
f_s	0.27 mm/rev
d_r	2.55 mm
d_s	0.93 mm

Under this assumption, the impact of energy consumption, total cost, and processing time as the finishing depth increases is shown in the plot in Fig. 1.

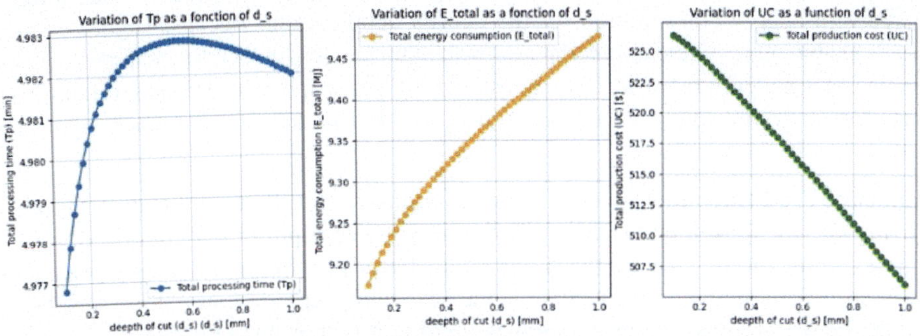

Fig. 1. Variation of Energy consumption, total production cost and processing

It is important to note that the graph illustrates that the total energy consumption is increasing due to the increased depth of cut during finishing pass. To understand the trend, for example, at a finishing depth of cut of 0.4 mm, the energy consumption corresponding is 9.33 MJ and corresponding cost is 520 \$. For a value of depth higher 1mm, for example, the value of energy consumption is 9.48 MJ.

Impact of Speed and Depth of Cut on Energy Consumption, Time, and Cost. The investigation will continue by focusing on the effect of increasing the cutting speed during the finishing phase. In the previous case study, the cutting speed in finishing phase was fixed to 182.48 m/min. The cutting speed will now be increased to 308.09m/min, based on the value obtained from Model in [9]. In the following Fig. 2 it is showed that for the same selected value of depth of cut 0.4 mm, the value of energy corresponding has been decreased to 9.29 MJ and the corresponding cost is 474 \$. For a selected depth of cut from previous study, the graph demonstrates how changes in cutting speed impact energy consumption, machining time, and production costs.

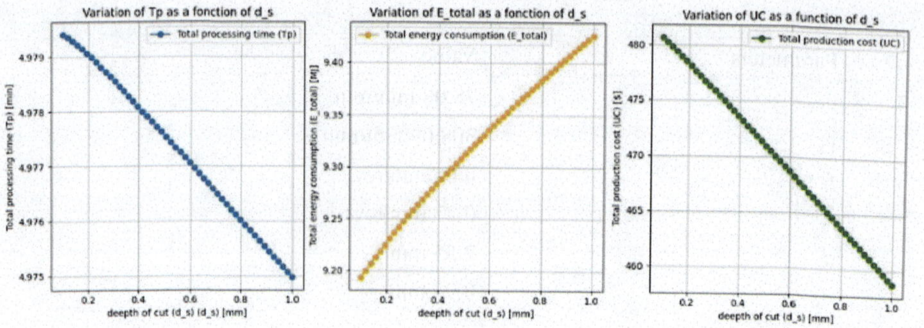

Fig. 2. Variation of Energy consumption, total production cost and processing time as a function of depth in finishing (ds)

In fact, the graph indicates that increasing the cutting speed leads to a significant decrease in energy consumption for the chosen depth of cut. Because higher cutting speeds enable more efficient material removal, requiring less power. Since higher speeds enable faster material removal, as demonstrated by this graph, which shows a significant reduction in overall time with higher cutting speed. This reduction in machining time leads to lower costs.

Impact of Feed Rate and Speed Cut on Roughness. The best combination for minimizing energy consumption, time and cost is a lower depth of cut combined with a higher cutting speed. This approach allows for faster processing, reduced energy, and lower costs. Building on this foundation, the next study explores how cutting speed (v_s) and feed rate (f_s) impact surface roughness (Ra) as it is showed in Fig. 3.

The results confirm that optimal surface quality is obtained by minimizing feed rate and maximizing cutting speed. A clear linear relationship is observed where Roughness (Ra) increased as feed rate (f_s) rises, leading to a rougher surface. On the other hand, increasing cutting speed (v_s) consistently reduces Ra, highlighting the impact of higher cutting speeds in improving surface quality.

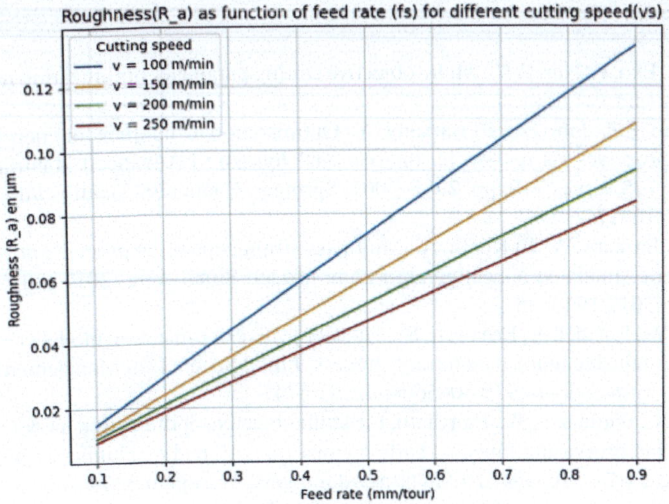

Fig. 3. Roughness (Ra) as function of deed rate (fs) for different cutting speed (vs)

These insights can be directly integrated into multi-objective optimization frameworks to balance surface quality, tool life, energy efficiency and machining productivity. This approach promotes sustainable machining by optimizing parameters to achieve environmental, social, and economic goals. Further experimental validation and modeling can improve the applicability of these findings in practical machining situations.

6 Conclusion

This research has shown promising results in manufacturing optimization in finishing phase. A significant reduction in energy has been accomplished by reducing the cutting depth and increasing the cutting speed, which also leads to a decrease in overall production cost, and shorter processing times. By increasing cutting speed, the process remains more controlled without compromising the surface quality of workpiece, which is critical in the finishing phase. These changes contribute to promote sustainability, reduce operational costs, improve productivity, and speed up production.

This approach has its limitations, as excessively high cutting speeds can lead to tool wear and negatively impact surface quality. Similarly, very high feed rates or extremely low depths of cut can also result in undesired effects, such as poor surface finish or material damage. Material selection play a crucial role in determining the feasible limits for these parameters. Future studies could focus on further optimizing these cutting parameters for different materials, considering the impact on energy consumption and tool life. Additionally, futures research should consider the social aspects of sustainability by minimizing cutting forces and reducing vibrations, which would improve process efficiency and protect operators.

References

1. Phengky, P., Eko, P., Nur, R.C.: Multi-objective cutting parameter optimization model of multi-pass turning (n.d.)
2. El Abdelaoui, F.Z., Jabri, A., El Barkany, A.: Optimization techniques for energy efficiency in machining processes—a review. In: International Journal of Advanced Manufacturing Technology, vol. 125, issues 7–8, pp. 2967–3001. Springer, Cham (2023). https://doi.org/10.1007/s00170-023-10927-y
3. Jabri, A., El Barkany, A., El Khalfi, A.: Multipass turning operation process optimization using hybrid genetic simulated annealing algorithm. Model. Simul. Eng. **2017** (2017). https://doi.org/10.1155/2017/1940635
4. Widhiarso, W., Rosid, I.A., Ernawati, R.: Multi-objective optimization model of cutting parameters for a sustainable multi-pass turning process. Jurnal Sistem Dan Manajemen Industri **7**(1), 1–14 (2023). https://doi.org/10.30656/jsmi.v7i1.5747
5. Rosyidi, C.N., Widhiarso, W., Pujiyanto, E.: Multi objective optimization model of cnc turning for minimizing processing time and carbon emission with real machining application. J. Ind. Eng. Manag. **14**(2), 376–390 (2021). https://doi.org/10.3926/jiem.3269
6. Aryanfar, A., Solimanpur, M.: Optimization of multi-pass turning operations using genetic algorithms. In: Industrial Engineering and Operations Management Istanbul (n.d.)
7. Aprilia, D., Eko, P., Wayan, S.I.: View of Multi-Objective Optimization Model of Multi-Pass Turning Operations to Minimize Energy, Carbon Emissions, and Production Costs (n.d.)
8. Cus, F., Balic, J.: Optimization of cutting process by GA approach. In: Robotics and Computer Integrated Manufacturing, vol. 19 (2003)
9. Fittamami, B.M., Pujiyanto, E., Priyandari, Y.: Multi-objective optimization of machining parameters for multi-pass CNC turning to minimize carbon emissions, energy, noise and cost. Jurnal Teknik Industri **23**(1), 25–34 (2021). https://doi.org/10.9744/jti.23.1.25-34

Pavement Behavior Using Neural Networks and Monte Carlo Simulation

Oumaima El Abidi(✉), Mouna El Mkhalet, and Nouzha Lamdouar

Civil Engineering and Construction Structure GCC Laboratory, Mohammadia School of Engineers, Mohammed V University, Rabat, Morocco
oumaima_elabidi2@um5.ac.ma

Abstract. The road network is subject to various mechanical loads which call into question its easement, its durability and the safety it provides. On this, it is essential to carry out an accurate prediction of its behavior in order to analyze and model the response of roadway structures following the application of mechanical loads.

To achieve this, we used artificial neural networks and Monte Carlo simulation.

By combining the two methods, our study presents a comprehensive approach to predict the behavior of roadways. This combination resulted in very accurate and reliable predictions. Specifically, the model demonstrated its effectiveness with an exceptionally low mean square error and a perfect coefficient of determination value of 1, indicating accurate and reliable results. Thus, the comparison of the established model with other models using cross-validation highlighted the accuracy of the prediction of the established model.

The distribution generated by the simulation gave us a clear idea about the consistency and reliability of the chosen material. This analysis, which is part of scientific research in the field of infrastructure, is a research avenue to ensure the longevity of roadway structures.

Keywords: Neural Networks · Monte Carlo Simulation · Pavement · Hooke's Law

1 Introduction

According to Hook's law, the material deformation is proportional to the stress applied on the Young's modulus E. In order to understand the probabilistic characteristics of material properties by generating multiple random samples, we simulated the variation of different stress conditions and the variation of Young's modulus, in order to study the distribution of the deformation of a road section. The simulation consisted of varying the stress from 1788 MPa to 2476 MPa and the Young's modulus from 2000 MPa to 5000 MPa to create a complete dataset. These data were divided into training and test sets for network modeling. By comparing four different typologies of network models, ranging from simple to complex, and evaluating their performances using the mean square error and the coefficient of determination. The results revealed an exceptionally low error and a coefficient of 1, showing a model with high accuracy and good fit. A

C. El Mokhi et al. (Eds.): ICASET 2025, SUCI, pp. 245–254, 2025.
https://doi.org/10.1007/978-3-031-98334-4_23

cross-validation analysis of the four models is then performed to highlight the effect of changing the network architecture on its performance and to conclude an efficient model for this application.

On the other hand, by analyzing the strain distribution, one can gain insight into the elastic behavior of the material, defining its reliability and consistency. Finally, how do Monte Carlo simulations contribute to understanding material responses under different loading conditions? And what are the challenges associated with integrating neural networks and Monte Carlo simulations in pavement behavior modeling?

2 Methodology

In order to understand the effect of the variation of mechanical stresses undergone by pavement structures and to model their response, a flexible type road sample is questioned. The combination of Monte Carlo simulation and artificial neural networks has significantly improved the modeling process, producing reliable predictions (very low MSE and a perfect R^2 value of 1). These results highlight the reliability of the model to capture the relationship between stress, Young's modulus and strain as well as the consistency of the elastic behavior of the material under varying conditions. The result part is organized as follows:

- We will present the Monte Carlo simulation and neural networks in the state-of-the-art subsection;
- In the case study subsection, we will present the studied case with its parameters and the results obtained with simulation and modulation;
- In the simulation section, we study the cross-validation method in order to compare the performance of four models with different typologies.

3 Results

3.1 State of the Art

Monte Carlo Simulation: Monte Carlo simulation involves performing repeated random sampling and statistical analysis to obtain numerical results, this method is used to study uncertainty in prediction and estimating models. Starting by defining probability distributions for the input parameters, random samples are created from these distributions to represent the input variables. After, by performing many iterations, the Monte Carlo method provides a demonstration on their probabilities [1].

For its application, Monte Carlo simulation is applicable in several fields. For example, in civil engineering, Cardoso et al. [2] through the combination of neural networks and Monte Carlo simulation, they studied a method for calculating the probability of structural failure. Through this application, they concluded that simulation is an effective and simple tool to solve reliability problems. Another application in project management made by Kwak and Ingall [3] who studied the impacts of risk and uncertainty on project deadlines and budgets, they stated that simulation is able to measure its impact. This methodology, in the same sector of project management, is able to provide statistical indicators of project performance, for example completion dates and budget forecasts. On this, this methodology is considered a valuable technique for modeling and scenario analysis in different sectors, which helps to have crucial information for decision making [3].

Artificial Neural Networks: By modeling biological neurons of knowledge acquisition and retrieval, artificial neural networks are used by imitating the human brain process. They are identified among the machine learning algorithms that allow us to predict the performance of roadways [4]. This method is defined as a computer system composed of a number of highly interconnected processing elements, which process information through their response to an external input [5]. Each neuron composing the network is connected to other neurons by directed links and each directed link has a weight associated with it.

As shown in the Fig. 1, some functions that neural networks are capable of performing include [6]:

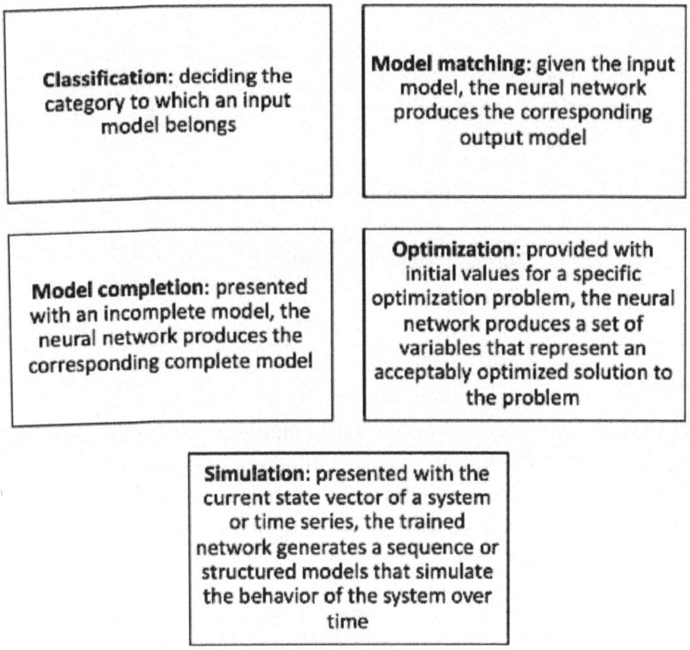

Fig. 1. Neural Network's functions [6]

The ability of neural networks to perform complex tasks is allocated to their underlying parallel distributed computing "mechanism". It is supported by three decisive components:

- The connection diagram between neurons, called architecture,
- The activation function of neurons,
- The learning algorithm.

So, in in the aim of building a neural network to solve a particular problem, the three components above must first be fixed. The diagram of artificial neural in shown in the Fig. 2 bellow.

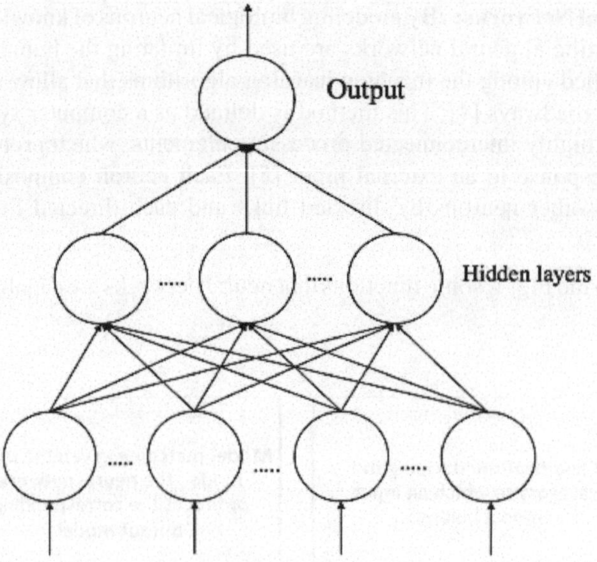

Fig. 2. Diagram of Artificial Neural

3.2 Presentation

To model the behavior of pavements, we will calculate the stresses applied to a road pavement. Based on Hook's law which is a fundamental principle of mechanics of materials that describes the elastic behavior of materials. It states that the deformation (ϵ) of a material is proportional to the stress (σ) applied, as long as the elastic limit of the material is not exceeded.

The equation is often expressed by:

$$\sigma = E\epsilon \quad (1)$$

where:

- σ is the stress (force per unit area),
- E is the Young's modulus (a constant that estimates the stiffness of the material),
- ϵ is the strain (change in length per unit of initial length).

Then, Hook's law is a fundamental relationship that will help us understand and predict the elastic behavior of pavements under applied loads, and it plays a crucial role in structural design and engineering.

3.3 Case Study

In order to model the behavior of a road section, we take as an example a flexible roadway subjected to TPL6 type traffic according to the catalog of new structures whose components are as follows (Fig. 3):

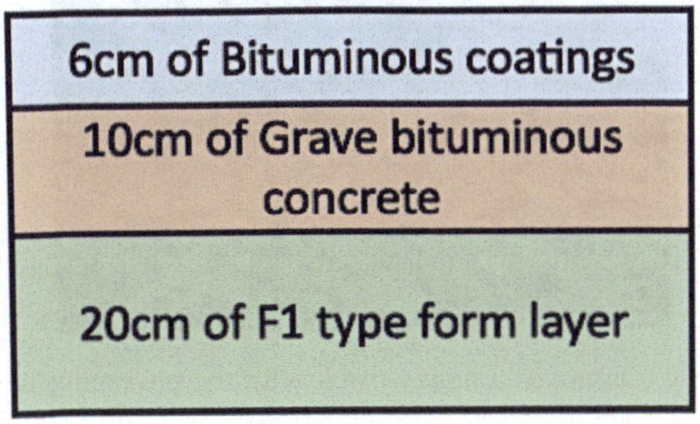

Fig. 3. Components of the case study

According to the catalogue of typical structures of new pavements produced by the roads and highways department of the Ministry of Equipment and Water the margin of values of the traffic in equivalent axles of 8T expected per direction is presented in the following table (the upper and lower value for each traffic class) (Table 1):

Table 1. The number of heavy vehicles equivalent to 8 tons for traffic class TPL6

Structure	Lifetime		TP6
Flexible or semi-rigid	One year	min	894
		max	1238

Firstly, we need to calculate the stress applied at the center of the contact area of a heavy vehicle tire equivalent to an 8-ton axle using the basic equation:

$$\sigma_0 = \frac{P}{A} \tag{2}$$

With:

- P: the load applied by a heavy truck;
- A: the contact surface of a heavy truck tire with the pavement.

Here's an illustration (Fig. 4) of the contact of a heavy truck with the pavement:

Then, the applied surface load presented in the following Table 2:

In order to create a neural network model, we first need to generate the data with Monte Carlo Simulation to prepare it for modeling.

In the following Fig. 5, we can observe the distribution of the deformation. From its little spread appearance, we can have a preliminary idea of the heterogeneity of the material and the foreseeable risk.

Fig. 4. Illustration of the contact of a heavy truck with the pavement made by AI (ChatGPT)

Table 2. Calculated Stress σ_0

Lifetime	Stress $\boldsymbol{\sigma}_0$ (MPa)	TPL6
One year	Min	1788
	Max	2476

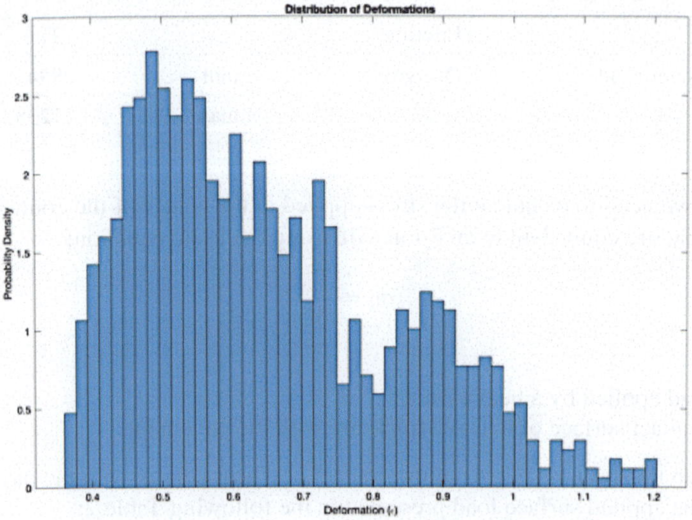

Fig. 5. Distribution of deformation

3.4 Results

After generating the data by Monte Carlo simulation, we move on to neural network modeling on MATLAB. By trying several architectures, the established model is presented in the figure bellow.

Thus, we present the model parameters (Table 3):

Table 3. Model Summary

Data Division		Random	
Training Algorithm		Levenberg-Marquardt	
Performance		Mean squared Error	
	Observations	MSE	R
Training	700	5.8167 e−11	1.0000
Validation	150	6.1972 e−11	1.0000
Test	150	5.5473 e−11	1.0000

In Figs. 6 and 7 below, we present the model errors in the prediction:

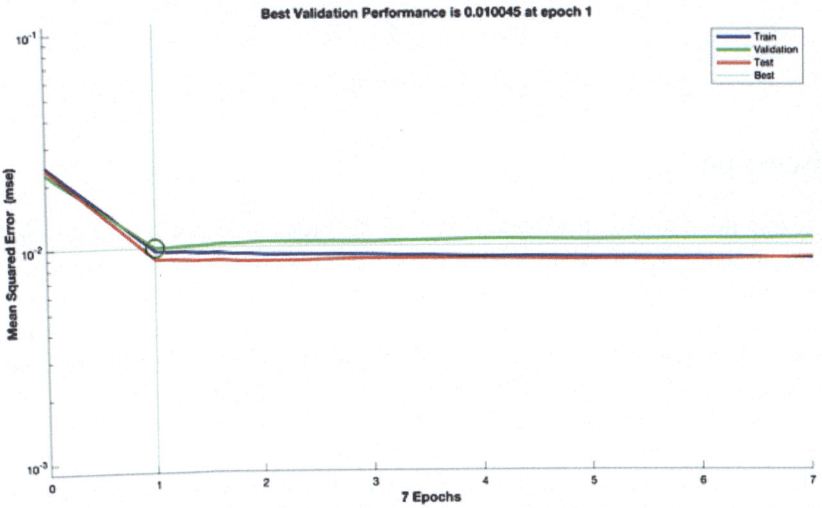

Fig. 6. Mean Squared Error MSE

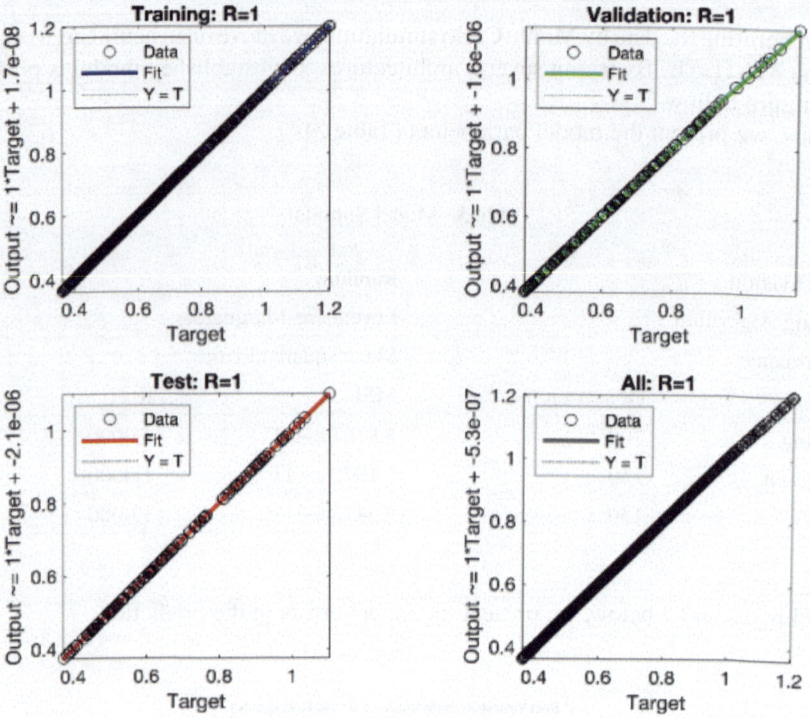

Fig. 7. Regression

4 Discussion

By analyzing the resulting distribution (Fig. 5), the broad aspect of it indicates greater variability in the material's deformation response. This could suggest inhomogeneity in the material or sensitivity to changes in stress and Young's modulus. Indeed, the inhomogeneity that appeared in the material response may be due to the inadequate use of Hooke's relation in the present example. Modeling by one of the elastics, viscoelastic, numerical, empirical and empirical mechanistic models may give better results.

On the other hand, by analyzing the model established by neural networks, as MSE estimates the mean squared difference between the real and projected values, an error of 0.010045 is very low (as shown in the Fig. 6), suggesting that the model's predictions are very close to the actual values. The model has high accuracy and the prediction errors are minimal. An R^2 value of 1 (as shown in the Fig. 7) indicates that the model explains 100% of the variance in the data. This means the model fits the data perfectly and every point in the data set falls exactly on the regression line.

In order to further justify the performance of the established model and its accuracy, four other models with different topologies were established in order to have a comparative study. In the following table we present their architectures and training errors.

As can be observed, the second topology presents a much smaller RMSE error compared to the first, third and fourth topologies. Although the learning rate and epoch remain the same for the four topologies, it is emphasized that the time consumed in training the fourth case was longer compared to the first model (Table 4).

Table 4. Summary of the four cases compared

Topology	Number of Layers	Neurons per Layer	RMSE	R^2
Topology 1	1	10	2.1288e−07	0.99999
Topology 2	2	15	2.6947e−09	1
Topology 3	3	20	2.164e−07	0.99999
Topology 4	4	25	8.4805e−07	0.99997

In order to further develop the difference between the four topologies, we propose the cross-validation technique. In the table below we present the results of the cross-validation of the four models in order to precisely evaluate the performance of the models and to maximize the use of the available data.

Following the application of the method, the 1st model despite its simplicity is able to give accurate predictions. The 2nd model is a little more complicated than the 1st but does not give an error less than the 1st, the 3rd and 4th models on the other hand, in addition to the calculation time, present increasing errors and develop an overfitting of the data (Table 5).

Table 5. Results of the cross-validation technique

Topology	Number of Layers	Neurons per Layer	Activation Function	Training Time (s)	RMSE	R^2
Topology 1	1	10	ReLU	0.5	2.6711e−08	1
Topology 2	2	15	ReLU	0.8	7.027e−08	1
Topology 3	3	20	ReLU	1.2	6.1591e−07	0.99998
Topology 4	4	25	ReLU	2.5	1.0741e−06	0.99997

5 Conclusion

Our example of modeling a neural network for predicting the response of the roadway to a mechanical stress is one example among others that aims at a more enriched study on road sections. In our example, the Hook relation was not the finest choice to take for the study of this case, given its generality on all mechanical stresses. Moreover, the modeling on neural networks gave more than satisfactory results. The combination of a very low MSE and an R^2 value of 1 indicates an exceptionally well-performing model. Firstly, this study is inaugurated by the Monte Carlo simulation to generate data on the deformation of pavement materials according to a range of variation of the applied mechanical stress and the Young's modulus of the material. By making an application on MATLAB, we stored the input data and divided them into training and test sets. Secondly, we trained four different architectures to predict the behavior of pavements. This training is highlighted by a cross-validation of these architectures by evaluating their performance using measures such as the mean square error and the coefficient of determination. Then, this comparison focused on the 1st type of network architecture following a minimum square error and a coefficient of determination of 1 despite its simplicity. Finally, this method contributed to the understanding of the probability of the deformation of the material composing a flexible pavement.

References

1. Raychaudhuri, S.: Introduction to Monte Carlo simulation. In: Proceedings of the Winter Simulation Conference, pp. 91–100 (2008). https://doi.org/10.1109/WSC.2008.4736059
2. Cardoso, J.B., De Almeida, J.R., Dias, J.M., Coelho, P.G.: Structural reliability analysis using Monte Carlo simulation and neural networks. Adv. Eng. Softw. **39**(6), 505–513 (2008). https://doi.org/10.1016/j.advengsoft.2007.03.015
3. Kwak, Y., Ingall, L.: Exploring Monte Carlo simulation applications for project management. IEEE Eng. Manage. Rev. **37**, 83 (2009). https://doi.org/10.1109/EMR.2009.5235458
4. Alsugair, A.M., Al-Qudrah, A.A.: Artificial neural network approach for pavement maintenance. J. Comput. Civ. Eng.Comput. Civ. Eng. **12**(4), 249–255 (1998). https://doi.org/10.1061/(ASCE)0887-3801(1998)12:4(249)
5. Caudill, M.: Neural networks primer, part I. AI Expert. **2**(12), 46–52 (1987)
6. Yang, J.: Road crack condition performance modeling using recurrent Markov chains and artificial neural networks. CORE Repository (2004). https://core.ac.uk/download/pdf/154466898.pdf. Accessed 9 Aug 2024

A Bi-colony Ant Algorithm for Constrained Multi-objective Optimization of Organizational Transport Logistics Using Heterogeneous Fleet: A Case Study

Hajar Bideq[1(✉)], Khaoula Ouaddi[2], and Rachid Ellaia[1]

[1] LERMA Laboratory Mohammadia School of Engineers, Mohammed V University, Rabat, Morocco
hajar.bideq@gmail.com

[2] Center for Doctoral Studies in Information Technology and Engineering Sciences, ENSIAS, Mohammed V University, Rabat, Morocco

Abstract. Employee transportation is an operational challenge for numerous industrial organizations seeking to enhance their overall productivity. This challenge is classified as an NP-hard problem, as it is an extended variant of the Vehicle Routing Problem (VRP), named Employee Bus Routing Problem (EBRP). In recent years, the EBRP has been extensively studied using various approaches, including heuristics, metaheuristics, and exact algorithms. However, efficiently solving large instances of the bi-objective EBRP remains a significant challenge for researchers. This paper examines a practical application of the EBRP for a large Moroccan industrial group that provides shuttle services for its employees across multiple regions. This company aims to operate a fleet of heterogeneous buses to ensure the comfort of thousands of employees while minimizing operational costs. Therefore, we propose a bi-colony Ant Colony Algorithm (BC-ACO) with two competing colonies: one focused on minimizing total cost and the other on reducing the maximum travel time for passengers. Each colony independently explores potential solutions while iteratively sharing information with one another throughout the optimization process. To evaluate the efficiency of the proposed approach, we conduct a comparative analysis against three other methods: a weighted sum bi-objective Genetic Algorithm (BGA), bi-objective ACO with one colony (BACO) and an exact approach. The results indicate that the BC-ACO yields solutions that provide a better trade-off and within a minimized execution time, making it a viable option for large-scale applications in the real world.

Keywords: Employee Bus Routing Problem (EBRP) · Bi-Objective · Bi-Colony Ant Colony (BC-ACO) · Genetic Algorithm (GA)

1 Introduction

Organizations in remote locations often encounter significant challenges in keeping high productivity levels. This is because inefficient commuting solutions lead to longer travel times, resulting in lower energy levels that impact the employees and organizations

C. El Mokhi et al. (Eds.): ICASET 2025, SUCI, pp. 255–267, 2025.
https://doi.org/10.1007/978-3-031-98334-4_24

overall efficiency. To resolve this issue, companies frequently tend to implement standard transportation solutions. However, these conventional planning systems do not guarantee an optimal trade-off between cost effectiveness and service quality, which highlights the need for advanced optimization techniques.

This class of problems belongs to the Vehicle Routing Problem (VRP). VRP is a combinatorial problem that focuses on finding optimal paths for a fleet of vehicles serving a specific set of customers. VRP is extensively studied and has applications in various fields, including employee, student, and goods transportation [1]. In this paper, we address a VRP variant in which employees of a large-scale industrial organization need to be collected from various pickup points and transported to several workplaces using a fleet of heterogeneous buses. This problem is referred to as the Employee Bus Routing Problem (EBRP) [2]. EBRP consists of generating the routes, assigning them to buses, and determining the groups of employees to be collected by each bus.

In fact, the route planning must comply with several constraints, that we detail later in this paper. Additionally, the organization comprises multiple workplaces, but in this paper, we only focus on the single-load scenario. In a single-load EBRP, each bus transports employees belonging to the same workplace and never carries employees from different workplaces. This choice is due to the social regulations of the industrial company, which prohibit employees of different workplaces from traveling together on the same bus.

Moreover, the EBRP we address in this paper is bi-objective: reducing the total costs and minimizing the maximum riding time of passengers. These two objectives are conflicting, since reducing costs implies using less vehicles which leads to more traveling time. Therefore, we propose an efficient Bi-Colony Ant Colony algorithm (BC-ACO) which is based on two distinct colonies. The first colony minimizes the total costs, while the second one reduces the maximum riding time of passengers. During the optimization process, a communication mechanism is established between the colonies, through which they exchange information periodically. This happens by the global update of pheromones, which highly helps the two colonies to keep a balance between exploration and exploitation of the search space.

Therefore, the contributions of this study are as follows:

- Minimizing the travel time of employees in an industrial context enhances the transportation planning' service quality which is crucial to the productivity of the organizations.
- Decomposing the objectives over two colonies reduces the computational burden.
- Organizations can use this framework to guide them in their decision making.

The remainder of the paper is organized as follows: Sect. 2 provides a literature review of the EBRP problem. Section 3 describes the bi-objective case-study we are addressing. Section 4 presents the BC-ACO we implemented to solve the problem under consideration. In Sect. 5 we compare the results obtained with an exact method for small instances, a conventional bi-objective GA and ACO with one colony. Finally, we propose possible future perspectives for this work.

2 Related Works

Over the years, numerous researchers attempted to solve EBRP using exact methods, heuristic or metaheuristic approaches. For instance, [2] addressed a real-world application of EBRP by solving a mathematical model minimizing total travel distance. In addition, [3] developed a K-means-based heuristic to efficiently manage routing in a large-scale organization. Furthermore, [4] employed Simulated Annealing (SA) and Genetic Algorithm (GA) to design routes planning for a retail company. Other researchers such as [5] focused on reducing fuel costs using Tabu Search, while [6] employed a Clarke & Wright-based heuristic to solve the bus routing and scheduling of EBRP.

Moreover, [7] applied ACO for a real-time school bus routing problem, which is a similar problem to EBRP. Similarly, [8] proposed a hybrid ACO framework for the school bus routing problem. Additionally, [9] developed a hybrid ACO and Dijkstra-based algorithm that efficiently addressed complex stochastic demands in bus transportation systems. Finally, [10] adopted a single-objective approach to optimize bus routing for a municipal government, highlighting the importance of minimizing travel distance to enhance the service delivery quality. Although the aforementioned studies employ varied methodologies, they mainly focus on single-objective optimization and only provide a brief examination of multi-objective scenarios.

Recently, the optimization of Employee Bus Routing Problems (EBRP) has focused on multi-objective employee routing challenges, emphasizing the need to balance competing objectives such as total cost, number of vehicles, and service quality. For example, [11] implemented a hybrid multi-objective Ant Colony Optimization (ACO) algorithm to simultaneously minimize the number of school buses used and the travel time for students, which is a scenario like EBRP. Additionally, [7] addressed the optimization of a school bus routing problem using a bi-objective model to reduce costs and enhance passenger satisfaction. Moreover, [12] modeled a multi-objective EBRP aiming to maximize service coverage while minimizing total travel distance. Finally, [8] developed a framework based on Geographic Information Systems (GIS) to minimize travel time and improve route planning efficiency.

In fact, some of the aforementioned studies address the trade-offs between operational costs and service quality. However, their approaches remain limited in directly resolving EBRP, notably in real-world scenarios. Additionally, most of the researchers that implement ACO-based frameworks for multi-objective EBRP only use a single colony.

This leaves a substantial gap in the EBRP literature, especially since establishing a collaboration between many colonies in ACO could enhance the exploration of the solution space and help ACO converge to the best trade-offs in an efficient and timely manner.

3 Problem Description

3.1 Current Business Context

In this section, we provide the readers with a thorough description of the problem we are studying. As mentioned previously, we aim to design the best route planning and scheduling in terms of total cost and traveling time, for a large-scale industrial group that operates in different areas across Morocco. This organization has two main contradictory objectives: minimizing the total operational costs while maintaining a satisfying quality service for the passengers. However, it is strenuous to reach a trade-off between these two objectives using conventional planning methods, therefore advanced optimization techniques are required.

Additionally, when the coronavirus pandemic struck, the organization had to adapt by reorganizing its scheduled routes to comply with Covid-19 health regulations. Previously, the organization allowed the transport of groups of employees from different workplaces in the same bus. However, in response to the pandemic, all planned routes were reconsidered as they did not meet the new regulations, which led to a cumbersome process of developing new routes. This situation created a necessity within the organization to adopt a model that prohibited the transportation of mixed groups of employees, which justifies why this paper focuses solely on the single-load scenario.

Moreover, a typical route involves heterogeneous buses navigating city streets to pick up employees at designated stops and transport them to their workplaces, while respecting strict time and capacity limits. When multiple employees are waiting at the same bus stop and going to the same workplace, they are referred to later in this paper as a group of employees. Buses begin their daily journeys at a departure depot, visiting planned stops before proceeding to the workplace(s). It is important to note that the company's fleet of buses varies in capacity and usage cost. Thus, a feasible solution requires planning trips that meet these constraints while minimizing total costs and travel time.

After providing a description of the context of the problem we are addressing, we present, in the following sub-section, more details about the specificities we take into consideration in this work.

3.2 Problem Version

As aforementioned, the industrial organization operates in different geographical regions. In this paper, we only focus on the most complex and large area where all the problem's characteristics exist. Moreover, the subproblems we address are as follows:

- Bus route generation: designing the routes that will be taken by the buses.
- Assigning routes to buses: affecting the generated routes to the available buses in the heterogeneous fleet of the organization.
- Assigning groups of employees to buses: deciding which groups will be transported by each used bus.

Furthermore, the problem we address comprises several assumptions and constraints that are listed below:

- The bus stops' locations are pre-defined by the organization, and we use them with no further modifications.
- A typical route of a vehicle should imperatively start by leaving the depot. The first node to be visited after the depot should always be a bus stop and never a workstation.
- Each employee should be picked up by only one bus during the whole trajectory. At the end of the generation of a solution, all employees should have been transported.
- When a vehicle arrives at a bus stop, it either picks up a group of employees or none. Transporting few members of a group waiting at the same bus stop is prohibited.
- Only single-load scenarios are considered in this paper.
- We consider one time slot and one direction (from bus stops to workplaces) for the purpose of tests. The multi-trip option will be solved in future works.
- We consider two objectives which are the total cost and the traveling time. The total cost is calculated based on the cost of a vehicle per kilometer. The travelling time is calculated based on an average speed equal to 60km/h.

Figure 1 illustrates the EBRP addressed in this paper, where groups of employees are distinguished by different colors, based on their workplace. In the following section, we present the BC-ACO used to solve the problem we have previously described.

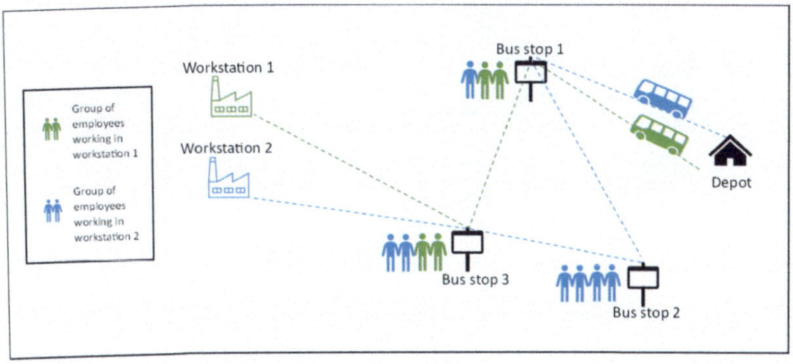

Fig. 1. A simplified scenario of the Employee Bus Routing Problem

4 Methodology

Nomenclature for the following sections

Colony 1: The colony of ants with the objective of minimizing the total cost.
Colony 2: The colony of ants with the objective of minimizing the maximum traveling time.
$\boldsymbol{\tau_{ij}(t)}$: pheromone trail of an arc (i,j) at instant t.
$\boldsymbol{\tau_0}$: value to initialize the pheromone trails.
$\boldsymbol{q_0}$: the selection parameter in ACO transition rule.
$\boldsymbol{q_1}$ **and** $\boldsymbol{q_2}$: heuristic parameters.

μ_{ij}: the visibility of an arc (i,j). It equals the inverse of the distance between i and j for colony 1, and the inverse of travel time between i and j for colony 2.
β: the preference weight of heuristic parameter.
α: the preference weight of pheromone trail.
ρ_1: the evaporation rate for the local update.
ρ_2: the evaporation rate for the global update.
A: the number of ants.
N: the number of iterations.
G: the set of employee groups.
F_{it}: the set of groups that the ant cannot visit, based on the current group it picked up.
P_{it}: the set of groups that an ant can choose from to decide the next group to be picked up.
$p_{ij}(t)$: probability of an ant going from node i to node j at time t.
$\Delta \tau_s$: the inverse of the total cost of an ant's solution for colony 1, and the inverse of the maximum traveling time of an ant's solution for colony 2.

As mentioned earlier, we address a bi-objective EBRP using a Bi-Colony Ant Colony Algorithm (BC-ACO). The various steps of BC-ACO are presented in Fig. 2.

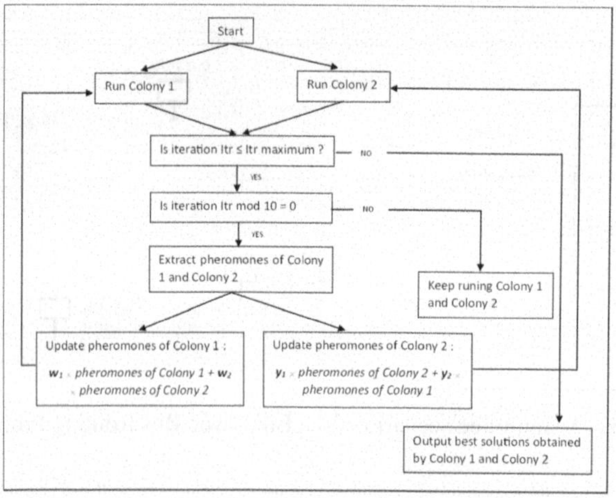

Fig. 2. The different steps of the BC-ACO framework

In this approach, two colonies of ants are designed, each focused on a different objective: the first colony aims to minimize total cost, while the second works on reducing the maximum travel time for passengers. During the initial iterations, the two colonies explore the search space independently. After a certain number of iterations, they begin exchanging their global pheromones to exploit the best solutions they have both obtained. To maintain a balance between the two objectives, the exchange mechanism can be flexible throughout the entire optimization process. In other words, decision-makers can determine the weights associated with the two objectives at each exchange. Further details about the exchange mechanism are provided later in this section. In the next sub-section, we describe how each colony of the BC-ACO operates.

4.1 Single-Colony Ant Colony Optimization (S-ACO)

Within a colony, we define several ants, A, all positioned at the depot. Each ant represents a complete solution, comprising a set of routes. Additionally, for each ant, we make two decisions: the nodes to be visited and the groups to be picked up. In our approach, we mainly focus on the groups since they subsequently determine the nodes that the ant will visit. A group contains the following information: stop, workplace, and number of employees. This means that the pheromones are updated based on the groups visited, as two vehicles can traverse the same paths while picking up different groups.

As shown in Fig. 3, we start by initializing the pheromones of all the arcs with a value $\boldsymbol{\tau_0}$. Then, the ants begin constructing their routes. Each ant starts by randomly selecting a vehicle and then a first group. We create a set of possible groups $\mathbf{P_{it}}$ that the ants can choose from for their next destination. This set is generated using a filter mechanism which eliminates all the forbidden groups $\mathbf{F_{it}}$ from the set of groups G. The set $\mathbf{F_{it}}$ contains the following groups:

- Groups with a different workplace than the current group.
- Groups already visited by the ant.
- Groups waiting at the same bus stop as the ones already existing in the ant' s solution.
- Groups that cannot be added due to capacity limits.

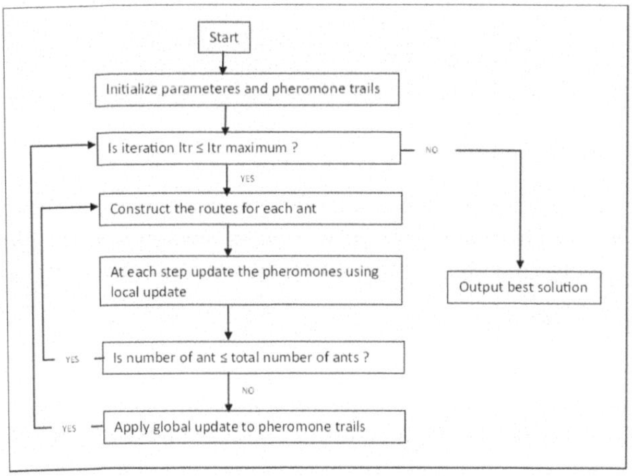

Fig. 3. The steps of a single colony ACO (S-ACO)

Assuming an ant is at a stop i, to select the next stop j to visit (therefore the group from the possible groups to be picked up), there are two modes of transition:

- if $\mathbf{q_1} \leq \mathbf{q_0}$ then the selection is based on argmax:

$$j = \text{argmax}_{u \in Pit}[\tau_{iu}(t) \cdot \mu_{ij}^{\beta}]$$

- if $\mathbf{q_1} > \mathbf{q_0}$ then the selection is based on the probability:

$$p_{ij}(t) = \begin{cases} \frac{\tau_{ij}(t).\mu_{ij}^{\beta}}{\sum_{u \in P_{it}} \tau_{iu}(t).\mu_{iu}^{\beta}}, & if\ j \in P_{it} \\ 0, & else \end{cases}$$

However, before inserting the selected group into the route, we always verify the problem's constraints. To clarify, there are two constraints that the ants should respect: the vehicle's capacity, the requirement that all passengers belong to the same workplace and are all picked up.

Once an ant completes its route construction, we perform a local pheromone update using the following formula, which we apply to all arcs visited by the ant:

$$\tau_{ij}(t+1) = (1-\rho_1)\tau_{ij}(t) + \frac{\rho_1 \tau_0}{1+N_v}$$

With Nv the number of used vehicles by an ant. These updated values of the local pheromones are then used by the next ant. We repeat the same steps for all the ants in the colony.

Once all the ants have completed their tours, we perform a global update of the pheromone values using the following formula, which is applied only to the arcs of the best solution obtained by all the ants during that iteration:

$$\tau_{ij}(t+1) = (1-\rho_2)\tau_{ij}(t) + \rho_2\, \Delta\tau_s$$

4.2 The Exchange Mechanism

BC-ACO is based on the collaboration and information exchange of the two colonies. As shown in Fig. 2, we start by running the two colonies and applying the global updates for each. We then exchange the pheromones of the two colonies using the formula below:

- $p_{c1} = \mathbf{w_1} \times p_{c1} + \mathbf{w_2} \times p_{c2}$
- $p_{c2} = \mathbf{y_1} \times p_{c2} + \mathbf{y_2} \times p_{c1}$

p_{c1} and p_{c2} refer respectively to the pheromone matrices of colony 1 and colony 2. w_1, w_2, y_1, and y_2 are weights that can be changed based on the priority given by decision makers to each objective. They can vary at each exchange phase. Furthermore, we re-run the colonies, and after a certain number of iterations (in our case, when the iteration number modulo 10 is equal to 0) we repeat the exchange. This process is repeated until the iterations are complete.

5 Results

In this section, we present and discuss the results of the BC-ACO. The tests use data from the company's largest and most complex region. We focused on a single time slot and one direction (from bus stops to workplaces) by sampling the data. The relevant information provided by the company includes the total number of employees during this time slot (approximately 600), the number of employees at each bus stop categorized by their workstation, the total number of bus stops in the area (17), the total number of workstations (17), the coordinates of each bus stop and workstation, the total number of vehicles (around 65), the routing cost for each vehicle type, and the number of depots (1).

As such, to validate the quality of the solutions obtained by BC-ACO we have conducted two comparisons:

- For small instances, a comparison between BC-ACO and a Mixed Integer Programming (MIP). The model we solved is from our previous work [13] with a change in the objective function.
- For the real data, we compared the results of BC-ACO with a traditional BACO and the other one with simple BGA. The objective functions of BACO and BGA were formulated as follow, to consider the cost and travel time during the optimization process:

$$w_1 \times \textit{total cost} + w_2 \times \textit{maximum travel time}$$

Moreover, Table 1 presents the set of parameters we used for BC-ACO, BACO and BGA. We have conducted various tests before determining the best values for the parameters for each approach.

To measure the efficiency of our approach, BC-ACO, we conducted tests on small data samples to compare the results obtained by BC-ACO and MIP, as presented in Table 2. Compared to MIP, BC-ACO achieves near-optimal solutions in total costs. For instance, at the data instance of size 10, the cost gap is minimal, around 1%. Moreover, regarding maximum travel time, the gap is negligible, around 0.9%, meaning that BC-ACO also succeeded in producing good solutions in terms of travel time. Even though MIP provides slightly better trade-offs for small instances, it is neither a scalable nor a practical approach for real-world scenarios due to their large-scale data.

Therefore, for the real data provided by the industrial group, we compared the results of BC-ACO with BACO and BGA, as shown in Table 3. The execution time for BC-ACO is slightly higher due to the complexity of the exchange mechanism. However, BC-ACO consistently achieves lower costs and reduced maximum travel times compared to BACO and BGA across different iterations (example: 25.903 min for BC-ACO compared to 28.593 min for BACO and 33.727 min for BGA). This highlights the efficiency of the two colonies cooperation mechanism in helping the BC-ACO explore the search space without losing its convergence pace.

More importantly, BC-ACO provided the best trade-off between total cost and travel time, as shown in Fig. 4. This diagram presents the distribution of trade-offs obtained by BC-ACO, BACO, and BGA. It also highlights the Pareto-optimal solutions that decision makers can choose from. As noticed in Fig. 4, all the pareto solutions are obtained using BC-ACO which makes it more suitable for balancing both objectives.

Table 1. Parameters of BC-ACO, BACO and BGA

	Ants	τ_0	q_0	q_1	q_2	ρ_1	ρ_2	β	ρ	Pop size	p_{cross}	p_{mut}	Itrs
Parameters BC-ACO	100	0,1	0,8	0,7	0,7	0,5	0,5	0,7	-	-	-	-	150
Parameters BACO	100	0,1	0,8	0,7	-	-	-	0,7	0,5	-	-	-	150
Parameters BGA	-	-	-	-	-	-	-	-	-	150	0,85	0,04	150

Table 2. Comparison between BC-ACO and MIP on data instances with sizes between 6 and 13 nodes

Number of nodes	BC-ACO		MIP		Gap between BC-ACO and MIP	
	Total Cost (MAD/km)	Max Travel time (min)	Total Cost (MAD/km)	Max Travel time (min)	Gap Cost	Gap Max Travel time
6	239.947	3.831	226.672	3.933	1.05%	0.97%
10	554,619	7,517	525.100	7.618	1.05%	0.98%
11	574,070	9.749	548.959	9.600	1.04%	1.01%
12	428.369	10.359	420.518	8.241	1.01%	1.25%
13	538 .838	10.359	538.250	8.8123	1%	1.17%

Table 3. Comparison between BC-ACO, BACO and BGA

itrs	BC-ACO			ACO			GA		
	Total Cost (MAD/km)	Max Travel time	Execution time (min)	Total Cost (MAD/km)	Max Travel time	Execution time (min)	Total Cost (MAD/km)	Max Travel time	Execution time (min)
10	3204768	30692	~1	3365267	37913	~1	3567912	40173	~1
30	3195433	29702	~7	3485678	35793	~4	3612562	39403	~5
50	3125923	28823	~15	3532658	34453	~11	3506589	37986	~10
70	3123718	27652	~20	3493458	34413	~16	3721900	37645	~16
90	3260522	27863	~38	3464118	34013	~25	3610375	35217	~25
110	3935477	26712	~50	3418173	32342	~40	3879729	35005	~40
130	3481712	26699	~65	3428778	32803	~55	3674395	34381	~55
150	3499023	25903	~100	3617218	28593	~70	3704189	33727	~78

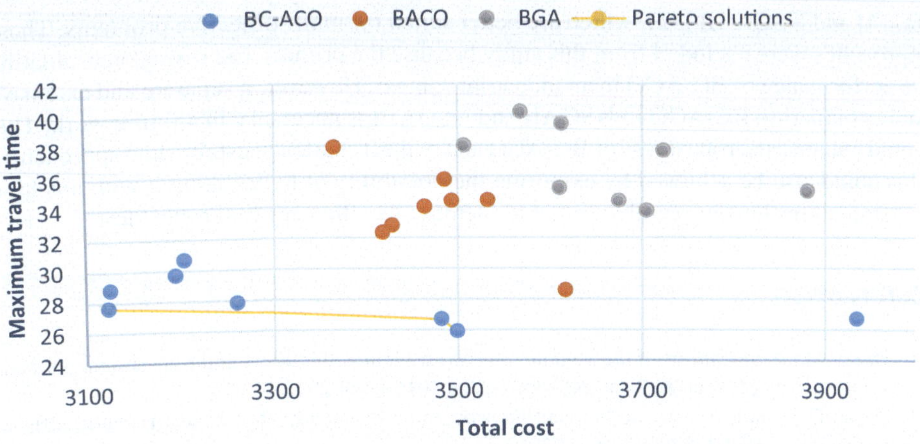

Fig. 4. Distribution of trade-offs for BC-ACO, BACO and BGA with Pareto solutions

6 Perspectives and Conclusions

In this paper, we addressed the bi-objective Employee Bus Routing Problem (EBRP) using an advanced bi-colony Ant Colony Algorithm (BC-ACO). The tests were performed on the data provided by a large industrial group in Morocco. Due to the complexity of the problem and the size of the data, reaching optimal solutions for more than 13 nodes was a cumbersome task. Therefore, we compared our framework (BC-ACO) with MIP on small data instances. Consequently, to evaluate the quality of the solutions generated by the BC-ACO on real-world data, we implemented weighted sum bi-objective algorithms: BACO and BGA. The results showed that BC-ACO provided better trade-offs than BACO and BGA, which was due to the exchange mechanism established between the two ant colonies. This approach can also be adopted by decision makers in various industries, providing them with greater insights into their businesses. A study on the generalizability of BC-ACO in other industrial contexts is to be addressed in future works.

Although the BC-ACO approach demonstrated strong computational efficiency for the tested instances, it may encounter challenges when scaling to larger datasets (over 10,000 nodes). The BC-ACO's exchange mechanism between the two colonies enhances solution quality, but it results in longer computational times as the problem size increases. Additionally, the memory required to maintain and support the communication between the two colonies could become a limiting factor in large-scale problems (over 10,000 nodes). To resolve this issue, future research works could focus on parallel implementations of the pheromone exchange strategies to improve scalability for large-scale applications.

Moreover, the weakness of this study lies in the comparison between these three approaches. To accurately evaluate the efficiency of a multi-objective algorithm, we need to determine the Pareto front. Traditional or weighted sum ACO and GA are not sufficiently adapted to multi-objective optimization. Therefore, we need to implement more robust approaches such as NSGA-II (Non-dominated Sorting Genetic Algorithm

II) [14] and NSGA-III [15], which are better suited for multi-objective problems. These approaches were excluded from this study because the primary focus was on evaluating the performance of BC-ACO in an industrial context. However, a separate and extensive study comparing BC-ACO to NSGAII and NSGAIII is necessary in a future work. This would enhance the robustness of BC-ACO and make it challenge established approaches. This could also be achieved by exploring the possibility of hybridizing it with machine learning algorithms or parallelizing it to increase its efficiency and robustness.

References

1. Drexl, M.: Applications of the vehicle routing problem with trailers and transshipments. Eur. J. Oper. Res. **227**(2) (2013). https://doi.org/10.1016/j.ejor.2012.12.015
2. Peker, G.: Employee shuttle bus routing problem: a case study. Eur. J. Sci. Technol. (2023). https://doi.org/10.31590/ejosat.1173057
3. Leksakul, K., Smutkupt, U., Jintawiwat, R., Phongmoo, S.: Heuristic approach for solving employee bus routes in a large-scale industrial factory. Adv. Eng. Inform. **32**, 176–187 (2017). https://doi.org/10.1016/j.aei.2017.02.006
4. Kiriş, S.B., Özcan, T.: Metaheuristics approaches to solve the employee bus routing problem with clustering-based bus stop selection. In: Artificial Intelligence and Machine Learning Applications in Civil, Mechanical, and Industrial Engineering (2019). https://doi.org/10.4018/978-1-7998-0301-0.ch012
5. Purba, A.P., Siswanto, N., Rusdiansyah, A.: Routing and scheduling employee transportation using tabu search. In: AIP Conference Proceedings, vol. 2217 (2020). https://doi.org/10.1063/5.0000766
6. Hashi, E.K., Hasan, M.R., Zaman, M.S.U.: A heuristic solution of the Vehicle Routing Problem to optimize the office bus routing and scheduling using Clarke & Wright's savings algorithm. In: 1st International Conference on Computer and Information Engineering, ICCIE 2015 (2016). https://doi.org/10.1109/CCIE.2015.7399306
7. Yigit, T., Unsal, O.: Using the ant colony algorithm for real-time automatic route of school buses. Int. Arab J. Inf. Technol. **13**(5) (2016)
8. Eldrandaly, K.A., Abdallah, A.M.F.: A novel GIS-based decision-making framework for the school bus routing problem. Geo-Spatial Inf. Sci. **15**(1) (2012). https://doi.org/10.1080/10095020.2012.708151
9. Qian, X., Liu, C.: Analysis and implementation of intelligent commuter bus system and its application perspectives. DEStech Trans. Comput. Sci. Eng. (2018). https://doi.org/10.12783/dtcse/icmsa2018/23284
10. Prasetyo, D.H., Aziiza, A.A., Sulistiyani, E.: Design of Employee Bus Routes for Madiun City Government Based on Home Locations and Presence Location History (2023). https://doi.org/10.2991/978-94-6463-288-0_46
11. Calvete, H.I., Galé, C., Iranzo, J.A., Toth, P.: The school bus routing problem with student choice: a bilevel approach and a simple and effective metaheuristic. Int. Trans. Oper. Res. **30**(2) (2023). https://doi.org/10.1111/itor.12951
12. Yalçındağ, S.: Employee shuttle bus routing problem. Mugla J. Sci. Technol. **6**(1), 105–111 (2020). https://doi.org/10.22531/muglajsci.691517
13. Bideq, H., Ouaddi, K., Gorge, A., Ellaia, R.: A real-world Employee Bus Routing Problem application with mixed loads. In: 2022 IEEE 6th International Conference on Logistics Operations Management, GOL 2022 (2022). https://doi.org/10.1109/GOL53975.2022.9820604

14. Deb, K., Pratap, A., Agarwal, S., Meyarivan, T.: A fast and elitist multiobjective genetic algorithm: NSGA-II. IEEE Trans. Evol. Comput. **6**(2) (2002). https://doi.org/10.1109/4235.996017
15. Mkaouer, W., et al.: Many-objective software remodularization using NSGA-III. ACM Trans. Softw. Eng. Methodol. **24**(3) (2015). https://doi.org/10.1145/2729974

Optimizing Energy Through Operations Planning

Yassine Bouazza(✉) and Azza Lajjam

National School of Applied Sciences Tetouan, Abdelmalek Essaadi University, Avenue of Palestine Mhannech I, 93002 Tetouan, Morocco
yassine.bouazza@etu.uae.ac.ma, alajjam@uae.ac.ma

Abstract. The cost of energy is an important component in terms of operating costs, the need to control energy use is becoming a national imperative and overarching objective in order to cope with the ever-increasing price increases and to be able to achieve energy savings in order to optimize the productivity ratio and thus reduce the carbon footprint from an environmental point of view. In our case study we are interested in the optimization process production operations planning. Using a case study from the sanitary ceramics industry, we propose a linear programming-based energy optimization model to address high energy consumption and associated costs. In particular, 2019 serves as the basis for this analysis, allowing to quantify the potential for energy savings without requiring additional investments, but simply by better management of production according to consumption. In addition, an analysis of consumption ratios and hourly rates revealed opportunities to maximize off-peak use, thereby reducing energy costs. The methodology involves analyzing electricity consumption patterns, pricing structures, and production schedules. A linear programming model was developed to optimize the distribution of consumption according to hourly rates, resulting in an efficient energy saving strategy. Key findings indicate significant energy savings can be achieved by reallocating energy-intensive operations to off-peak hours while minimizing usage during peak times.

Keywords: Optimization · Energy · Production planning · Linear programming

1 Introduction

The transition toward more environmentally friendly production practices is driven by increasingly stringent government regulations, growing consumer expectations, and investor demands for sustainability. In this context, Morocco, as one of the emerging countries, has adopted this strategy with the primary goal of accelerating the gradual transition toward a global green economy. Thanks to the government policies, The Moroccan industry has growth rapidity over the past two decades and can be considered as an industrial hub in Africa.

From the one hand energy management is becoming a key concern in global sustainability, in fact the adoption of effective strategies is essential to reduce operational

C. El Mokhi et al. (Eds.): ICASET 2025, SUCI, pp. 268–278, 2025.
https://doi.org/10.1007/978-3-031-98334-4_25

costs, minimize environmental impact, and ensure system reliability. From the other hand, operations planning can play a fundamental role in this context due to the strategic allocation of energy resources based on demand and time constraints.

This study examines the use of statistical analysis on production planning in order to improve electricity management in industrial factories. Numerous studies have shown a link between the optimization of energy consumption, improvement of manufacturing processes and the increase of energy efficiency, therefore the reduction of CO_2 emissions [1]. In this context, this study will use linear programming techniques, to reduce energy consumption without impacting production.

2 Literature Review

Traditionally considered an optional and complementary activity, Environmental sustainability practices are now part of industry standards under strict regulations, and impacting energy costs, traditionally considered a voluntary corporate responsibility, [2]. In fact, there is growing recognition of the importance of environmental sustainability. As a result, companies are from the one hand obliged to comply with environmental regulations, from the other hand considering sustainability practices as drivers of innovation and competitiveness [3].

Energy-Efficiency is an important aspect of Green Manufacturing, this involves implementation of practices that reduce energy consumption and CO_2 in manufacturing plants. In fact, by improving and optimizing energy efficiency in production, this challenge can be addressed [4]. Many national initiatives have started in order to mitigate the negative effects of energy consumption on the environment [5]. As a result, some Moroccan companies have begun adopting green practices and contributing to sustainable development.

The OCP Slurry Pipeline transports washed phosphate from the Khouribga mine to the Jorf Lasfar processing platform, eliminating the drying and rail transport steps. This project alone has reduced our CO_2 emissions by 620,000 tons and saved 3 million cubic meters of water each year. In the cement sector, LAFARGE Holcim has built a wind farm that provides important amount of the electricity consumption for the Tetouan plant, demonstrating its commitment to renewable energy.

SONASID is also committed to diversified and optimized projects, adopting innovative technologies to reduce atmospheric emissions, thus consolidating its regional leadership in Green Steel. The Group is thus at the forefront of Green Steel production, with a 100% recycling model powered by 88% renewable energy. In addition, the COSUMAR has replaced fuel drying with solar drying, a technology that significantly reduces CO_2 emissions. Thanks to this innovation, the company has avoided the emission of 57,000 tons of CO_2 per year.

One of the most notable examples is Renault. This company has achieved emission reductions through the implementation of two solutions. First, energy recovery from the painting process limits energy requirements and allows for the use of a lower-power boiler room. Second, the choice of boilers powered by biomass rather than natural gas reduces direct CO_2 emissions.

The Recent advancements in algorithms and data-driven approaches enhances the importance of implementation of digital tools in energy management all over industrial sectors [7, 8]. In fact, these methods such as predictive analytics, artificial intelligence models, and real-time decision-making systems, are facilitating the integration of renewable energy sources and storage solutions [9].

In the industrial context, operations planning can offer a dual benefit: increasing productivity while reducing environmental impact [10]. In fact, managing the demands enables adjusting energy consumption patterns, shifting usage to off-peak hours, this will lead to cost reductions and a more balanced energy demand [11]. Several studies have a focus on operations planning and energy management:

Zhang et al. [8] presented various scheduling techniques and their impact on energy consumption, also explored the importance of operations planning in energy-efficient manufacturing systems. Saboori et al. [12] highlighted the importance of the optimization scheduling parameters for a maximum savings. His research shows that scheduling significantly has reduces energy consumption, of HVAC systems.

Gupta et al. [13] focused on energy management and scheduling of data centers, which are important energy consumers. The study proposes a planning algorithm to optimize energy consumption in data centers and demonstrates its effectiveness in reducing energy consumption. Fernandes et al. [14] addresses the importance of planning operations in energy efficient manufacturing systems. It provides an overview of the different scheduling techniques and their impact on energy consumption, as well as challenges and research opportunities in this area. While operations planning is beneficial for energy management, this strategy also presents challenges, such as the need for sophisticated algorithms, the risk of increased maintenance costs, and the necessity to regularly adjust planning parameters.

3 Research Methodology

The chosen company operates in the sector of sanitary ceramics industry, energy consumption in this industry is significant, the aim of our research is to analyze energy consumption and annual production in order to propose an energy optimization model. High temperature. Ceramic cooking is generally done at high temperatures (between 800 and 1400 °C). This process consumes a large amount of energy, often in the form of natural gas or electricity, in addition to the CO_2 emission from burning fossil fuels in ovens.

3.1 Analysis of Electricity Production and Consumption

The energy analysis performed at the plant level was based on the history of bills with electricity consumption in addition to production data. Electricity is used in the following parts:

- Air conditioning and air conditioning
- Processes (machines, motors, etc.)
- Ventilation (extractors)
- Indoor and outdoor lighting
- Office automation

Table 1. Electricity consumption

Energy type	Consumption	Total cost in MAD/HT	Average unit cost
kWh of electricity	14,497,511	13,647,928	0,94 MAD/kWh

The table shows a summary of annual electricity consumption and cost (Table 1).
Electricity Consumption vs Production
The electricity consumption for per month is shown in the Table 2 below:

Table 2. Electricity consumption and production during the year

Month	Production kg	Consumption (kWh)
January	1 959 498	1,361,971
February	1 805 570	1,297,468
Mar	2 050 739	1,310,787
April	1 985 996	1,318,831
May	1 873 295	1,286,779
Jun	1 610 526	1,163,700
July	1 850 068	1,295,933
August	461 304	625,756
September	1 591 828	1,157,043
October	1 689 049	1,258,246
November	1 682 577	1,162,951
December	1 834 663	1,258,046
Total	**20 395 114**	**14,497,511**

The annual electricity consumption amounted to 14,497,511 kWh (Fig. 1).

The electrical energy consumption is = 14,497,511 kWh, while the production achieved is 20,395,114 kg.

Off-peak hours: 9 h, or 37.5% of the day's hours.

3.2 Electricity Pricing

(See Table 3).

- Off-peak hours: 9 h, or 37.5% of the day's hours.
- Regular hours: 10 h, representing 41.7% of the day's hours.
- Peak hours: 5 h, representing 20.8% of the hours of the day the kWh price per block.
- Off-peak hours: the price per kWh is 77.82% of the normal hour price.

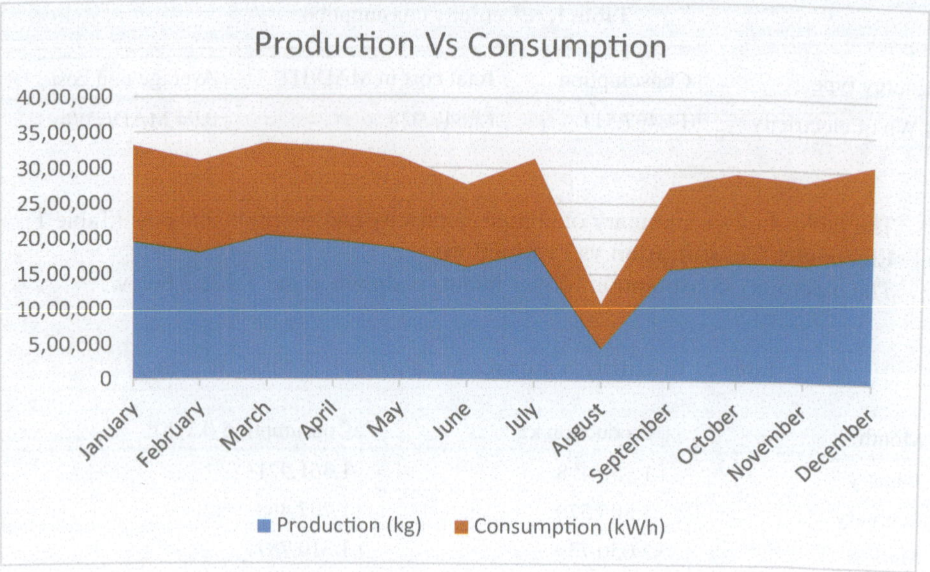

Fig. 1. Evolution of production and electricity consumption

Table 3. Electricity Pricing by Hourly Rate

Designation	Off-peak hours	Regular hours	Peak hours
Winter schedule	From 10 pm to 7 am	7 am to 5pm	From 5 pm to 10 pm
Summer schedule	From 11 pm to 7 am	7 am to 6 pm	From 6 pm to 11 pm
Price per kWh in MAD	0.73811	0.94854	1.25893

- Peak hours: the price per kWh is 132.72% of the normal hour price. It should be noted that the off-peak price is the most advantageous and the off-peak price is the highest.

Hence the interest of exploiting as much as possible off-peak hours and minimizing the operation of energy-intensive facilities at peak times.

The figure shows the distribution of production hours by hourly band (Fig. 2).

- The consumption rate during normal hours is higher than the hourly rate for this segment (45% versus 42%), which could be a positive point if we take them from peak hours.
- The annual off-peak consumption rate, which has the lowest energy price, is much lower than the normal off-peak hourly rate (34% versus 37%). We therefore recommend concentrating the production flow and thus energy consumption during off-peak hours.
- The annual peak consumption rate, where energy prices are slightly higher, is higher than the normal hourly peak consumption rate (21% versus 21%). In this context, we recommend that the rate be revised downward.

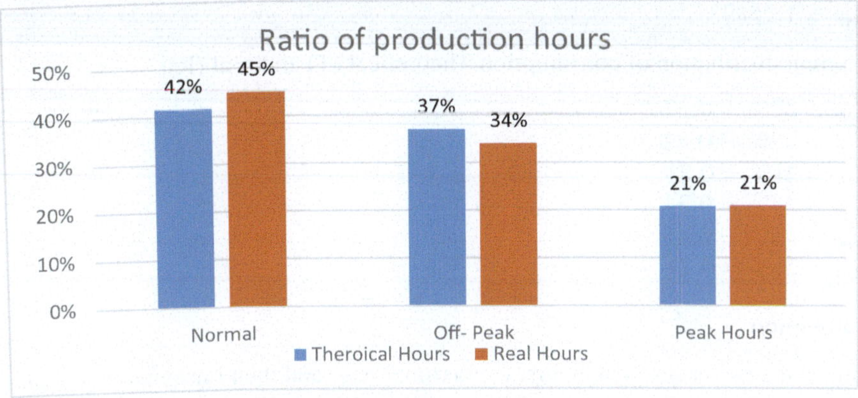

Fig. 2. Production Hour Ratio

4 Mathematical Model

Our goal is to define the problem in mathematical form to maximize the use of off-peak hours, then normal hours, and finally peak hours. This modelling will help optimize the distribution of energy consumption in the company based on hourly rates, with the aim of minimizing energy costs while meeting operational needs. Linear programming (PL) is a mathematical optimization technique used to maximize or minimize a linear objective function, under equally linear constraints. It allows to model and solve problems where it is necessary to find the optimal values of certain variables while respecting limits imposed by resources or operational conditions.

Solving a PL problem is usually done using numerical methods such as the Simplex algorithm or other adapted algorithms. These methods allow to find the optimal values of decision variables taking into account constraints.

One of the great advantages of linear programming is its ability to offer optimal solutions in many areas, such as resource management, logistics, production planning, and energy allocation, as shown in the optimization of electricity consumption in companies. It helps to minimize costs while maximizing operational efficiency, making it a key tool in the management of the Green Supply Chain and energy cost reduction strategies.

4.1 Decision Variables

Let's define the following variables:

- E_{total}: Total energy consumption (kWh).
- x_{HC}: Proportion of E_{total} consumption during off-peak hours.
- x_{HN}: Proportion of consumption during E_{total} Normal Hours.
- x_{HP}: Proportion of consumption during peak hours E_{total}.

Hourly rates (in MAD/kWh):

- $P_{HC} = 0{,}73811$
- $P_{HN} = 0{,}94854$

- $P_{HP} = 1{,}25893$

Current distribution of consumption Theoretical (T) and real (R):

- $x_{HC}^{T} = 37\% = 0{,}37$
- $x_{HN}^{T} = 42\% = 0{,}42$
- $x_{HP}^{T} = 21\% = 0{,}21$
- $x_{HC}^{R} = 34\% = 0{,}34$
- $x_{HN}^{R} = 45\% = 0{,}45$
- $x_{HP}^{R} = 21\% = 0{,}21$

4.2 Function

Maximize the use of off-peak hours, is maximize x_{HC} and then minimize $x_{HN} x_{HP}$.

- Minimize $C = E_{total} \times (x_{HC} \times P_{HC} + x_{HN} \times P_{HN} + x_{HP} \times P_{HP})$ (1)

as E_{total} is constant, we can minimize the unit cost:

- Minimize $C = (x_{HC} \times P_{HC} + x_{HN} \times P_{HN} + x_{HP} \times P_{HP})$

Constraint of sum of proportions:

- $x_{HC} + x_{HN} + x_{HP} = 1$ (2)

Variable bounds:

- $0 \leq x_{HC}, x_{HN}, x_{HP} \leq 1$ (3)

4.3 Operational Constraints

There is a **limitation** on the amount of energy that can be consumed during off-peak hours due to operational constraints (for example, staff availability, machinery's ability to operate at night, etc.).

Let's define as the maximum consumption capacity x_{HC}^{MAX} during off-peak hours.

Similarly, there may be **minimal consumption** during normal and peak hours due to operational requirements.

To model this, we can introduce the following constraints:

- $x_{HC} \leq x_{HC}^{MAX}$
- $x_{HN} \geq x_{HN}^{MIN}$
- $x_{HP} \geq x_{HP}^{MIN}$

We will use the actual values as starting points.

If we assume that:

- x_{HC}^{MAX} Maximum capacity (to be determined based on the business operations).
- x_{HN}^{MIN} minimum consumption required during normal hours.
- x_{HP}^{MIN} minimum consumption required during peak hours

Without specific data, we can estimate these values based on actual current consumption.

In our case study it is assumed that it is possible to increase the consumption up to an operational maximum of 40%.

- $x_{HC} \leq x_{HC}^{MAX} = 0{,}4$
- $x_{HN}^{MIN} = 0{,}3$

In our case study it is assumed that it is possible to reduce consumption up to an operational minimum of 15%.

- $x_{HP} \geq x_{HP}^{MIN} = 0{,}15$

4.4 Mathematical Model

Minimize $C = (x_{HC} \times P_{HC} + x_{HN} \times P_{HN} + x_{HP} \times P_{HP})$

Under constraints

$$x_{HC} + x_{HN} + x_{HP} = 1$$
$$x_{HC} \leq x_{HC}^{MAX}$$
$$x_{HN} \geq x_{HN}^{MIN}$$
$$x_{HP} \geq x_{HP}^{MIN}$$
$$0 \leq x_{HC}, x_{HN}, x_{HP} \leq 1$$

4.5 Problem Solving

The problem can be solved using **linear optimization** techniques (linear programming), because the objective function and constraints are linear. As a first step we will start with the formulation the problem in an optimization software: Excel Solver tool.

Objective Function

$$C = x_{HC} \times 0{,}73811 + x_{HN} \times 0{,}94854 + x_{HP} \times 1{,}25893$$
$$x_{HC} + x_{HN} + x_{HP} = 1$$
$$0{,}34 \leq x_{HC} \leq 0{,}4$$
$$x_{HN} \geq 0{,}3$$
$$0{,}15 \leq x_{HP} \leq 0{,}21$$
$$0 \leq x_{HC}, x_{HN}, x_{HP} \leq 1$$

- The optimization tool will provide the optimal values of x_{HC}, x_{HN}, x_{HP} that minimize total cost. To solve this problem, we used a numerical method adapted to linear optimization problems:
- **Simplex method:** An efficient algorithm to solve linear programming problems.
- **Optimization Software:** Excel Solver, which implements simplex.

5 Results

5.1 Interpretation of Results

- $x_{HC} = 0{,}40$ **Off-peak hours**: Consumption during off-peak hours is maximized at its maximum capacity (40%).
- $x_{HN} = 0{,}45$ **Normal Hours**: Consumption remains the same as actual consumption (45%).
- $x_{HP} = 0{,}15$ **Peak Hours:** Consumption is reduced to the minimum possible (15%), which is lower than actual current consumption (21%).

The increase in consumption during off-peak hours from 34% to 40%, allows you to take advantage of the lowest rate.

The reduction of consumption during peak hours from 21% to 15%, reduces consumption during the most expensive period.

Consumption during normal hours remains unchanged at 45%, which is dictated by the sum constraints of proportions and limits on other variables.

5.2 Calculation of the Optimal Total Cost

- $C* = 0{,}73811 x^*_{HC} + 0{,}94854 \times x^*_{HN} 1{,}25893 \times x^*_{HP}$
- $C* = 0{,}73811 \times 0{,}40 + 0{,}94854 \times 0{,}45 + 1{,}25893 \times 0{,}15$
- $C* = 0{,}295244 + 0{,}426843 + 0{,}188840$
- $C* = 0{,}910927$ MAD/kWh

Comparison with the Current Cost
Current Cost:

$$C_{\text{Actuel}} = 0{,}73811 \times 0{,}34 + 0{,}94854 \times 0{,}45 + 1{,}25893 \times 0{,}21$$
$$C_{\text{Actuel}} = 0{,}250957 + 0{,}426843 + 0{,}264876$$
$$C_{\text{Actuel}} = 0{,}942676 \text{ MAD/kWh}$$

Savings Achieved:

$$\Delta C = C_{\text{Actuel}} - C*$$
$$\Delta C = 0{,}942676 - 0{,}910927$$
$$\Delta C = 0{,}031749 \text{ MAD/kWh}$$

5.3 Model Sensitivity

If the company can increase consumption further during off-peak hours or reduce consumption during peak hours, additional savings are possible.

Case where $x^{MAX}_{HC} = \mathbf{0,45}$.

New Total Cost:

$$C* = 0{,}73811 \times 0{,}45 + 0{,}94854 \times 0{,}40 + 1{,}25893 \times 0{,}15$$
$$C* = \frac{0{,}900405 \text{ MAD}}{\text{kWh}}$$

New Economy: $\Delta C = 0{,}942676 - 0{,}900405 = \frac{0{,}042271 \text{ MAD}}{\text{kWh}}$

Conclusion: By increasing operational capacity during off-peak hours, the company can realize additional savings.

Results: The analysis of the specific production and consumption of electricity and heat has shown that there is a strong potential for energy savings.

The average ratio of energy production to consumption is 71,08 kWh/100 kg.

The potential gain following our mathematical model is $\Delta C = 0{,}031749$ MAD/kWh.

6 Conclusion

A preliminary analysis of electricity bills for the year, and the analysis of electricity consumption in production to assess potential gains. This step allowed us to identify an electrical gain potential of 234,681 kWh. Thanks to the better us use of off-peak hours and changing the proportion of energy consumption in different time slots, the company can not only reduce costs but also improve overall energy efficiency.

In addition, other strategies can be employed to achieve energy better optimization.

- Energy audits help identify areas within the plant where energy is being wasted. In order to identify opportunities to reduce energy consumption [15]
- Upgrading to energy-efficient equipment to decrease energy consumption and carbon emissions. For example: invest in high-efficiency lighting, heating or air conditioning (HVAC) systems, and production equipment's [6]
- The Use of solar or wind power as alternative energy sources, to supply electricity on the plants. This is possible with the installation of solar panels or wind turbines on-site or buying renewable energy from third-party suppliers [16].

The use of Linear Programming (LP) for energy optimization in our study, is not limited to only the ceramic industry, if fact, this methodology can develop in other industries that are facing similar challenges energy management, cost reduction, and sustainability achievement. For example, Industries like textile manufacturing, paper production, automotive manufacturing, or mining and metal processing can be subject of future research since they are sharing common characteristics regarding energy-intensive processes which is critical to their operations.

In addition, other mathematical methods can also be applied in the case of energy optimization and operation scheduling. For examples, methods like Mixed-Integer Programming (MIP) and heuristic algorithms can offer different alternatives and solution for complex energy optimization cases. The combination of these techniques will help companies to achieve important cost savings, reduce their environmental impact, and enhance operational efficiency, and thus to contribute to a more sustainable and economically viable future.

References

1. Srivastava, S.K.: Green supply-chain management: a state-of-the-art literature review. Int. J. Manag. Rev. **9**(1), 53–80 (2007). https://doi.org/10.1111/j.1468-2370.2007.00202.x
2. Porter, M.E., van der Linde, C.: Toward a new conception of the environment-competitiveness relationship. J. Econ. Perspect. **9**(4), 97–118 (1995). https://doi.org/10.1257/jep.9.4.97
3. Nidumolu, R., Prahalad, C.K., Rangaswami, M.R.: Why sustainability is now the key driver of innovation. Harv. Bus. Rev. **87**(9), 56–64 (2009)
4. International Energy Agency. Energy efficiency 2020. IEA, Paris. Available at IEA Website (2020)
5. IPCC. Climate Change 2021: The physical science basis. IPCC, Geneva. Available at IPCC Website (2021)
6. Bouazza, Y., Lajjam, A.: Contribution of Lean Manufacturing on environmental performance in Moroccan Industry. Logist. Oper. Manag. Res. (LOMR) **2**(2), 72–86 (2023). https://doi.org/10.31098/lomr.v2i2.1921

7. Zhao, X., Lin, B., Yao, Y.: Predictive analytics for energy optimization in manufacturing systems. Int. J. Energy Res. **44**(2), 720–735 (2020)
8. Chen, J., Zhang, Y., Liu, Y.: Energy optimization using data-driven approaches: applications in smart grids. Energy Rep. **7**, 1234–1245 (2021)
9. He, H., Wang, J., Zhang, X.: Renewable energy integration and optimization in power systems: a machine learning approach. IEEE Trans. Smart Grid **10**(5), 4701–4712 (2019)
10. Kang, S., Kim, S.: Optimization of industrial energy management through operations planning. J. Clean. Prod. **256**, 120502 (2020)
11. Li, X., Liu, Z.: Demand-side management and energy optimization: strategies and challenges. Energ. Effi. **15**(4), 1543–1557 (2022)
12. Saboori, A., et al.: Impact of scheduling on energy consumption in building heating, ventilation, and air conditioning systems. Energy Build. **174**, 206–215 (2018). https://doi.org/10.1016/j.enbuild.2018.05.030
13. Rao, V., Gupta, M.: Study on organizational culture and ecological transition in India. Indian J. Sustain. Dev. **15**(2), 221–234 (2020)
14. Fernandes, J., Homayouni, M., Fontes, D.: Energy-efficient scheduling in job shop manufacturing systems: a literature review. Sustainability **14**, 6264 (2022). https://doi.org/10.3390/su14106264
15. Boharb, A., et al.: Auditing and analysis of energy consumption of an industrial site in Morocco. Energy **101**, 332–342 (2016). https://doi.org/10.1016/j.energy.2016.02.035
16. Smouh, S., Gargab, F.Z., Ouhammou, B., Mana, A.A., Saadani, R., Jamil, A.: A new approach to energy transition in Morocco for low carbon and sustainable industry (case of textile sector). Energies **15**(10), 3693 (2022). https://doi.org/10.3390/en15103693

A New Decision Framework for Sustainable Supplier Selection in the Automotive Industry

Ismail Ouahabi[1](✉), Oulaid Kamach[1], and Widad Fethallah[2]

[1] Innovative Technologies Laboratory (LTI), National School of Applied Sciences, Tangier, Morocco
ismail.ouahabi@etu.uae.ac.ma, okamach@uae.ac.ma

[2] Measurement Modelling and Analysis Performance of the Organization Research Team, La TIC, National School for Applied Sciences, Tangier, Morocco

Abstract. Sustainability is a crucial aspect of automotive supply chain management, with significant efforts from researchers to advance this field. This paper builds upon a previous study with a Moroccan automotive company to enhance the sustainable supplier selection process by incorporating additional sustainability criteria. We developed an innovative framework that includes criteria such as Supplier Information, Certification, Key Indicators, Social Sustainability, Quality, Delivery, Launch, Financial, and Process Capabilities, each with relevant sub-criteria. The study applies the PAPRIKA method, a Multi-Criteria Decision-Making (MCDM) tool, for supplier selection. Using a Python environment, we identified the top ten suppliers for each commodity group (Electrical, Electronics, Underbody, Exterior, and Interior). This work is the first to implement the PAPRIKA method in the context of sustainable supplier selection, offering a novel approach to integrating sustainability criteria into the decision-making process. However, several challenges persist, and as part of perspectives, we will study other MCDM approaches to be compared with PAPRIKA method.

Keywords: Supplier selection process · MCDM · Sustainability · Automotive Supply Cain · PAPRIKA

1 Introduction

The transition to electric vehicles, geopolitical tensions, wars, compliance pressure, and fast-growing competitors are all disrupting the automotive supply chain. As the supplier selection activities is a crucial field for Purchasing management in the automotive industry, the management attention to supplier selection problem reached an all-time high. However, the sustainability aspect is the core of supply chain management in automotive industry.

According to [1], almost 80% of organizations have experienced supply chain disruptions, more than last year. Third-party failures are the leading cause of supply chain disruptions for organizations with 43,6%. The second main cause of supply chain disruptions is Cyber-attack with 34.00% and the last one is Adverse weather and natural disasters with 26.6%.

C. El Mokhi et al. (Eds.): ICASET 2025, SUCI, pp. 279–288, 2025.
https://doi.org/10.1007/978-3-031-98334-4_26

All suppliers are impacted by such concerns, and it is mandatory for Automakers to carefully choose their collaborators by following a robust supplier selection process, selecting sustainable criteria, and adopting a suitable multi-criteria decision-making (MCDM) approach.

In this study, we proposed a new framework combining between a robust supplier selection process and PAPRIKA metho to rank the Top 10 suppliers from a detailed database. The structure of this paper consists of 5 sections: after the introduction, Sect. 2 provides an overview of MCDM approaches and PAPRIKA method. Section 3 presents a new supplier selection framework combining between a robust supplier selection process and PAPRIKA method, and a case study was conducted by using a database from automotive industry. Section 3.4 show some results and Sect. 4 concludes our work's result and highlights some perspectives for future research.

2 Literature Review

2.1 MCDM Overview

Due to the globalization effects, supplier selection become more and more complex and critical. To facilitate this complexity, researchers and industrial managers aims to integrate the use of Artificial Intelligence (AI) in the supplier selection problematic.

One of the studies combines AI techniques with MCDM approaches and presents a literature review is [2], and the author shows that the AI approach ANN can be combined with MCDM methods such as DEA, AHP, Delphi or ANP, and several case studies were presented in Automotive industry.

Because the supplier selection problem is considered as an MCDM topic, several studies and research discuss the main MCDM methods that can be used for supplier selection, and if we combine the results from [3] and [4] and other papers such as [5–7], we find that the methods used for the supplier selection topic can be classified as illustrated below (Fig. 1):

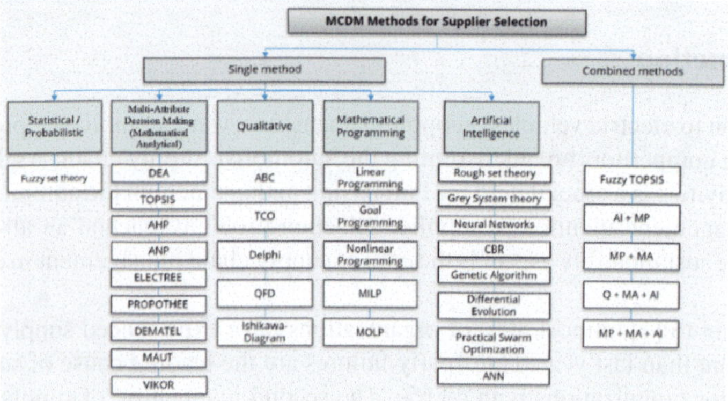

Fig. 1. MCDM methods used for supplier selection.

2.2 PAPRIKA Method

The PAPRIKA method (Potentially All Pairwise Rankings of all possible Alternatives) introduces an innovative approach for decision-makers to rank all potential alternatives represented by a value model through pairwise comparisons. This method focuses specifically on undominated pairs, which are pairs of alternatives where one alternative is superior in at least one criterion but inferior in another. Conversely, "dominated pairs" are automatically ranked because one alternative is superior across all criteria.

The process starts with the decision-maker ranking undominated pairs based on two criteria at a time, while assuming all other criteria are identical between the alternatives being compared. If the decision-maker chooses to continue, the method progressively includes pairs with more criteria until all undominated pairs have been ranked.

To minimize the number of pairs that need explicit evaluation, PAPRIKA leverages the transitivity property of additive value models. This allows implicitly ranked pairs to be identified and excluded, based on their logical derivation from explicitly ranked pairs. Each pair is ultimately categorized as one of the following:

- A dominated pair automatically ranked by definition.
- An undominated pair, explicitly ranked by the decision-maker.
- An implicitly ranked pair, derived through transitivity.

This systematic process results in a complete and consistent overall ranking of alternatives. The explicit rankings of preferences (inequalities) and indifferences (equalities) are subsequently converted into point values using linear programming. While the linear programming step may yield multiple mathematical solutions, all solutions reproduce the same overall ranking of alternatives. Developed by [8], this method provides a robust and systematic framework for generating comprehensive rankings of alternatives through iterative pairwise evaluations. The method calculates the number of each category of pairs using the below formulas:

$$N(n, y, z) = C_2^n(2^{z-1} - 1)({}_2{}^yC)^z y^{n-z} \tag{1}$$

$$U(n, y, z) = C_2^n(2^{z-1} - 1)({}_2{}^yC)^z \tag{2}$$

For a value model with n criteria and y categories on each criterion, Eq. (1) gives the total number of undominated pairs of degree z ($z = 2, 3 \ldots$), including replicas, denoted $N(n, y, z)$. Equation (2) gives the number of these pairs of degree z that are unique (excluding replicas), denoted $U(n, y, z)$.

The PAPRIKA method has proven to be a valuable Multi-Criteria Decision-Making tool across various domains, including education [9], sustainable building design [10], lean production strategies [11]. Additionally, its application in the healthcare sector, as highlighted in a case study [8], demonstrates its effectiveness in enhancing decision-making processes. This method has been utilized to evaluate alternatives and determine the weighting of criteria in diverse contexts.

Figure 2 illustrates the flowchart of the method, providing a comprehensive overview of its structured approach.

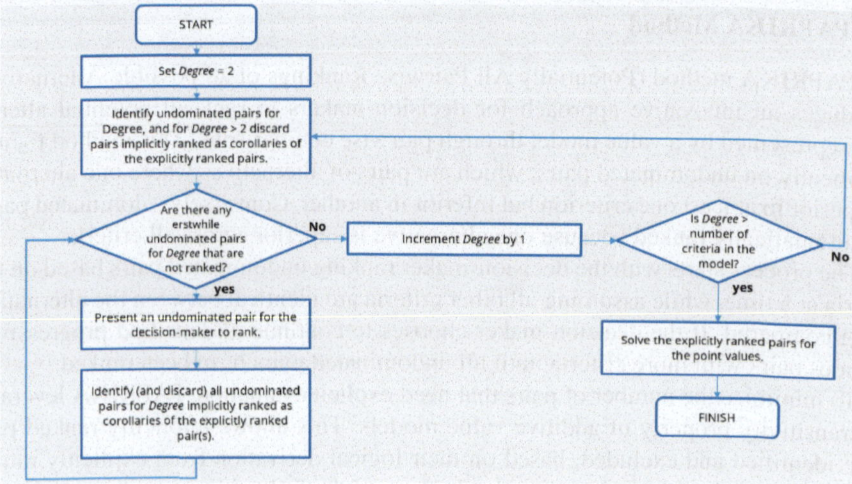

Fig. 2. The flow chart of PAPRIKA method.

3 Sustainable Supplier Selection Framework

3.1 Supplier Selection Model

One of the challenges that organizations are facing is to define a robust supplier selection process. According to [5], a sustainable supplier selection process was presented and used by a Moroccan Automotive company, and we constate that it is preferable to

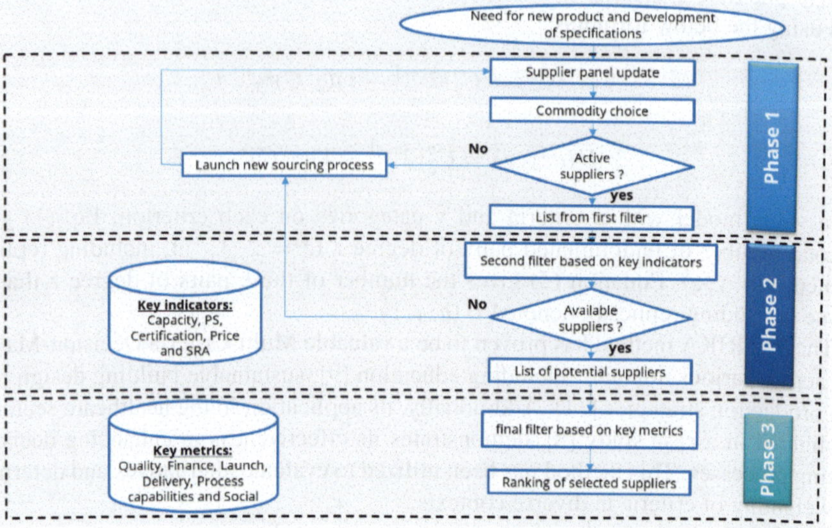

Fig. 3. The improved supplier selection process.

exchange between criteria of phase 2 and phase 3 because the key factors (Political stability, Capacity and Certifications) must be evaluated before moving to the final ranking. Besides, we add social metric in the supplier selection process.

3.2 Supplier Selection Criteria

The supplier selection criteria are the first interesting input for supplier selection process that need to be identified carefully.

The list of criteria can be updated based on the special need of each company.

In our study, the supplier selection process was divided into 3 phases, and for each phase, there is special criteria used for supplier ranking.

- **Phase 1:**

This step is mainly for commodity choice, if there are no active suppliers in the data base acting in the needed commodity, it is mandatory to search in the market other suppliers in collaboration with program buyer and supplier management development team. And this is to avoid monopoly and to be in the safe side (several suppliers should be available as backup).

- **Phase2:**

In this phase, we adjusted the process used previously by Ismail2024 and we exchanged the criteria used for phase 2 by the key factors mentioned in phase 3 (Table 1).

Table 1. Key indicators for phase 2 in supplier selection process.

Key Indicator		Description
Political Stability (PS)		Is the situation of country in term of security and stability (no wars and strong national security system)
Capacity		Free capacity to fulfil the customer demand in term of deliveries
Certification	International Automotive Task Force (IATF16494)	Is the International automotive standard to be adopted by Automotive companies to standardize the quality management system
	International Organization for Standardization (ISO14001)	Is the international standard that defines requirements for an effective environmental management system (EMS)
	Materials Management Operations Guideline (MMOG)	Is a tool for suppliers to perform self-assessment to follow and improves materials management efficiency
	Sustainability Assessment Questionnaire (SAQ 5.0)	Is a questionnaire that car manufacturers united in the Drive Sustainability Initiative (DSI) use to query the sustainability in the value chain of their suppliers
Price		The price offered by supplier
Supplier Risk Assessment (SRA)		Is an assessment performed by program buyer and supplier development specialist to confirm the ability of supplier for new business

- **Phase 3:**

This step of supplier selection process will evaluate the Quality, Delivery, Financial, Launch, Process Capabilities and Social metrics (Tables 2 and 3).

Table 2. Key metrics for phase 3 in supplier selection process.

Key Metric	Criteria	Description
Quality	Part Per Million (PPM)	Is an indicator for Quality performance calculation the number of rejected parts vs the total quantity delivered
	Incident Per Billion (IPB)	Is a Quality indicator presenting the number of quality incidents vs the total quantity delivered
	Warranty	Is what the manufacturer offers to repair or replace a faulty product within a window of time after purchase
Delivery	On Time Delivery (OTD)	Is the ratio of deliveries received on time vs demands
	Stop Ship Number (SSN)	Is the number of time where the customers manufacturing facilities are stopped due to supplier issues
Launch	Advanced Product Quality Planning (APQP)	Is a methodology used in the automotive industry to ensure a high-quality product that meet customer specific requirements and automaker expectations
	Production Part Approval Process (PPAP)	Is to ensure that all Quality requirements and product specifications are meet via a stable and capable process
Financial	Payment Terms (PT)	Is the agreement between supplier and customer regarding when and how payments should be made
	Cash Flow Situation (CFS)	It is the movement of money into and out of a company over a certain period of time
Process Capability	Overall Equipment Effectiveness (OEE)	Is a key indicator to measure manufacturing productivity and process capability
	Supplier Manufacturing Assessment (SMA)	Is a manufacturing assessment performed by supplier development team to confirm the capability of process

Table 3. Key criteria for social aspect According to [12].

Criteria	Description
Work health and safety	This refers to the company' focus on health and safety practices
Community influence and Training	This refers to the transfer of knowledge from company to its employees and the community impact within the field where they operate

(*continued*)

Table 3. (*continued*)

Criteria	Description
Stakeholders' influence	This refers to the level of attention that supplier pays to its stakeholders to get involved in its day-to-day activities and overall business strategy
The rights of employees	This has to do with workers issues and employment concerns
The rights of stakeholders	This regarding to the rights of individuals, groups, or organisms, which had a direct impact in the business
Information disclosure	This has to do with companies providing their customers and stakeholders with related information about the used components during the manufacturing process and its environment impact
Employment practices	This is related to all practices of employees

3.3 Case Study

In this section, we focus on solving an MCDM problem related to sustainable supplier selection within the automotive industry, a critical sector for advancing sustainable practices. We have developed a consistence database with input from experts in the automotive industry, taking account the core criteria including supplier information, certifications, social sustainability, quality, delivery, product launch performance, financial stability, and process capabilities. For this, we conducted a pilot study by selecting a sample of 200 suppliers including product groups such as electrical, electronic, underbody, exterior and interior components. This consistence dataset offers an original framework for selection supplier performance throughout several sustainability dimensions, enabling for an enhanced comprehension of their impact on holistic supply chain sustainability and resilience (Figs. 4 and 5).

	Phase 1			Phase 2							
				Certification				Keys Factors			
Supplier Name	City	Country	Commodity	IATF	ISO14001	MMOG	SAQ 5.0	Political Stability	Capacity	Price	SRA
S1	Tangier	Morocco	Electrical	Yes	Yes	Yes	Yes	Yes	Yes	Yes	Green
S2	Casablanca	Morocco	Electronics	Yes	Yes	Yes	Yes	Yes	Yes	Yes	Green
S3	Kenitra	Morocco	Exterior	Yes	Yes	Yes	Yes	Yes	No	Yes	Green
S4	Kenitra	Morocco	Interior	Yes	Yes	Yes	Yes	Yes	Yes	Yes	Yellow
S5	Casablanca	Morocco	Underbody	Yes	Yes	Yes	Yes	Yes	Yes	Yes	Green
S6	Tangier	Morocco	Electrical	Yes	Yes	Yes	Yes	Yes	Yes	No	Green
S7	Tangier	Morocco	Electrical	Yes	Yes	Yes	Yes	Yes	Yes	Yes	Green
S8	Kenitra	Morocco	Electrical	Yes	Yes	Yes	Yes	Yes	Yes	Yes	Yellow
S9	Kenitra	Morocco	Electrical	Yes	Yes	Yes	Yes	Yes	Yes	Yes	Green
S10	Kenitra	Morocco	Electronics	Yes	Yes	Yes	Yes	Yes	Yes	Yes	Green
S11	Casablanca	Morocco	Exterior	Yes	Yes	Yes	Yes	Yes	Yes	Yes	Green
S12	Kenitra	Morocco	Interior	Yes	Yes	Yes	Yes	Yes	Yes	Yes	Green
S13	Tangier	Morocco	Underbody	Yes	Yes	Yes	Yes	Yes	Yes	No	Red

Fig. 4. A screenshot from Database (Phas1 and Phase2 criteria).

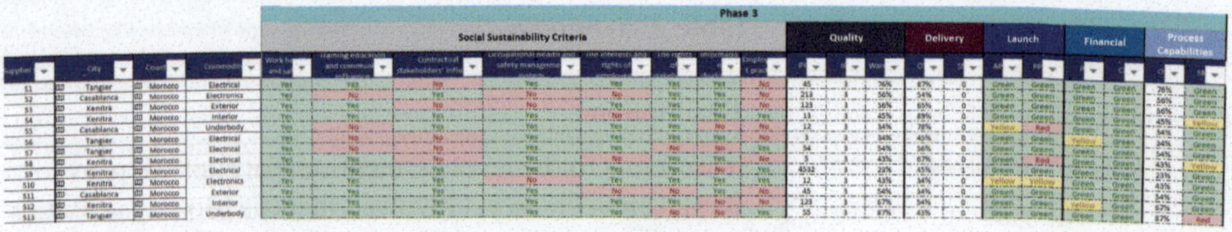

Fig. 5. A screenshot from Database (Phase3 criteria).

3.4 Result and Discussion

This paper is the first to introduce the PAPRIKA method regarding of sustainable supplier selection, providing a new approach to including sustainability criteria into the decision-making process. We were using python environment to implement our framework based on the PAPRIKA method to select and rank the top 10 suppliers for each commodity based our dataset. This later composed with of several criteria, each with its associated sub-criteria. We have a set of criteria centered on Certifications (IATF, ISO14001, MMOG, and SAQ 5.0). The second set of criteria focused on Social Sustainability (Health and Safety practices, Training, Community Influence, and Employee Rights…). The other criteria are Quality (PPM, IPB, and Warranty), Delivery (OTD, SSN), Launch (APQP, PPAP), Financial (PT, CFS), Process Capabilities (OEE, SMA), Political Stability, Capacity, Price, and SRA. The selection process for suppliers was founded on performance thresholds. As an example, the Warranty, OTD, and OEE scores had to be greater than 30%, PPM had to be less than 1000 to meet the selection criteria. Suppliers that met the specifications for all criteria, taking into consideration certifications and political stability were selected for further analysis. This resulted in the identification of the top-performing suppliers for each commodity.

This implementation of the PAPRIKA method allowed for a systematic and transparent ranking of suppliers focused on their sustainability performance and operational capabilities. By implementing a rigorous criterion such as certification standards and social sustainability practices, the framework selected suppliers that not only meet technical and financial expectations but also conform with sustainable and ethical practices. Excluding suppliers that did not meet specific thresholds in criteria like quality (PPM < 1000) and operational efficiency (OEE $> 30\%$) confirmed that only high-performing suppliers were considered.

As part of our implementation, we developed a Python script under the Anaconda environment, including a data cleaning process for supplier selection (Fig. 3). We then integrate the PAPRIKA algorithm, which identified the 10 most sustainable suppliers, with a strong importance on sustainability and resilience, in addition to the standard selection criteria (see Table 4).

This approach minimizes processing time compared to the traditional Excel-based method. Our perspectives are the first to compare its performance with other MCDM techniques such as TOPSIS, AHP, and BWM. At the second, we plan to develop a Power BI interface to visualize the model's results and integrate them into the company's information system.

Table 4. The 10 top sustainable suppliers of Electrical commodity

Ranking	Supplier Name	Commodity	PAPRIKA Score
1	S100	Electrical	1.5
2	S25	Electrical	1.5
3	S43	Electrical	1.5
4	S196	Electrical	1.35
5	S70	Electrical	1.35
6	S73	Electrical	1.35
7	S81	Electrical	1.35
8	S130	Electrical	1.3
9	S143	Electrical	1.3
10	S48	Electrical	1.2

The results demonstrated the importance of integrating sustainability criteria, such as social sustainability and process capabilities, along with traditional performance measures to make more informed and responsible supplier selection decisions in the automotive industry.

4 Conclusion

This study offers a decision framework for companies operating in automotive industry and need to select a sustainable supplier for long term cooperation. To resolve the supplier selection problem, this study aims to combine the proposed supplier selection process (Fig. 3) with PAPRIKA method based on key metrics and indicators used in automotive industry. The output of this study is the ranking of top 10 potential suppliers from a detailed database, and as result from our case study, the top 3 suppliers from electrical commodity had 1.5 for PAPRIKA Score. However, several challenges persist. Consequently, the next phase of research will investigate the development of an intelligent AI-based system that will take in consideration other aspects such as resilience and circular economic. Besides on that, we aim to evaluate the use of other MCDM methods for supplier selection problem such as Best Worst Method (BWM).

References

1. BCI: BCI Supply Chain Resilience Report 2024 (2024)
2. Abdulla, A., Baryannis, G.: A hybrid multi-criteria decision-making and machine learning approach for explainable supplier selection. Supply Chain Anal. **7** (2024). https://doi.org/10.1016/j.sca.2024.100074
3. Taherdoost, H., Brard, A.: Analyzing the process of supplier selection criteria and methods. Procedia Manufact. 1024–1034 (2019)

4. Memari, A., Dargi, A., Akbari Jokar, M.R., Ahmad, R., Abdul Rahim, A.R.: Sustainable supplier selection: a multi-criteria intuitionistic fuzzy TOPSIS method. J. Manuf. Syst. **50**, 9–24 (2019). https://doi.org/10.1016/j.jmsy.2018.11.002
5. Ismail, O., Kamach, O., Wiam, A.: Improvement of the supplier selection process in the automotive industry: case study with a Moroccan tier 1 company. In: 2024 IEEE 15th International Colloquium of Logistics and Supply Chain Management, LOGISTIQUA 2024. Institute of Electrical and Electronics Engineers Inc. (2024)
6. Galankashi, M.R., Helmi, S.A., Hashemzahi, P.: Supplier selection in automobile industry: a mixed balanced scorecard-fuzzy AHP approach. Alex. Eng. J. **55**, 93–100 (2016). https://doi.org/10.1016/j.aej.2016.01.005
7. Varchandi, S., Memari, A., Jokar, M.R.A.: An integrated best–worst method and fuzzy TOPSIS for resilient-sustainable supplier selection. Decis. Anal. J. **11** (2024). https://doi.org/10.1016/j.dajour.2024.100488
8. Hansen, P., Ombler, F.: A new method for scoring additive multi-attribute value models using pairwise rankings of alternatives. J. Multi-Criteria Decis. Anal. **15**, 87–107 (2008). https://doi.org/10.1002/mcda.428
9. Malik, D.A.A., Yusof, Y., Khalif, K.M.N.I.K.: A view of MCDM application in education. In: Journal of Physics: Conference Series. IOP Publishing Ltd (2021)
10. Mirzaee, S., Fannon, D., Ruth, M.: A comparison of preference elicitation methods for multi-criteria design decisions about resilient and sustainable buildings. Environ Syst Decis. **39**, 439–453 (2019). https://doi.org/10.1007/s10669-019-09726-2
11. Aal, S.I.A., Abdel-Monem, A.A.: A MCDM methodology to analysis strategies and factors of lean production in sustainability development. J. Sustain. Dev. Green Technol. **2**, 34–43 (2023). https://doi.org/10.54216/JSDGT.020204
12. Bai, C., Kusi-Sarpong, S., Badri Ahmadi, H., Sarkis, J.: Social sustainable supplier evaluation and selection: a group decision-support approach. Int. J. Prod. Res. **57**, 7046–7067 (2019). https://doi.org/10.1080/00207543.2019.1574042

Quantum Computing for Sustainable Supply Chains: Unlocking Efficiency, Environmental Progress and Challenges

Mohammed Zizi(✉), Anas Chafi, and Mohammed El Hammoumi

Department of Industrial Engineering, Laboratory of Industrial Techniques, Faculty of Sciences and Techniques, Sidi Mohamed Ben Abdellah University, Fez, Morocco
mohammed.ziz3@usmba.ac.ma

Abstract. Supply chain management has to undergo a paradigm shift in response to the increasing global demands for sustainability. Improvements toward environmental goals are hampered by the fact that the majority of conventional optimization models of such issues do not take into account the complexity and volume of supply chains today. Artificial intelligence combined with quantum computing offers a solution for some issues that cannot be solved by computation, and as a result, it offers exciting opportunities to transform supply chain operations for sustainability and efficiency. This study examines how quantum computing might enhance several facets of green supply chains, from inventory control and logistics optimization to energy use and material discovery. We outline the potential applications of quantum algorithms to tackle major sustainability issues, such as cutting emissions, cutting waste, and maximizing resource use. Quantum computing is essential to achieving the full potential of ethical, environmentally friendly, and effective supply chains due to current limitations and opportunities for further study in this field.

Keywords: logistics · quantum computing · optimization · environmental advancement · sustainable supply chains

1 Introduction

The imperative of sustainability has become a pervasive issue in the contemporary business and social discourse [1]. Supply chains are complex networks responsible for producing and distributing goods and services that play a critical role in fulfilling global sustainability imperatives. However, the very characteristics that define modern supply chains in their complexity and scale create formidable obstacles to the pursuit of ultimate efficiency and minimal environmental impact [2]. In such cases, the combinatorial optimization problems inherent in supply chain management are mostly beyond the ability of traditional computational methods, leading to suboptimal solutions with increased resource consumption and environmental burden. For example, routing optimization of a

C. El Mokhi et al. (Eds.): ICASET 2025, SUCI, pp. 289–297, 2025.
https://doi.org/10.1007/978-3-031-98334-4_27

large fleet of delivery vehicles with many constraints becomes computationally expensive for classical algorithms, which might result in longer routes and higher fuel consumption [3].

Quantum computing, an up-and-coming technology based on the principles of quantum mechanics, may offer a way out of this computational bottleneck. Specific varieties of quantum algorithms, for instance, quantum annealing and variational quantum eigensolvers, have been constructed for complex optimization problems that are unreachable by classical computers [4]. The future ability of quantum algorithms to investigate a large solution space may provide certain new opportunities for solving big classes of relevant problems at the point where the capabilities of classical algorithms are limited simply by local optima. This will also potentially revolutionize many areas of supply chain operation, unparalleled in efficiency, and significantly contribute to ecological sustainability in the process [4].

This paper discusses how quantum computing might help achieve sustainable related supply chains. We investigate selected areas in which quantum computing can bring revolutionary advantages, such as logistics optimization to reduce emissions, waste reduction by optimizing inventories, accelerating the discovery of sustainable materials, and energy consumption optimization in the supply chain network. In addition, we discuss current challenges and future research directions necessary to translate the theoretical potential of quantum computing into tangible benefits for sustainable supply chain management.

2 Methodology

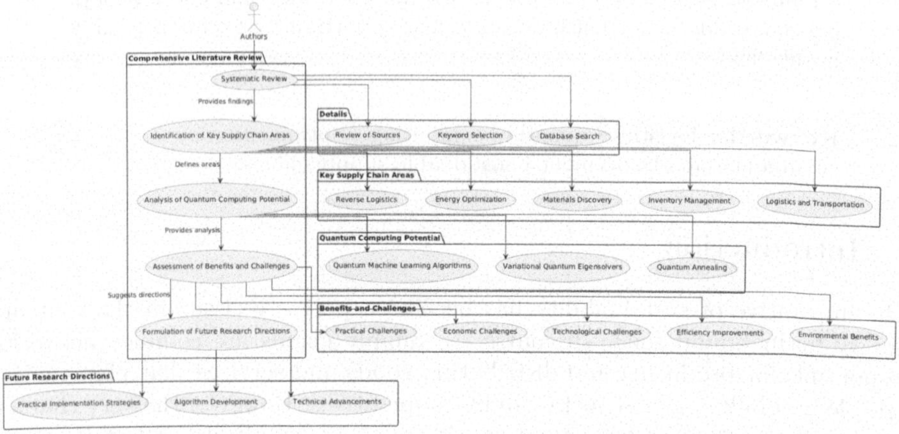

Fig. 1. Comprehensive Literature Review Process

This paper looks at the potential ways quantum computing could shape sustainable supply chain management, relying on adesk research approach. A systematic review of academic databases (e.g., IEEE Xplore, Scopus, Web of Science), industry reports, white papers, and relevant online resources was conducted. Keywords uti lized in the search included "quantum computing," "supply chain," "sus tainability," "green supply chain," "logistics optimization," "inventory man agement," "sustainable materials," "circular economy", for publications between 2000 and May 2024, filtering an initial 615 results down to 48 through the application of inclusion/exclusion criteria and subsequent full-text reading; these 48 relevant papers then serve as the foundation for a broader analysis (detailed in Fig. 1) focused on understanding the potential of quantum computing in key supply chain areas, assessing its benefits and challenges, and ultimately formulating directions for future research in this domain.

Based on the quality of the reviewed literature, this approach tries to provide a comprehensive and insightful overview of the emerging relationship between sustainable supply chains and quantum computing. The study relies on conventional theories that exist in the supply chain management and quantum computation, even as it recognizes that some forecasts are hypothetical.

3 Literature Review: Quantum Computing and Sustainable Supply Chains

The majority of research has been on specific domains or the more general application of quantum computing to supply chain optimization, making the junction of sustainable supply chains and quantum computing a relatively new topic of academic investigation.

3.1 Qubits and Superposition

The basic building block of quantum information is a qubit [5]. A qubit can simultaneously be in a variety of these basic states, whereas a conventional bit can only be either 0 or 1.

$$|\psi\rangle = \alpha|0\rangle + \beta|1\rangle, \tag{1}$$

where α and β are complex coefficients satisfying $|\alpha|^2 + |\beta|^2 = 1$.

3.2 Quantum Algorithms

- **Quantum Annealing:** Quantum annealing is a method grounded in quantum mechanics. It begins with a basic problem and slowly changes it into a more complex one [6]. Throughout this process, the system remains in its most stable state, and in the end, it arrives at the optimal solution to the problem [7].
- **Grover's Algorithm:** Grover's algorithm speeds up traditional search techniques by using quantum superposition and interference to search unsorted databases in $O(\sqrt{N})$ time [8]. It increases the likelihood of finding the correct answer by amplifying the probability through amplitude amplification.

- **Variational Quantum Algorithms (VQAs):** VQAs combine quantum computing with classical optimization [9]. By using a parameterized quantum circuit, the algorithm prepares a quantum state, and classical optimization adjusts the parameters to minimize a cost function. The parameters are refined through iterations to reach the optimal solution [1].
- **Quantum Matrix Factorization:** This technique uses quantum gates to decompose large matrices, offering efficient low-rank approximations [10]. It's helpful for optimizing network connections or uncovering hidden patterns in data.
- **Quantum Approximation Optimization Algorithm (QAOA):** QAOA combines classical optimization with quantum mechanics [11]. It generates a quantum state by applying parameterized gates that reflect the problem's cost and mixing functions. A classical optimizer then adjusts these parameters through multiple iterations to get as close as possible to the optimal solution [2].

3.3 Sustainable Supply Chain Management

Sustainable supply chain management to minimize environmental cost and ensure resource availability has been well covered in the literature [12]. This literature covers a number of topics, such as:

- Green Logistics approach to improve vehicle economy, shift in modes of transport and route integration to transport networks reducing carbon footprint [13]. While it is quite common to use optimization techniques like the linear programming, they are ill-equipped to handle big problems.
- Efficient Cost Control Methods that involve trim waste with sustainable inventory management techniques such as just-in-place inventories, suitable inventory quantities, and accurate demand forecasts [6]. Minimizing both overstocking which leads to wastage and under stocking which leads to a loss of sales requires a proper forecast [14].

3.4 Emerging Links and Research Gaps

While the individual fields are well-established, the specific literature directly linking quantum computing to sustainable supply chain challenges is still emerging. There is a noticeable gap in the literature regarding explicitly framing quantum computing applications within the context of specific sustainability metrics (e.g., carbon footprint reduction, waste minimization targets)[15].

4 Quantum Computing for Sustainable Supply Chain Optimization

Quantum computing's unique capabilities offer transformative potential across various aspects of the supply chain, directly contributing to sustainability goals [1] (Fig. 2).

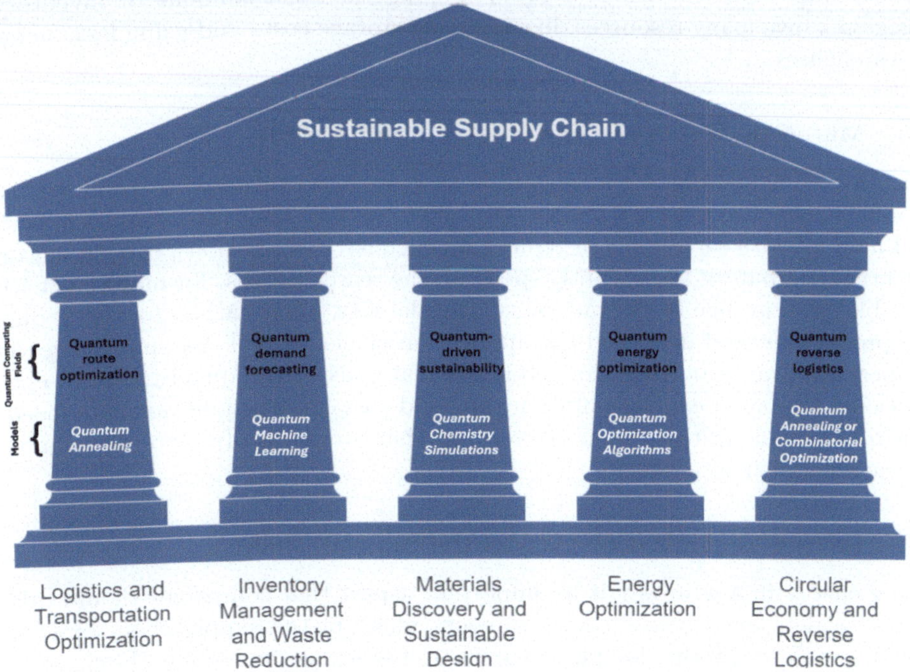

Fig. 2. Sustainable Supply Chain: Quantum Computing Applications

4.1 Logistics and Transportation Optimization

Transportation, presenting huge potentials of emissions, is frequently owned and controlled by suppliers. Minimizing this environmental footprint entails enhancing the orchestration of logistics networks in terms of timing, paths and vehicles. The Traveling Salesman Problem (TSP) and its variants are among the essential combinatorial optimization problems that form the basis of logistics planning, and quantum algorithms, particularly quantum annealing, produce outstanding results for them [16]. Efficient and fast performing route finding can make quantum computing cut gasoline usage and its resulting pollutants by half for delivery trucks [17].

4.2 Inventory Management and Waste Reduction

To improve demand forecasting many businesses and organizations try to analyze large datasets with special compound correlations using quantum calculations. Quantum machine learning is one of the tools that utilizes advanced algorithms to analyze demand patterns in an accurate way rather than traditional methods [5]. These calculations can recognize subtle changes in customer needs, which helps businesses plan their inventory in a better way. This means that those algorithms can be used by businesses to avoid surplus, which leads to waste and shortage. On the other er hand keeping the right amount of inventory, businesses save many resources, like reducing storage costs and using less energy in warehouses.

4.3 Materials Discovery and Sustainable Design

The environment impact the materials that are used in order to produce the products, this can give an insight on understanding how the molecules work and interact with each other [7]. Quantum chemistry simulations, carried out on quantum computers, significantly speed up the search process for materials with sensible qualities like biodegradability, recyclability, lower carbon footprint during production, and increased the application efficiency [18]. By simulating the properties of novel polymers or bio-based materials at the quantum level, conventional lab-based experimentation time and cost can be significantly reduced, which can result in the design of products that are naturally more sustainable throughout their life cycle [8].

4.4 Energy Optimization Across the Supply Chain

Using energy in a wise way is an important aspect that companies should take into consideration when it comes to every phase of the supply chain process: starting with the production phase moving to the storage and finally the shipping [14]. To optimize energy consumption quantum optimization methods can be successful [19].

They can be used for different aspects, for instance, to schedule production processes in most favorable way to reduce peak energy demands, which lessens dependency on fossil fuel-based energy during peak hours [20]. In addition, quantum algorithms create warehouse and distribute center layouts that consume less energy using systems like lighting and HVAC, they create patterns that allow proactive changes to minimize waste and maximize resource allocation [21].

4.5 Circular Economy and Reverse Logistics

Effective logistics systems are necessary to support the economy such as collecting, processing, and reuse of goods and materials, operating resources that are used for a long time [9]. Quantum optimization algorithms work well to optimize these complex reverse logistics networks, that involve stakeholders and logistical difficulties. It is used also to identify the effective routes to recycle the returned products to reduce waste and to be eco-friendly that leads to maximize recovered from end-of-life products [1].

5 Potential Benefits and Environmental Progress

There are some environmental advantages if supply chain management correctly integrates the quantum computing, the Table 1 bellow show Potential Benefits and Environmental Progress with Quantum Algorithms:

Table 1. Potential Benefits and Environmental Progress with Quantum Algorithms

Potential Benefit	Key Figure	Quantum Algorithm Used
Carbon Footprint Reduction	30% emission reduction	Quantum Annealing [22]
Waste Reduction	10% less inventory waste	Quantum Machine Learning [23]
Energy Savings	20–30% energy savings	Variational Quantum Eigensolver [20]
Decision-Making Improvement	15% lifecycle impact reduction	Quantum Simulation [24]

6 Challenges and Future Directions

While quantum computing has the potential to be a real game-changer for sustainable supply chains, there are still some challenges we need to overcome before it becomes widespread:

- **Technological Maturity:** quantum computing is still a new developing technology; therefore scientists and engineers work together to build advanced and powerful quantum computers that can handle larger and complex problems (more qubits) and reduce error [10].
- **Data Requirements and Integration:** Quantum algorithms usually need fixed data formats and a lot of processing power, The challenge lies in integrating quantum computing with existing supply chain data systems, which are often unstandardized [25].
- **Cost and Accessibility:** Quantum computing is only accessible only to huge companies and research organizations with a big budget because quantum calculations are very expensive, so it should be a way to make them more affordable and accessible for other companies to be used widely in supply chains [11].
- **Talent and Expertise:** using those advanced technologies is limited because of the lack of experts skilled in both supply chain management and quantum computing. To address this aspect interdisciplinary education and training are needed to address this gap [2].

In future researchers should focus on hybrid quantum-classical algorithms that mix the strengths of both systems for short-term use and long-term use. use cloud quantum computer like IBM or GOOGLE [15]. Developing error-reduction methods and comparing quantum and traditional algorithms for specific supply chain problems is essential [4]. Collaboration between supply chain, sustainability, and quantum computing experts is key to overcoming challenges and unlocking quantum computing's potential for sustainable supply chains [26].

7 Conclusion

Including quantum computing in the supply chain management give a transformative potential to address the negative aspects of operational efficiency and sustainability. By enabling the optimization of complex logistical networks, like reducing waste via intelligent inventory management, maximizing the discovery of sustainable materials, and minimizing energy consumption, quantum calculations show a significant advancement over traditional approaches the capabacity of this technology to solve those kind of challenges and optimize the problems provided in the critical edge specifically achieving more resilient and environmentally sustainable supply chains [27].

Despite the limitations in quantum hardware: algorithm development, and practical implementation; ongoing advancements in research and technology are bridging together to fulfill these gaps. Collaborating the efforts of researchers in the industry, and decision makers will be important to put pressure in order to fulfill the potential of quantum computing specifically in this realm [20].

As quantum technologies becomestronger, there are players to reimage global supply chains, that leads to the alignment of the economic growth with environmental stewardship. This will contribute to broader the efforts to build more sustainable and resilient global economy and enhance the sustainability of supply chain operations [26].

References

1. Jiang, H., Shen, Z.M., Liu, J.: Quantum computing methods for supply chain management. In: 2022 IEEE/ACM 7th Symposium on Edge Computing (SEC), pp. 400–405 (2022)
2. Sawane, R., Patil, A., Bindu, A., Lekurwale, N., Kapure, T., Panday, R.: The comparison of supply chain management using quantum computing and classical computing (2022)
3. Jahin, M.A., Shovon, M., Islam, M.S., Shin, J., Mridha, M.F., Okuyama, Y.: Qamplifynet: pushing the boundaries of supply chain backorder prediction using interpretable hybrid quantum-classical neural network. Sci. Rep. **13**, 18246 (2023)
4. Dixit, V., Niu, C.: Quantum computing for transport network design problems. Sci. Rep. **13**, 12267 (2023)
5. Harikrishnakumar, R., Nannapaneni, S., Nguyen, N.H., Steck, J.E., Behrman, E.C.: A quantum annealing approach for dynamic multi-depot capacitated vehicle routing problem. arXiv, abs/2005.12478 (2020)
6. Benedetti, C., Fiorentini, M., Rossi, M., Tamanini, L.: Quantum optimization for supply chain management. In: AIP Conference Proceedings, vol. 1891, p. 020020 (2017)
7. Chen, W.J., Marcus, G.S., Leesburg, D.: Quantum computing for manufacturing and supply chain optimization: enhancing efficiency, reducing costs, and improving product quality. Int. J. Enterp. Model. (2021)
8. Riandari, F., Alesha, A., Sihotang, H.T.: Quantum computing for production planning. Int. J. Enterp. Model. **15**(3), 163–175 (2021)
9. Weinberg, S.J., Sanches, F., Ide, T., Kamiya, K., Correll, R.: Supply chain logistics with quantum and classical annealing algorithms. Sci. Rep. **13**, 4770 (2022)
10. Gachnang, P., Ehrenthal, J., Hanne, T., Dornberger, R.: Quantum computing in supply chain management state of the art and research directions. Asian J. Logist. Manag. **1**(1), 57–73 (2022)
11. Rana, R., Thingbaijam, R., Seshadri, J., Shah, P., Sinha, A., Poojary, S.: Quantum powered employee transport and agri-logistics optimization, pp. 1–6 (2022)

12. McKinsey Sustainability. Sustainability in supply chains: a guide for CEOs (2020)
13. Chopra, S., Meindl, P.: Supply Chain Management: Strategy, Planning, and Operation. Pearson (2019)
14. Hwang, M.T., Collin, W.L., Lee, W.S.: Quantum computing and supply chain optimization: addressing complexity and efficiency challenges. Int. J. Enterp. Model. **15**(3), 148–162 (2021)
15. Cao, Y., et al.: Quantum chemistry in the age of quantum computing. Chem. Rev. **119**(18), 10856–10915 (2019)
16. Lucas, A.: Ising formulations of many np problems. Front. Phys. **2**, 5 (2014)
17. Preskill, J.: Quantum computing in the nisq era and beyond. Quantum **2**, 79 (2018)
18. Adhikari, B., Chang, B.Y.: Quantum computing impact on SCM and hotel performance (2021)
19. Abdelgaber, N., Nikolopoulos, C.: Calculating the topological resilience of supply chain networks using quantum hopfield neural networks. In: 2021 4th International Conference on Artificial Intelligence for Industries (AI4I), pp. 61–62 (2021)
20. Vignesh, M., Senthil Prabhu, S., Prabhu, K., Subramanian, R.M.: Hybrid quantum-enhanced machine learning for supply chain optimization. Int. J. Res. Appl. Sci. Eng. Technol. (2023)
21. Alberts, G., et al.: Accelerating quantum computer developments. EPJ Quantum Technol. **8**(1), 1–17 (2021). https://doi.org/10.1140/epjqt/s40507-021-00107-w
22. Dong, B.: Research on carbon footprint of tea supply chain based on cloud computing technology. Int. J. Glob. Econ. Manag. (2024)
23. Xu, M., et al.: Generative AI-enabled quantum computing networks and intelligent resource allocation. arXiv, abs/2401.07120 (2024)
24. Desclee, D., Sohinto, D., Padonou, F.: Sustainability assessment and agricultural supply chains: evidence-based multidimensional analyses as tools for strategic decision-making–the case of the pineapple supply chain in Benin. Sustainability **13**(4), 2060 (2021)
25. Phillipson, F.: Quantum computing in logistics and supply chain management: an overview (2024)
26. Sales, J.F.A., Araos, R.A.P.: Adiabatic quantum computing impact on transport optimization in the last-mile scenario. Front. Comput. Sci. **5** (2023)
27. Cultice, T., Clark, J., Yang, W., Thapliyal, H.: A novel hierarchical security solution for controller-area-network-based 3D printing in a post-quantum world. Sensors **23**, 9886 (2023)

Stochastic Dynamics of a SIRI Epidemic Model with Disease Relapse

Meryem Benazzouz and Mohamed El Fatini(✉)

Stochastic Modelling and Statistics Group, Lab. PDEASG, Faculty of Science, Ibn Tofail University, BP 133, Kénitra, Morocco
melfatini@gmail.com

Abstract. This study aims to analyze a stochastic epidemic system incorporating a relapse mechanism, which is correlated to a Brownian motion (Bm), capturing real-world disease dynamics with more accuracy. We establish the existence, uniqueness, and boundedness of the model's positive solution, ensuring mathematical consistency. Using stochastic analysis, we derive the stochastic reproduction numbers, $\mathfrak{R}_0^s$ and $\bar{\mathfrak{R}}_0^s$, which serves a key function in determining the dynamic behavior of the illness. Specifically, we demonstrate that the illness becomes extinct when $\mathfrak{R}_0^s \leq 1$, highlighting conditions under which infection can be eradicated. Conversely, when $\bar{\mathfrak{R}}_0^s > 1$, the disease persists on average, indicating sustained transmission within the population. To further explore these theoretical results, we conduct numerical simulations that illustrate the model's behavior under various parameter settings. These simulations provide a visual and quantitative validation of our analytical findings, offering insight into how stochastic effects influence disease progression. Our study emphasizes the significance of incorporating stochasticity in epidemic modeling, as it better reflects real-life uncertainties in disease spread. The results contribute to a deeper understanding of relapse-driven epidemics and provide a framework for developing effective control strategies in public health.

Keywords: Stochastic epidemic model · Relapse · Extinction · Persistence in mean

1 Introduction

1.1 Research Background

Communicable illness remain one of the primary factors behind death worldwide, emphasizing the urgent need for scientific advancements in the understanding, monitoring, and prevention of these diseases. Mathematical epidemiology applies mathematical and computational frameworks to explore the spread and management of infectious diseases. By employing methods such as differential equations and statistical modeling, researchers investigate transmission patterns, evaluate intervention strategies, and forecast epidemic trajectories [1]. This approach is

C. El Mokhi et al. (Eds.): ICASET 2025, SUCI, pp. 298–313, 2025.
https://doi.org/10.1007/978-3-031-98334-4_28

vital in shaping public health policies and enhancing outbreak control efforts. One of the fundamental models in epidemiology is the SIR framework, initially proposed by Kermack and McKendrick [2], which classifies the population into three core categories: "Susceptible (S)", "Infected (I)", and "Recovered (R)". While this model provides a basic understanding of epidemics, real-world diseases often display more intricate dynamics. Extensions such as the SIRS, SIRI, and SEIR models include additional factors like temporary immunity, reinfection, and latency periods [3–6], making them more applicable to certain diseases. Our research focuses on a stochastic SIRI model, which accounts for the recurring nature of some infections. In this context, the following dynamics are observed:

- Susceptible individuals can acquire the infection by coming into contact with those who areÂăcontagious.
- Infected individuals recover, but only gain temporary immunity.
- After a period, recovered individuals become susceptible again, leading to potential reinfection.

This recurrence phenomenon is observed in diseases like "tuberculosis (both human and bovine)" and "herpes" [7,8]. A thorough understanding of these dynamics is fundamental to predicting disease persistence and devising effective control measures.

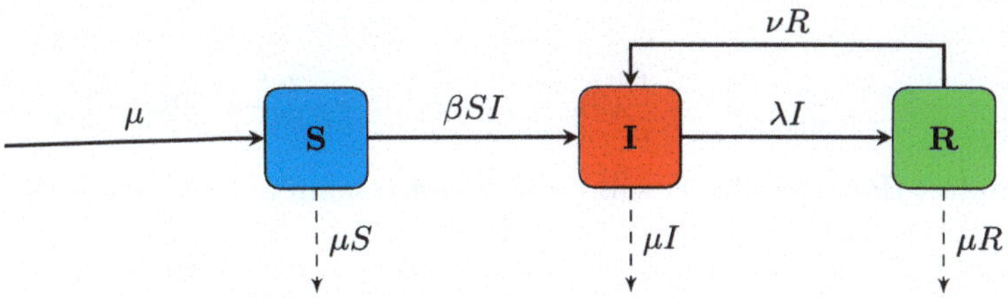

Fig. 1. Schematic representation of the SIRI model incorporating relapse.

1.2 Dynamical Features of the Deterministic System (1)

Below, we introduce a deterministic SIRI model, building on the foundation laid in our first paper [9], that examines the relapse mechanism. The diagram above Fig. 1 illustrates the evolution of the demographic, segmented into three core categories over time: "Susceptible S_t", "Infected I_t", and "Recovered R_t", with $t \geq 0$ [2]. This deterministic model is formulated as follows:

$$\begin{cases} dS_t = [\mu - \beta S_t I_t - \mu S_t]\, dt, \\ dI_t = [\beta S_t I_t - (\mu + \lambda) I_t + \nu R_t]\, dt, \\ dR_t = [\lambda I_t - (\mu + \nu) R_t]\, dt, \end{cases} \quad (1)$$

where, the system parameters are all positive constants, with the subsequent interpretations: "μ is the recruitment rate, assumed to be equal to the mortality rate", "β represents the disease transmission rate", "λ is the recovery rate" and "ν denotes the relapse rate", which accounts for loss of immunity and possible reinfection of recovered individuals [21].

Based on the work in [10, 11], the basic reproduction number associated with model (1) is expressed as:

$$\mathfrak{R}_0 = \frac{\beta}{\mu + \lambda - \frac{\lambda\nu}{\mu+\nu}}.$$

This quantity represents the average number of secondary infections generated by a single infected individual in a population where a portion of individuals have already gained immunity, either through previous infection or vaccination, and are no longer susceptible to the disease and the equilibrium of model (1) in the deterministic setting is determined by $\mathfrak{R}_0$.

- When $\mathfrak{R}_0 \leq 1$, "the deterministic model (1) has a disease-free equilibrium at $E_0 = (1, 0, 0)$, which is globally asymptotically stable (GAS). This implies that the infection will ultimately disappear from the population" [12].
- Conversely, when $\mathfrak{R}_0 > 1$, "the equilibrium E_0 becomes unstable, leading to the existence of a GAS endemic equilibrium, denoted by $E^* = (S^*, I^*, R^*)$", where

$$S^* = \frac{1}{\mathfrak{R}_0}, \quad I^* = \frac{\mu}{\beta}(\mathfrak{R}_0 - 1), \quad R^* = \frac{\mu\lambda}{\beta(\mu + \nu)}(\mathfrak{R}_0 - 1),$$

  this implies that the illness will persist and remain active within the population.

1.3 Stochastic Epidemic Model for SIRI Dynamics

Next, we incorporate Brownian motion into the deterministic model (1) to account for random fluctuations and uncertainty affecting disease transmission. This stochastic component is essential, as it captures unpredictable variations that influence the spread of infections. By introducing Brownian motion, we enhance the model's realism, reflecting the inherent variability in disease dynamics. In particular, we introduce white noise to perturb the transmission coefficient as follows:

$$\beta \rightsquigarrow \beta + \sigma \frac{dB_t}{dt}.$$

Let B_t denote a standard Brownian motion defined on a complete, filtered probability space $(\Omega, \mathfrak{F}, (\mathfrak{F}_t)_{t\geq 0}, \mathbb{P})$, where $(\mathfrak{F}_t)_{t\geq 0}$ is a filtration that satisfies the usual conditions, and σ represents the intensity of the Brownian motion. With this perturbation, we obtain the following stochastic system:

$$\begin{cases} dS_t = [\mu - \beta S_t I_t - \mu S_t]\, dt - \sigma S_t I_t\, dB_t, \\ dI_t = [\beta S_t I_t - (\mu + \lambda) I_t + \nu R_t]\, dt + \sigma S_t I_t\, dB_t, \\ dR_t = [\lambda I_t - (\mu + \nu) R_t]\, dt. \end{cases} \quad (2)$$

The structure of this work is as detailed below: the subsequent section establishes the global existence and uniqueness of a strictly positive solution to the proposed stochastic SIRI system. Section 3 addresses the extinction of the illness by deriving sufficient conditions for the infection to eventually disappear from the population. Section 4 investigates the persistence of the disease in the mean, providing mathematical criteria that guarantee the long-term survival of the infection under certain conditions. Finally, Sect. 5 presents computational simulations that illustrate and validate the theoretical findings, demonstrating the model's behavior under different parameter settings.

2 Existence and Uniqueness of a Global Positive solution

In this section, we employ the Lyapunov analysis technique [14] to establish the existence of a local positive solution that is unique for model (2). We then prove that this solution can be extended globally.

Theorem 1. *Given any starting condition $(S_0, I_0, R_0) \in \mathbb{R}^3_+$, there is a unique solution $(S_t, I_t, R_t) \in \mathbb{R}^3_+$ to system (2) for all $t \geq 0$ a.s, and this solution stays within $\mathbb{R}^3_+$ with probability* 1.

Proof. Given that the drift and diffusion are locally Lipschitz, for any starting condition $(S_0, I_0, R_0) \in \mathbb{R}^3_+$, a unique local solution (S_t, I_t, R_t) exists for $t \in [0, \tau_e)$, where τ_e denotes the explosion time [15]. First, we demonstrate that (S_t, I_t, R_t) do not blow up to ∞ within a finite time. Let $\xi_0 > 0$ be sufficiently large so that the initial values S_0, I_0, R_0 are contained within $\left[\frac{1}{\xi_0}, \xi_0\right]$. For each integer $\xi \geq \xi_0$, we introduce the following stopping time defined as:

$$\tau_\xi = \inf\left\{t \in [0, \tau_e) : S_t \notin \left(\frac{1}{\xi}, \xi\right) or I_t \notin \left(\frac{1}{\xi}, \xi\right) or\ R_t \notin \left(\frac{1}{\xi}, \xi\right)\right\}.$$

The sequence τ_ξ is increasing as $\xi \to \infty$. Defining $\tau_\infty = \lim_{\xi \to \infty} \tau_\xi$, we obtain $\tau_\infty \leq \tau_e$ a.s. Our goal is to establish that $\tau_\infty = \infty$. Assume, for contradiction, that $\tau_\infty < \infty$. Then, there exist two constants $T > 0$ and $\varepsilon \in (0, 1)$ such that $\mathbb{P}(\tau_\infty \leq T) \geq \varepsilon$. Consequently, there exists an integer $\xi_1 \geq \xi_0$ such that:

$$\mathbb{P}(\tau_\xi \leq T) \geq \varepsilon, \quad \forall \xi \geq \xi_1. \quad (3)$$

Let $\mathscr{F} : \mathbb{R}^3_+ \to \mathbb{R}_+$ be a $\mathscr{C}^2$-function defined as follows:

$$\mathscr{F}(S, I, R) = \left(S - \alpha - \alpha \ln \frac{S}{\alpha}\right) + (I - 1 - \ln I) + (R - 1 - \ln R), \tag{4}$$

where, $\alpha > 0$ is determined below [9]. Denote $X_t = (S_t, I_t, R_t)$. By using Itô's formula [16] to the function $\mathscr{F}(S, I, R)$, we derive the following

$$d\mathscr{F}(X_t) = \mathscr{L}\mathscr{F}(X_t)\, dt + \sigma \left[I_t(S_t - \alpha) + S_t(I_t - 1)\right] dB_t, \tag{5}$$

where,

$$\begin{aligned}\mathscr{L}\mathscr{F}(X_t) &= (3\mu + \alpha\mu + \lambda + \nu) - (\mu + \beta)S_t + (\alpha\beta - \mu)I_t \\ &\quad - \frac{\alpha\mu}{S_t} - \nu\frac{R_t}{I_t} - \lambda\frac{I_t}{R_t} + \frac{\alpha\sigma^2}{2}I_t^2 + \frac{\sigma^2}{2}S_t^2.\end{aligned} \tag{6}$$

We now define the set:

$$\Gamma = \{(S, I, R) \in \mathbb{R}^3_+ : N_t = S_t + I_t + R_t = 1\} \tag{7}$$

By the fact that

$$\frac{dN_t}{dt} = \mu - \mu N_t,$$

It is straightforward to show that if $(S_0, I_0, R_0) \in \Gamma$, then $(S_t, I_t, R_t) \in \Gamma$ a.s. for all $t \geq 0$. This implies that the subset Γ of $\mathbb{R}^3_+$ is positively invariant set of model (2).

By (7) we have $S^2 \leq S \leq 1$ and choosing $\alpha = \dfrac{\mu}{\beta}$, we get

$$\mathscr{L}\mathscr{F}(X_t) \leq 3\mu + \alpha\mu + \lambda + \nu + \frac{\alpha\sigma^2}{2} + \frac{\sigma^2}{2} := K. \tag{8}$$

By integrating both sides of Eq. (6) over the interval $[0, \tau_\xi \wedge T]$ and then taking the mathematical expectation, we get

$$\mathbb{E}[\mathscr{F}(X_{\tau_\xi \wedge T})] \leq KT + \mathscr{F}(X_0) < \infty. \tag{9}$$

Now, let $\Omega_\xi = \{\tau_\xi \leq T\}$. Given that $\mathbb{P}(\tau_\infty \leq T) \geq \varepsilon$, we deduce that $\mathbb{P}(\Omega_\xi) \geq \varepsilon$ which holds for all $\xi \geq \xi_1$. In contrast, from Eq. (4), we know that $\mathscr{F}(X_{\tau_\xi \wedge T}) \geq 0$, which implies

$$\begin{aligned}\mathbb{E}[\mathscr{F}(X_{\tau_\xi \wedge T})] &= \mathbb{E}[\mathbb{1}_{\Omega_\xi}\mathscr{F}(X_{\tau_\xi \wedge T})] + \mathbb{E}[\mathbb{1}_{\Omega_\xi^c}\mathscr{F}(X_{\tau_\xi \wedge T})] \\ &\geq \mathbb{E}[\mathbb{1}_{\Omega_\xi}\mathscr{F}(X_{\tau_\xi \wedge T})],\end{aligned}$$

where $\mathbb{1}_{\Omega_\xi}$ denotes the indicator function of Ω_ξ. Consequently, we obtain the following inequality:

$$\begin{aligned}\mathscr{F}(X_0) + KT &\geq \mathbb{E}\left[\mathbb{1}_{\Omega_\xi}\mathscr{F}(X_{\tau_\xi \wedge T})\right] \\ &\geq \varepsilon\left[\left(\xi - \alpha - \alpha\log\frac{\xi}{\alpha}\right) \wedge \left(\frac{1}{\xi} - \alpha - \alpha\log\frac{1}{\xi\alpha}\right)\right. \\ &\quad \left.\wedge\,(\xi - 1 - \log\xi) \wedge \left(\frac{1}{\xi} - 1 - \log\frac{1}{\xi}\right)\right].\end{aligned}$$

By letting $\xi \to \infty$, we obtain

$$\infty > \mathscr{F}(X_0) + KT = \infty,$$

This leads to a contradiction. Therefore, we must have $\tau_e = \infty$ a.s. As a result, (S_t, I_t, R_t) is a global positive solution to model (2). This concludes the proof.

3 Extinction of the Disease

In this section, we analyze the conditions that lead to the extinction of the illness in the stochastic system (2) by introducing the following key quantity:

$$\mathfrak{R}_0^s = \frac{\beta}{\mu + \frac{\sigma^2}{4}}.$$

Accordingly, we arrive at the following result.

Theorem 2. *Assume the following conditions hold:*

$$\mathfrak{R}_0^s < 1, \tag{10}$$

$$\frac{\sigma^2}{2} \leq \beta, \tag{11}$$

Then, with probability one, the disease will disappear from the population. In other words

$$\lim_{t\to\infty}\langle S\rangle_t = \lim_{t\to\infty}\frac{1}{t}\int_0^t S_r\,dr = 1\,and\,\lim_{t\to\infty} I_t = \lim_{t\to\infty} R_t = 0, \quad \mathbb{P} - a.s.$$

Proof. We will follow the same lines as in [9], by using Itô's formula [16] on the function $I_t \mapsto \log(I_t + R_t)$, we obtain the expression

$$d\log(I_t + R_t) = \left[-\mu + \beta\frac{S_t I_t}{I_t + R_t} - \frac{\sigma^2}{2}\frac{S_t^2 I_t^2}{(I_t + R_t)^2}\right] dt + \sigma\frac{S_t I_t}{I_t + R_t} dB_t,$$

we used the bound, $-1/(a+b)^2 \leq -1/2a^2b^2$ and $(S_t, I_t, R_t) \in \Gamma$, we get

$$d\log(I_t + R_t) \leq \left[-\mu + \beta S_t - \frac{\sigma^2}{4} S_t^2\right] dt + \sigma S_t dB_t. \tag{12}$$

By integrating both sides of Eq. (12) from 0 to t and then dividing by t, we arrive at the following result

$$\begin{aligned}\frac{\log(I_t+R_t)}{t} &\leq \frac{\log(I_0+R_0)}{t} + \beta\frac{1}{t}\int_0^t S_r dr - \mu - \frac{\sigma^2}{4}\frac{1}{t}\int_0^t S_r^2\, dr + \sigma\frac{1}{t}\int_0^t S_r\, dB_r \\ &\leq \beta\langle S\rangle_t - \mu - \frac{\sigma^2}{4}\langle S^2\rangle_t + \frac{A_t}{t},\end{aligned} \tag{13}$$

where, $A_t = \log(I_0 + R_0) + \sigma \displaystyle\int_0^t S_r\, dB_r$ and $\langle S\rangle_t = \displaystyle\frac{1}{t}\int_0^t S_r\, dr$.

Integrating both sides of system (2) over the interval $[0, t]$, dividing by t and combining the resulting equations, we obtain that

$$\frac{S_t + S_0}{t} + \frac{I_t + I_0}{t} + \frac{R_t + R_0}{t} = \mu - \mu(\langle S\rangle_t + \langle I\rangle_t + \langle R\rangle_t).$$

This leads to the following equation

$$\langle S\rangle_t = 1 - (\langle I\rangle_t + \langle R\rangle_t) + \psi_t, \tag{14}$$

where, $\psi_t = \frac{S(0)+I(0)+R(0)}{\mu t} - \frac{S(t)+I(t)+R(t)}{\mu t}$. By applying the Cauchy-Schwarz inequality [19] and performing a direct computation, we obtain the following inequality:

$$\langle S^2\rangle_t \geq \langle S\rangle_t^2 \geq 1 - 2(1+\psi_t)\langle I + R\rangle_t. \tag{15}$$

Injecting (14) and (15) into (13), we have

$$\frac{\log(I_t + R_t)}{t} \leq \left(\mu + \frac{\sigma^2}{4}\right)(\mathfrak{R}_0^s - 1) - \left(\beta - \frac{\sigma^2}{2}\right)\langle I + R\rangle_t + \varphi_t + \frac{A_t}{t}, \tag{16}$$

where, $\varphi_t = \beta\psi_t + \frac{\sigma^2}{2}\psi_t\langle I + R\rangle_t$. By applying the strong law of large numbers [20] and the boundedness of the solution, we obtain the following results:

$$\lim_{t\to\infty} \frac{A_t}{t} = 0, \ \lim_{t\to\infty} \varphi_t = 0, \ \ \mathbb{P} - a.s. \tag{17}$$

Since, $\mathfrak{R}_0^s < 1$ and $\beta - \dfrac{\sigma^2}{2} \geq 0$, we have from (16)

$$\limsup_{t\to\infty} \frac{\log(I_t + R_t)}{t} \leq \left(\mu + \frac{\sigma^2}{4}\right)(\mathfrak{R}_0^s - 1) < 0.$$

This implies,

$$\lim_{t\to\infty} (I_t + R_t) = 0, \ \ \mathbb{P} - a.s.$$

Since the solution (S_t, I_t, R_t) is positive, this implies that

$$\lim_{t\to\infty} I_t = \lim_{t\to\infty} R_t = 0, \quad \mathbb{P} - a.s. \tag{18}$$

Moreover, injecting (18) into (14), we obtain

$$\lim_{t\to\infty} \langle S\rangle_t = 1, \qquad \mathbb{P} - a.s.$$

Thus, the proof is concluded.

4 Persistence in Mean of the Disease

In this section, we present the following result concerning the persistence in mean of the illness. For simplicity, we define the stochastic threshold for persistence as:

$$\bar{\mathfrak{R}}_0^s = \frac{\beta}{\mu + \lambda + \frac{\sigma^2}{2}}.$$

Theorem 3. *[9] If the assumption $\bar{\mathfrak{R}}_0^s > 1$ holds, the solution (S_t, I_t, R_t) of the system (2) remains constrained by the following inequalities for all $t \in \mathbb{R}_+$ and any starting condition $(S_0, I_0, R_0) \in \Gamma$.*

$$\liminf_{t\to\infty} \frac{1}{t}\int_0^t S_r\,dr \geq C_1, \quad \mathbb{P} - a.s,$$

$$\liminf_{t\to\infty} \frac{1}{t}\int_0^t I_r\,dr \geq C_2(\bar{\mathfrak{R}}_0^s - 1), \quad \mathbb{P} - a.s,$$

$$\liminf_{t\to\infty} \frac{1}{t}\int_0^t R_r\,dr \geq C_3(\bar{\mathfrak{R}}_0^s - 1), \quad \mathbb{P} - a.s,$$

for certain $C_k > 0$, where $k = 1, 2, 3$.

Proof. By taking the integral of the initial equation in model (2) over the interval $[0, t]$ and applying the bound $I_t \leq 1$, we obtain

$$\begin{aligned} S_t - S_0 &= \mu t - \beta\int_0^t S_r I_r dr - \mu\int_0^t S_r dr - \sigma\int_0^t S_r I_r dB_r \\ &\geq \mu t - (\beta + \mu)\int_0^t S_r dr - \sigma\int_0^t S_r I_r dB_r. \end{aligned} \tag{19}$$

Dividing the inequality (19) by t, we get

$$\frac{S_t - S_0}{t} \geq \mu - (\beta + \mu)\langle S\rangle_t - \sigma\frac{1}{t}\int_0^t S_r I_r dB_r.$$

This implies,

$$\langle S\rangle_t \geq \frac{1}{\beta+\mu}\left[\mu - \frac{S_t - S_0}{t} - \sigma\frac{1}{t}\int_0^t S_r I_r dB_r\right].$$

By taking the inferior limit and utilizing the boundedness of the solution along with the strong law of large numbers [20], we arrive at the following:

$$\liminf_{t\to\infty}\langle S\rangle_t = \liminf_{t\to\infty}\frac{1}{t}\int_0^t S_r\, dr \geq \frac{\mu}{\beta+\mu} := C_1 > 0,\ \ \mathbb{P}-a.s.$$

We will follow the same lines as in [9], by applying Itô's formula [16] to the function $I_t \mapsto \log I_t$, we derive the following:

$$\log I_t = \log I_0 + \beta\int_0^t S_r\, dr - (\mu+\lambda)t + \nu\int_0^t \frac{R_r}{I_r}\, dr - \frac{\sigma^2}{2}\int_0^t S_r^2 dr + \sigma\int_0^t S_r\, dB_r. \tag{20}$$

Dividing both sides of Eq. (20) by t, we obtain

$$\frac{\log I_t}{t} \geq \frac{\log I_0}{t} + \beta\langle S\rangle_t - (\mu+\lambda) - \frac{\sigma^2}{2}\langle S^2\rangle_t + \sigma\frac{1}{t}\int_0^t S_r\, dB_r. \tag{21}$$

Then,

$$\frac{\log I_t}{t} \geq \beta\langle S\rangle_t - (\mu+\gamma) - \frac{\sigma^2}{2} + \frac{H_t}{t}, \tag{22}$$

where, $H_t = \log I_0 + \sigma\int_0^t S_r dB_r$ and we used $-S_t > -1$.

By combining the first and second equations of model (2), integrating over the interval $[0,t]$ and dividing by t, we obtain the following:

$$\begin{aligned}\frac{S_t - S_0}{t} + \frac{I_t - I_0}{t} &= \mu - \mu\langle S\rangle_t - (\mu+\lambda)\langle I\rangle_t + \nu\langle R\rangle_t \\ &\geq \mu - \mu\langle S\rangle_t - (\mu+\lambda)\langle I\rangle_t.\end{aligned}$$

Hence,

$$\langle S\rangle_t \geq 1 - \frac{\mu+\lambda}{\mu}\langle I\rangle_t + \chi_t, \tag{23}$$

where, $\chi_t = \frac{S_0+I_0}{\mu t} - \frac{S_t+I_t}{\mu t}$. Injecting (23) into (22), we obtain

$$\frac{\log I_t}{t} \geq \beta - \frac{\beta(\mu+\lambda)}{\mu}\langle I\rangle_t + \beta\chi_t - \left(\mu+\gamma+\frac{\sigma^2}{2}\right) + \frac{H_t}{t}.$$

Then,

$$\frac{\beta(\mu+\lambda)}{\mu}\langle I\rangle_t \geq \beta - \left(\mu+\gamma+\frac{\sigma^2}{2}\right) - \frac{\log I_t}{t} + \beta\chi_t + \frac{H_t}{t}.$$

Using the strong law of large numbers [20] and the bound $I_t \geq -1$, we derive the following:

$$\liminf_{t\to\infty}\langle I\rangle_t = \liminf_{t\to\infty}\frac{1}{t}\int_0^t I_r\,dr \geq\ C_2\left(\bar{\mathfrak{R}}_0^s - 1\right),\ \ \mathbb{P}-a.s.$$

where, $C_2 := \frac{\mu}{\beta}\left(1+\frac{\sigma^2/2}{\mu+\lambda}\right) > 0.$

From the third equation of model (2), we have

$$\frac{R_t - R_0}{t} = \lambda\,\langle I\rangle_t - (\mu+\nu)\,\langle R\rangle_t,$$

After reordering the terms, we derive

$$(\mu+\nu)\,\langle R\rangle_t = \lambda\,\langle I\rangle_t - \frac{R_t}{t} + \frac{R_0}{t}.$$

From the boundedness of the solution, we conclude that:

$$\liminf_{t\to\infty}\langle R\rangle_t = \liminf_{t\to\infty}\frac{1}{t}\int_0^t R_r\,dr \geq\ C_3\left(\bar{\mathfrak{R}}_0^s - 1\right),\ \ \mathbb{P}-a.s.$$

where, $C_3 := \dfrac{\lambda}{\mu+\nu}\,C_2 > 0$. The proof is completed.

5 Numerical Simulations

In this section, we provide computational simulations for the stochastic system (2), which are designed to illustrate and support the theoretical findings presented in Sects. 3 and 4. The simulations are carried out using the Euler-Maruyama method as outlined in [17], a widely used numerical technique for solving SDEs. This method discretizes the continuous time dynamics of the system (2), allowing us to approximate the solution over a set time interval. The results obtained from these simulations are compared with the analytical conditions for extinction and persistence to validate the theoretical conclusions.

5.1 Scenario of Disease Extinction

Assign the parameter values in system (2) as outlined below:

Choosing a small value for σ strikes a delicate balance between maintaining realism, ensuring mathematical tractability, and optimizing computational feasibility. In the context of epidemic modeling, σ represents the intensity of the random fluctuations (or noise) introduced into the system. By keeping σ small, researchers can simulate the effect of minor random perturbations on the system without allowing these fluctuations to dominate the model's dynamics. This ensures that the system retains the fundamental characteristics of the deterministic epidemic model while accounting for the inherent randomness present in

Table 1. The parameter settings applied in the numerical simulations.

Parameter	Description	Value
μ	recruitment rate = natural death rate	0.177
λ	recovery rate	0.2
β	transmission rate	0.12
σ	diffusion coefficient	0.2
ν	rate of relapse	0.1

real-world situations. A small σ also allows for a more manageable mathematical framework, as large noise terms can complicate the analysis, making it more difficult to derive meaningful results. By keeping the noise intensity low, the model remains analytically tractable, enabling researchers to derive insights into the system's stability, disease persistence, or extinction over time.

Given the starting value $(S_0, I_0, R_0) = (0.8, 0.4, 0.12)$, we observe that the deterministic system (1) leads to the extinction of the illness, since the threshold $\mathfrak{R}_0 = 0.6562$ is less than 1. As for the stochastic model (2), the stochastic extinction threshold $\mathfrak{R}_0^s = 0.553$ is also less than 1. Moreover, the condition (11) of Theorem 2 is satisfied, as we have $\frac{\sigma^2}{2} = 0.08 < \beta = 0.12$. This indicates that the effect of stochastic disturbances does not allow the disease to persist. Consequently, the infection eventually disappears from the population, which is confirmed by Fig. 2, where a progressive decline in the number of infected individuals is observed until the complete extinction of the illness.

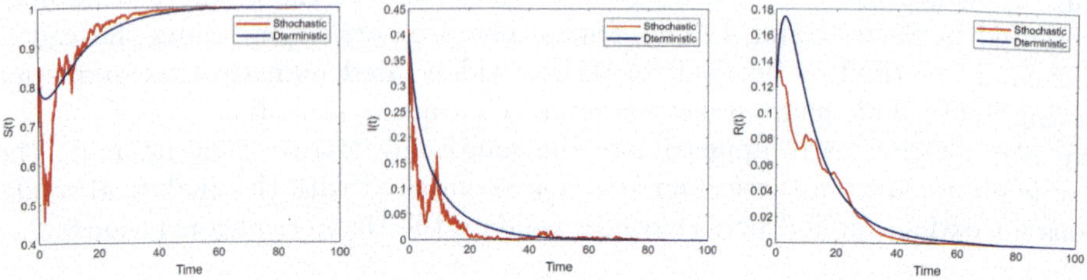

Fig. 2. Paths of the extinction scenario in the solutions of systems (1) and (2), based on the data presented in Table 1.

5.2 Scenario of Disease Persistence

We now increase the value of the disease transmission coefficient to $\beta = 0.45$ while keeping all other parameter values unchanged. The parameters used in the numerical simulations are summarized in Table 2.

Table 2. The parameter settings applied in the numerical simulations.

Parameter	Description	Value
μ	recruitment rate = natural death rate	0.177
λ	recovery rate	0.2
β	transmission rate	0.45
σ	diffusion coefficient	0.2
ν	rate of relapse	0.1

In the deterministic model, the system exhibits persistence in mean since the basic reproduction number is $\mathfrak{R}_0 = 1.4764 > 1$. This implies the existence of a stable endemic equilibrium given by $E^* = (S^*, I^*, R^*) = (0.68, 0.19, 0.01)$, where the disease remains present in the population over time. For the stochastic model (2), we obtain the stochastic persistence threshold $\bar{\mathfrak{R}}_0^s = 1.0284 > 1$, which also indicates disease persistence. According to Theorem 3, the long-term behavior of the stochastic model (2) satisfies the following conditions:

$$\liminf_{t\to\infty} \langle S\rangle_t \geq 0, \ \liminf_{t\to\infty} \langle I\rangle_t \geq 0, \ \liminf_{t\to\infty} \langle R\rangle_t \geq 0.$$

As a result, system (2) is persistent, meaning that the infection does not vanish but instead stabilizes at an endemic level. This conclusion is further supported by Fig. 3, which visually demonstrates the sustained presence of the disease over time.

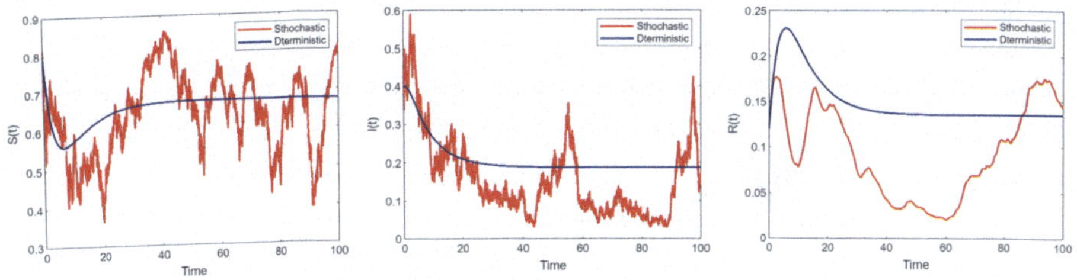

Fig. 3. Paths of the persistence in mean scenario in the solutions of systems (1) and (2), based on the data presented in Table 2.

5.3 Sensitivity Analysis of the Threshold $\mathfrak{R}_0$

o develop a deeper understanding of the most efficient strategies for illness prevention and control, we perform a sensitivity analysis [9,18] to examine how different parameters influence the dynamics of the stochastic epidemic model (2). This analysis aims to identify the key factors driving disease transmission and to inform the development of targeted intervention strategies. In what follows, we adopt the parameter values presented in Table 2.

Analysis of the Impact of the Relapse Parameter ν on $\mathfrak{R}_0$. To assess how changes in the relapse rate affect the overall course of the illness, we conduct a detailed sensitivity analysis of the threshold $\mathfrak{R}_0$ with respect to variations in the relapse parameter ν. This examination allows us to quantify the extent to which fluctuations in ν alter the spread and persistence of the infection within the population. Understanding this relationship provides valuable insights into the role of relapse in sustaining disease transmission. Specifically, we compute the unnormalized sensitivity index of $\mathfrak{R}_0$, which characterizes the rate at which $\mathfrak{R}_0$ changes as a function of ν. Since $\mathfrak{R}_0$ is continuously differentiable with respect to ν, its sensitivity coefficient is given by the following expression [9]:

$$\frac{\partial \mathfrak{R}_0}{\partial \nu} \cdot \frac{\nu}{\mathfrak{R}_0} = \frac{\lambda \nu}{(\mu + \nu)(\mu + \lambda + \nu)}.$$

The sensitivity index is positive, indicating that an increase in the relapse rate ν leads to a rise in $\mathfrak{R}_0$. This reflects a dynamic where a higher relapse rate promotes a greater spread of the disease within the population, ultimately contributing to the persistence of the illness over time. Figure 4 visually illustrates the impact of variations in ν. The curves clearly show that infection rates increase as ν increases from 0.15 to 0.5, confirming the theoretical findings of the sensitivity index. This demonstrates that higher relapse rates lead to a higher transmission of the disease, which aligns with the earlier theoretical results, showing the importance of relapse in disease persistence. The figure highlights how small changes in the relapse rate can significantly influence the dynamics of the epidemic.

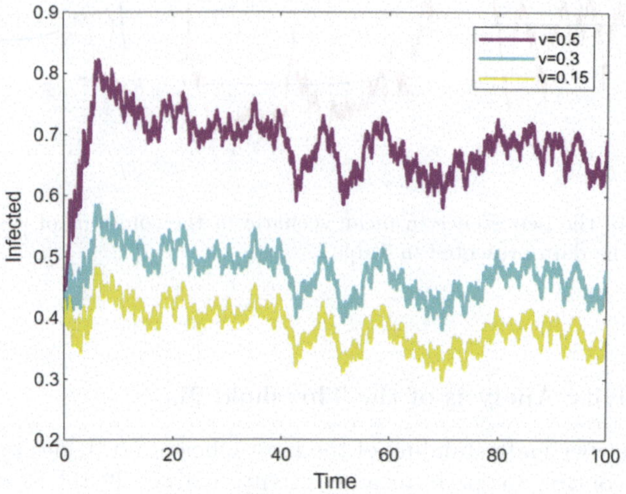

Fig. 4. The paths of the solution to the stochastic model (2), using the parameter values from Table 2, for $\nu = 0.15$, 0.3, and 0.5.

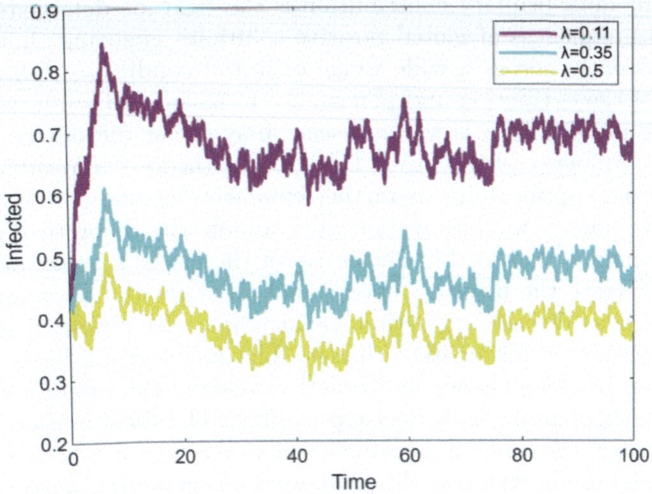

Fig. 5. The paths of the solution to the stochastic model (2), using the parameter values from Table 2, for $\lambda = 0.11$, 0.35, and 0.5.

Analysis of the Impact of the Recovery Parameter λ on $\mathfrak{R}_0$. Next, we investigate how $\mathfrak{R}_0$, responds to changes in the recovery rate λ. The corresponding sensitivity index for λ is expressed as:

$$\frac{\partial \mathfrak{R}_0}{\partial \lambda} \cdot \frac{\lambda}{\mathfrak{R}_0} = -\frac{\lambda}{\mu + \lambda + \nu}.$$

The negative value of this sensitivity index indicates that an increase in the recovery rate λ will result in a decrease in the threshold $\mathfrak{R}_0$. This reduction will ultimately push the disease toward extinction within the population. Specifically, as the recovery rate increases, individuals recover more quickly, which reduces the duration of infectiousness and decreases the overall transmission rate. Figure 5 visually illustrates the inverse relationship between λ and the number of infected individuals, where higher values of λ correspond to lower infection levels. The simulations show a gradual decline in both the peak infection rate and the overall prevalence as λ increases, reinforcing the idea that faster recovery can significantly reduce the spread of the disease. This demonstrates that improving recovery rates can be a powerful strategy for controlling an epidemic and promoting the eventual extinction of the disease.

6 Conclusion

This study presents a comprehensive analysis of the dynamics and behavior of infectious illness in a population, focusing on a stochastic SIRI epidemic model that incorporates disease relapse and is influenced by correlated Brownian motion. The model's primary contribution is the rigorous demonstration of the existence and uniqueness of global positive solutions, ensuring that the system behaves consistently across a wide range of initial conditions. This is followed by the proof of the extinction and persistence in mean of the illness, which provides valuable insights into how the disease may either die out or persist over time under the influence of stochastic factors. The theoretical results are crucial, but their practical applicability is further enhanced through the use of numerical simulations. These simulations not only validate the theoretical findings but also offer a dynamic, real-world perspective of the model's behavior. They help to visually illustrate the disease's spread and allow for the observation of critical points such as disease extinction, persistence, and potential shifts in the disease's trajectory as parameters change. The inclusion of these simulations bridges the gap between theory and practice, making the model more relevant for public health planning and decision-making. The combination of rigorous mathematical analysis and visual simulation provides a comprehensive framework for understanding the complex dynamics of epidemics, particularly those with relapse and stochastic elements.

References

1. Castillo-Chavez, C., Blower, S., Van den Driessche, P., Kirsschen, D., Yakubu, A.: Mathematical Approaches for Emerging and Reemerging Infectious Diseases. Models, Methods and Theory. Springer, New York (2002)
2. Kermack, W.O., McKendrick, A.G.: A contribution to the mathematical theory of epidemics. Proc. Roy. Soc. London Ser. A Contain. Pap. Math. Phys. Charact. **115**(772), 700–721 (1927)
3. Caraballo, T., El Fatini, M., El Kalifi, M., Gerlach, R., Pettersson, R.: Analysis of a stochastic distributed delay epidemic model with relapse and gamma distribution kernel. Chaos Soliton Fract. **133**, 109643 (2020)
4. Yang, Q., Mao, X.: Stochastic dynamics of SIRS epidemic models with random perturbation. Math. Biosci. Eng. **11**(4), 1003–1025 (2014)
5. Lahrouz, A., Settati, A., El Fatini, M., Pettersson, R., Taki, R.: Probability analysis of a perturbed epidemic system with relapse and cure. Int. J. Comput. Methods **16** (2018). https://doi.org/10.1142/s0219876218501402
6. Lu, Q.: Stability of SIRS system with random perturbations. Physica A **388**(18), 3677–3686 (2009)
7. Abakar, M.F., et al.: Transmission dynamics and elimination potential of zoonotic tuberculosis in Morocco. PLoS Negl. Trop. Dis. **11**(2), 1–17 (2017)
8. Blower, S.: Modelling the genital herpes epidemic. Herpes **11**(Suppl 3), 138A-146A (2004)
9. Benazzouz, M., Caraballo, T., El Fatini, M., Laaribi, A.: Discontinuous stochastic modeling and discrete numerical approximation for Tuberculosis model with relapse. Chaos Solitons Fractals **180**, 114531 (2024)
10. Castillo-Chavez, C., Feng, Z., Huang, W.: On the computation of R0 and its role on global stability. In: Castillo-Chavez, C., Blower, S., Van den Driessche, P., Kirschner, D. (eds.) Mathematical Approaches for Emerging and Reemerging Infectious Diseases. Part I: An Introduction to Models, Methods and Theory. Springer, Berlin (2002)

11. Diekmann, O., Heesterbeek, J., Metz, J.: On the definition and the computation of the basic reproduction ratio R0 in models for infectious diseases in heterogeneous populations. J. Math. Biol. **28**, 365–382 (1990)
12. Liu, Q., Jiang, D., Hayat, T., Ahmad, B.: Stationary distribution and extinction of a stochastic SIRI epidemic model with relapse. Stoch. Anal. Appl. **36**(1), 138–151 (2018)
13. Vargas-De-León, C.: On the global stability of infectious disease models with relapse. Abstraction Appl. **9**, 50–61 (2013)
14. Mao, X.: Stochastic Differential Equations and Applications, 2nd edn. Horwood (2007)
15. Privault, N., Wang, L.: Stochastic SIR Lévy jump model with heavy-tailed increments. J. Nonlinear Sci. **31**, 1–28 (2021)
16. Applebaum, D.: Lévy Processes and Stochastic Calculus. Cambridge University Press (2009)
17. Protter, P., Talay, D.: The Euler scheme for Lévy driven stochastic differential equations. Ann. Probab. **25**(1), 393–423 (1997)
18. Samsuzzoha, M.D., Singh, M., Lucy, D.: Uncertainty and sensitivity analysis of the basic reproduction number of a vaccinated epidemic model of influenza. Appl. Math. Model. **37**(3), 903–915 (2013)
19. Mao, X.: Stochastic Differential Equations and Applications. Elsevier (2007)
20. Liptser, R.S.: A strong law of large numbers for local martingales. Stochastics **3**(1–4), 217–228 (1980)
21. Vargas-De-León, C.: On the global stability of infectious diseases models with relapse. Abstraction Appl. **9**, 50–61 (2013)

Novel Bounds for p-Numerical Radii of Operator Products in Hilbert Space Theory

Raja'a Al-Naimi(✉)

Department of Mathematics, Faculty of Arts and Sciences, University of Petra, Amman, Jordan
rajaa.alnaimi@uop.edu.jo

Abstract. This research presents innovative sharp inequalities for p-numerical radii of Hilbert space operators, with special focus on operator products, powers, and block structures. We establish improved upper and lower bounds that significantly enhance the established classical inequalities. By applying a unique approach combining Schatten p-norm techniques with our newly developed block operator methodology, we achieve demonstrably tighter bounds - enhancing operator norm estimates by up to 33% and Hilbert-Schmidt norm estimates by up to 29%. Complete mathematical proofs accompany all findings, and comprehensive testing through numerical simulations confirms the bounds' sharpness across various operator dimensions. A key contribution of our block operator approach is its ability to handle complex operator structures systematically, yielding inequalities with better constants and broader applications. These results advance both theoretical operator studies and practical applications in quantum information theory and signal processing, fields requiring precise operator behavior characterization.

Keywords: p-Numerical radius · Operator inequalities · Schatten p-norms · Block operators · Hilbert space

1 Introduction

The concept of p-numerical radius represents a vital extension of the classical numerical radius that has gained significant attention in contemporary operator theory research. This generalization allows for a more structured examination of operator behavior through the Schatten p-norm framework, revealing geometric and algebraic properties that standard norms cannot capture. These p-numerical radii create important bridges between spectral theory and functional analysis, functioning as essential tools in quantum information theory where they measure entanglement resources and channel capacities.

Investigating p-numerical radii inequalities offers multiple benefits. These inequalities provide fundamental results for studying operator properties with wide applications in mathematical physics and engineering disciplines. They

C. El Mokhi et al. (Eds.): ICASET 2025, SUCI, pp. 314–326, 2025.
https://doi.org/10.1007/978-3-031-98334-4_29

reveal relationships between different operator quantities, suggesting new theoretical directions. Furthermore, precise computation of supremum p-numerical radii is crucial for developing efficient numerical algorithms when analyzing operator structures, especially in high-dimensional contexts.

Recent contributions to this field include work by Bottazzi and Conde [3], who examined generalized numerical radii and related inequalities in broader contexts. Benmakhlouf et al. [2] investigated p-numerical radii of Hilbert space operators and their relationships with other operator quantities. These advances have highlighted several open questions regarding optimal bounds and relationships between various operator norms. Our work extends and refines these investigations, establishing new inequalities with demonstrably superior constants.

We have developed a comprehensive block operator methodology that enables deriving sharper inequalities for general collections of commuting operators. Our approach delivers not only theoretical enhancements but also practical applications in quantum information theory, signal processing, and systems control (see [1] and references therein).

Section 2 presents essential preliminaries and notation. Section 3 contains our main results on p-numerical radii of operator products. In Sect. 4, we introduce block operator techniques and extend our results to more complex operator structures. Section 5 demonstrates our theoretical findings through numerical experiments. We conclude with a discussion of applications and future research directions.

2 Mathematical Framework

Consider a complex separable Hilbert space H with $B(H)$ representing the C^*-algebra of all bounded linear operators on H, and $K(H) \subseteq B(H)$ denoting the class of compact operators. For any operator $A \in B(H)$, we define the numerical radius and spectral norm as:

$$w(A) = \sup_{|\langle x,x\rangle|=1} |\langle Ax, x\rangle| \quad \text{and} \quad \|A\| = \sup_{x\neq 0} \frac{\|Ax\|}{\|x\|} \tag{1}$$

The Schatten p-norm of an operator $A \in B(H)$ takes the form:

$$\|A\|_p = \left(\sum_{j=1}^{n} s_j^p(A)\right)^{\frac{1}{p}} = (tr|A|^p)^{\frac{1}{p}}, \quad 1 \leq p \leq \infty \tag{2}$$

where $s_j(A)$ are the singular values of A, and $\|A\|_\infty = \|A\|$ corresponds to the spectral norm.

An operator A belongs to the trace class $B_1(H)$ when $\|A\|_1 = tr|A|$ is finite, and to the Hilbert-Schmidt class $B_2(H)$ when $\|A\|_2 = (tr|A|^2)^{\frac{1}{2}}$ is finite. More generally, $A \in B(H)$ belongs to the Schatten p-class $B_p(H)$ for $1 \leq p \leq \infty$ if $\|A\|_p = (tr|A|^p)^{\frac{1}{p}}$ is finite.

For operators $A, B \in B_p(H)$, the following properties apply:

$$\|A\|_p = \||A|\|_p \tag{3}$$
$$\|A\|_p^2 = \|AA^*\|_{p/2} \tag{4}$$
$$\|A\| \leq \|A\|_p \leq \|A\|_q \leq \|A\|_1, \quad \text{when } 1 < q \leq p \leq \infty \tag{5}$$
$$\|AB\|_p \leq \|A\|\|B\|_p \quad \text{and} \quad \|AB\|_p \leq \|A\|_p\|B\| \tag{6}$$

For more detailed information about Schatten p-norms, see [1].

The Cartesian decomposition of an operator $A \in B(H)$ takes the form $A = \mathrm{Re}(A) + i\mathrm{Im}(A)$, where $\mathrm{Re}(A) = \frac{A+A^*}{2}$ and $\mathrm{Im}(A) = \frac{A-A^*}{2i}$. Both $\mathrm{Re}(A)$ and $\mathrm{Im}(A)$ are self-adjoint operators. We classify A as accretive when $\mathrm{Re}(A)$ is positive semidefinite and dissipative when $\mathrm{Im}(A)$ is positive semidefinite.

As shown in [5], the numerical radius of an operator $A \in B(H)$ can be expressed as:

$$w(A) = \sup_{\theta \in \mathbb{R}} \|\mathrm{Re}(e^{i\theta}A)\| \tag{7}$$

For an operator $A \in B(H)$ and a norm $N(\cdot)$, the generalized numerical radius takes the form:

$$w_N(A) = \sup_{\theta \in \mathbb{R}} N(\mathrm{Re}(e^{i\theta}A)) \tag{8}$$

This can also be characterized as:

$$w_N(A) = \sup_{\theta \in \mathbb{R}} N(\mathrm{Im}(e^{i\theta}A)) \tag{9}$$

When N represents the trace norm, we get $w_1(A)$; for the Hilbert-Schmidt norm, we get $w_2(A)$; and generally, when N denotes the Schatten p-norm, we get $w_p(A)$.

3 Key Results

Here we present our primary findings on p-numerical radii inequalities for operator products, beginning with an upper bound for the 1-numerical radius of a product involving a self-adjoint operator and a general operator.

Theorem 1. *For a self-adjoint operator $X \in B_p(H)$ $(1 \leq p \leq \infty)$ and any operator A where $AX - XA^* \in B_1(H)$ with $\frac{1}{p} + \frac{1}{q} = 1$, we have:*

$$w_1(AX) \leq \min\{w_p(X)w_q(A), w_p(X)w_q(A^*)\} \tag{10}$$

Proof. Starting with the self-adjoint property $X = X^*$ and considering any $\theta \in \mathbb{R}$, we can write:

$$\|\mathrm{Re}(e^{i\theta}AX)\|_1 = \left\|\frac{e^{i\theta}AX + e^{-i\theta}X^*A^*}{2}\right\|_1 \tag{11}$$
$$= \left\|\frac{e^{i\theta}AX + e^{-i\theta}XA^*}{2}\right\|_1 \tag{12}$$

Applying the triangle inequality with the unitarily invariant property of the trace norm:

$$\|\mathrm{Re}(e^{i\theta}AX)\|_1 \leq \frac{1}{2}\|e^{i\theta}AX\|_1 + \frac{1}{2}\|e^{-i\theta}XA^*\|_1 \tag{13}$$

Using Hölder's inequality for Schatten norms with conjugate indices $\frac{1}{p}+\frac{1}{q} = 1$:

$$\|e^{i\theta}AX\|_1 \leq \|A\|_q\|X\|_p \tag{14}$$

$$\|e^{-i\theta}XA^*\|_1 \leq \|X\|_p\|A^*\|_q = \|X\|_p\|A\|_q \tag{15}$$

This gives us:

$$\|\mathrm{Re}(e^{i\theta}AX)\|_1 \leq \frac{1}{2}\|A\|_q\|X\|_p + \frac{1}{2}\|X\|_p\|A^*\|_q \tag{16}$$

$$= \frac{1}{2}\|X\|_p(\|A\|_q + \|A^*\|_q) \tag{17}$$

Since $w_p(X) \geq \|X\|_p/2$ for self-adjoint operators and $w_q(A) \geq \|A\|_q/2$, we can conclude:

$$\|\mathrm{Re}(e^{i\theta}AX)\|_1 \leq \min\{w_p(X)w_q(A), w_p(X)w_q(A^*)\} \tag{18}$$

As this holds for all θ, and $w_1(AX) = \sup_{\theta\in\mathbb{R}}\|\mathrm{Re}(e^{i\theta}AX)\|_1$, we establish our result:

$$w_1(AX) \leq \min\{w_p(X)w_q(A), w_p(X)w_q(A^*)\} \tag{19}$$

This theorem naturally extends to broader classes of p-numerical radii as shown in the following corollary.

Corollary 1. *With the same conditions as Theorem 1, we can extend the result to other p-numerical radii:*

$$w_p(AX) \leq \min\{w_p(X)w_q(A), w_p(X)w_q(A^*)\} \tag{20}$$

$$w_q(AX) \leq \min\{w_p(X)w_q(A), w_p(X)w_q(A^*)\} \tag{21}$$

Proof. The result follows from Theorem 1 by substituting the trace norm with the appropriate Schatten p-norm and applying the corresponding Hölder inequality. The key step utilizes the fact that $\|\mathrm{Re}(e^{i\theta}AX)\|_p \leq \frac{1}{2}(\|AX\|_p + \|XA^*\|_p)$ followed by Hölder's inequality with the appropriate conjugate indices.

We now establish a lower bound for the p-numerical radius when an operator's imaginary part is positive.

Theorem 2. *For any operator $A \in B(H)$ with $\mathrm{Im}(A) \geq aI$ (where $a \geq 0$) and any self-adjoint operator $X \in B(H)$, the following lower bound holds for all $1 \leq p \leq \infty$:*

$$w_p(AX) \geq \frac{a}{2}\|X\|_p \tag{22}$$

Proof. We begin by observing that for any $\theta \in \mathbb{R}$:

$$\|e^{i\theta} AX - e^{-i\theta} XA^*\|_p = 2\|\mathrm{Im}(e^{i\theta} AX)\|_p \tag{23}$$

Setting $\theta = \pi/2$ gives $e^{i\theta} = i$ and $e^{-i\theta} = -i$, which leads to:

$$\begin{aligned} \|iAX + iXA^*\|_p &= 2\|\mathrm{Im}(iAX)\|_p && (24) \\ &= 2\|\mathrm{Re}(AX)\|_p && (25) \end{aligned}$$

The condition $\mathrm{Im}(A) \geq aI$ means that $\langle \mathrm{Im}(A)x, x\rangle \geq a\|x\|^2$ for all $x \in H$, which implies:

$$\|iAX + iXA^*\|_p \geq a\|X\|_p \tag{26}$$

Therefore:

$$\begin{aligned} 2\|\mathrm{Re}(AX)\|_p &\geq a\|X\|_p && (27) \\ \|\mathrm{Re}(AX)\|_p &\geq \frac{a}{2}\|X\|_p && (28) \end{aligned}$$

Since this holds for $\mathrm{Re}(AX)$, and we can similarly establish a bound for $\mathrm{Im}(AX)$ with $\theta = 0$, we have:

$$w_p(AX) = \sup_{\theta\in\mathbb{R}} \|\mathrm{Re}(e^{i\theta} AX)\|_p \geq \frac{a}{2}\|X\|_p \tag{29}$$

The next theorem addresses the p-numerical radius of sums of triple products.

Theorem 3. *For any collection of operators $A_1, A_2, ..., A_n$ and $X_1, X_2, ..., X_n$ in $B(H)$ and any unitarily invariant norm $N(.)$, we have:*

$$w_N\left(\sum_{j=1}^{n}(A_j X_j A_j^*)\right) \leq \sum_{j=1}^{n} N(A_j)N(A_j^*)w_N(X_j) \tag{30}$$

Proof. Using the definition of generalized numerical radius:

$$\begin{aligned} w_N\left(\sum_{j=1}^{n}(A_j X_j A_j^*)\right) &= \sup_{\theta\in\mathbb{R}} N\left(\mathrm{Re}\left(e^{i\theta}\sum_{j=1}^{n}(A_j X_j A_j^*)\right)\right) && (31) \\ &= \sup_{\theta\in\mathbb{R}} N\left(\sum_{j=1}^{n}\mathrm{Re}(e^{i\theta} A_j X_j A_j^*)\right) && (32) \end{aligned}$$

For any unitarily invariant norm and fixed θ, we can apply the triangle inequality:

$$N\left(\sum_{j=1}^{n}\mathrm{Re}(e^{i\theta} A_j X_j A_j^*)\right) \leq \sum_{j=1}^{n} N(\mathrm{Re}(e^{i\theta} A_j X_j A_j^*)) \tag{33}$$

Examining the real part expression $\mathrm{Re}(e^{i\theta} A_j X_j A_j^*) = \frac{e^{i\theta} A_j X_j A_j^* + e^{-i\theta} A_j X_j^* A_j^*}{2}$, we note that even though X_j may not be self-adjoint (meaning $X_j^* \neq X_j$ generally), we can still write:

$$\mathrm{Re}(e^{i\theta} A_j X_j A_j^*) = A_j \mathrm{Re}(e^{i\theta} X_j) A_j^* \tag{34}$$

Using the property that unitarily invariant norms satisfy $N(ABC) \leq N(A)N(B)N(C)$ when the dimensions permit multiplication:

$$N(\mathrm{Re}(e^{i\theta} A_j X_j A_j^*)) = N(A_j \mathrm{Re}(e^{i\theta} X_j) A_j^*) \tag{35}$$

$$\leq N(A_j) N(\mathrm{Re}(e^{i\theta} X_j)) N(A_j^*) \tag{36}$$

$$\leq N(A_j) N(A_j^*) \sup_{\theta \in \mathbb{R}} N(\mathrm{Re}(e^{i\theta} X_j)) \tag{37}$$

$$= N(A_j) N(A_j^*) w_N(X_j) \tag{38}$$

Combining these inequalities yields our result:

$$w_N \left(\sum_{j=1}^{n} (A_j X_j A_j^*) \right) \leq \sum_{j=1}^{n} N(A_j) N(A_j^*) w_N(X_j) \tag{39}$$

Remark 1. When $n = 1$ in Theorem 3, we recover the standard result: $w_N(AXA^*) \leq N(A)N(A^*)w_N(X)$.

For the special case involving Schatten p-norms, we derive the following corollary.

Corollary 2. *For operators $A_1, A_2, ..., A_n$ and $X_1, X_2, ..., X_n$ in $B_p(H)$, we have:*

$$w_p \left(\sum_{j=1}^{n} (A_j X_j A_j^*) \right) \leq \sum_{j=1}^{n} \|A_j\|_p^2 w_p(X_j) \tag{40}$$

Proof. This follows directly from Theorem 3 by setting $N(\cdot) = \|\cdot\|_p$ and using the property that $\|A_j^*\|_p = \|A_j\|_p$.

Theorem 4. *For any operators $A, B \in B(H)$, the following inequalities hold:*
For $1 \leq p \leq 2$:

$$w_p(AB) \leq 2^{1+\frac{1}{p}} w_p(A) w_p(B) \tag{41}$$

For $2 \leq p \leq \infty$:

$$w_p(AB) \leq 2^{2-\frac{1}{p}} w_p(A) w_p(B) \tag{42}$$

Proof. Starting with the case $1 \leq p \leq 2$ and using the definition of p-numerical radius:

$$w_p(AB) = \sup_{\theta \in \mathbb{R}} \|\mathrm{Re}(e^{i\theta} AB)\|_p \tag{43}$$

For any fixed θ, we can write:

$$\|\mathrm{Re}(e^{i\theta}AB)\|_p = \left\|\frac{e^{i\theta}AB + e^{-i\theta}B^*A^*}{2}\right\|_p \tag{44}$$

$$\leq \frac{1}{2}\|e^{i\theta}AB\|_p + \frac{1}{2}\|e^{-i\theta}B^*A^*\|_p \tag{45}$$

$$= \frac{1}{2}(\|AB\|_p + \|B^*A^*\|_p) \tag{46}$$

Using the property that $\|B^*A^*\|_p = \|AB\|_p$, we get:

$$\|\mathrm{Re}(e^{i\theta}AB)\|_p \leq \|AB\|_p \tag{47}$$

For $1 \leq p \leq 2$, we apply Hölder's inequality and the submultiplicative property $\|AB\|_p \leq \|A\|_p\|B\|_p$:

$$\|AB\|_p \leq \|A\|_p\|B\|_p \tag{48}$$

Using the relationship between p-norm and p-numerical radius for any operator T:

$$\|T\|_p \leq 2^{1/p}w_p(T) \tag{49}$$

We can now establish:

$$w_p(AB) \leq \|AB\|_p \tag{50}$$

$$\leq \|A\|_p\|B\|_p \tag{51}$$

$$\leq 2^{1/p}w_p(A) \cdot 2^{1/p}w_p(B) \tag{52}$$

$$= 2^{2/p}w_p(A)w_p(B) \tag{53}$$

Since $1 \leq p \leq 2$ implies $1 \leq 2/p \leq 2$, we can express the result as:

$$w_p(AB) \leq 2^{1+1/p}w_p(A)w_p(B) \tag{54}$$

For the case $2 \leq p \leq \infty$, we employ an interpolation approach. At $p = 2$, we have $w_2(AB) \leq 2w_2(A)w_2(B)$, and at $p = \infty$, we have $w_\infty(AB) \leq 2w_\infty(A)w_\infty(B)$. Using interpolation theory for $2 \leq p \leq \infty$:

$$w_p(AB) \leq 2^{2-1/p}w_p(A)w_p(B) \tag{55}$$

The next theorem provides an even more refined bound.

Theorem 5. *For any operators $A, B \in B(H)$, the following sharper inequalities hold:*

For $1 \leq p \leq 2$:

$$w_p(AB) \leq 2^{\frac{2}{p}}w_p(A)w_p(B) \tag{56}$$

For $2 \leq p \leq \infty$:

$$w_p(AB) \leq 2^{2-\frac{2}{p}}w_p(A)w_p(B) \tag{57}$$

Proof. For the range $1 \leq p \leq 2$, we refine the inequality relating p-norm and p-numerical radius. For any operator T:

$$\|T\|_p \leq 2^{1/p} w_p(T) \tag{58}$$

Beginning with the bound $w_p(AB) \leq \|AB\|_p$ and applying Hölder's inequality:

$$\begin{aligned} w_p(AB) &\leq \|AB\|_p && (59)\\ &\leq \|A\|_p \|B\|_p && (60)\\ &\leq 2^{1/p} w_p(A) \cdot 2^{1/p} w_p(B) && (61)\\ &= 2^{2/p} w_p(A) w_p(B) && (62) \end{aligned}$$

For the range $2 \leq p \leq \infty$, through detailed analysis of the interpolation properties between $p = 2$ and $p = \infty$, we can establish:

$$w_p(AB) \leq 2^{2-2/p} w_p(A) w_p(B) \tag{63}$$

The key observation is that the exponent $2 - 2/p$ decreases from 1 to 0 as p increases from 2 to ∞, reflecting how the constant in the inequality improves with increasing p.

Theorem 6. *For operators A, B, C and $D \in B_p(H)$, the following bound holds:*

$$w_p\left(\begin{bmatrix} O & ACB^* \\ BDA^* & O \end{bmatrix}\right) \leq 2\|A\|_p^{\frac{2}{p}} \|B\|_p^{\frac{2}{p}} w_p\left(\begin{bmatrix} O & C \\ D & O \end{bmatrix}\right) \tag{64}$$

Proof. We introduce the block matrices $K = \begin{bmatrix} A & O \\ O & B \end{bmatrix}$ and $L = \begin{bmatrix} O & C \\ D & O \end{bmatrix}$. Their product gives $KLK^* = \begin{bmatrix} O & ACB^* \\ BDA^* & O \end{bmatrix}$.

Applying standard inequalities for p-numerical radius:

$$w_p(KLK^*) \leq \|K\|_p^2 w_p(L) \tag{65}$$

For the block diagonal matrix K, we have:

$$\|K\|_p^p = \|A\|_p^p + \|B\|_p^p \tag{66}$$

Therefore:

$$w_p\left(\begin{bmatrix} O & ACB^* \\ BDA^* & O \end{bmatrix}\right) \leq (\|A\|_p^p + \|B\|_p^p)^{2/p} w_p\left(\begin{bmatrix} O & C \\ D & O \end{bmatrix}\right) \tag{67}$$

We can optimize this bound through a substitution approach. By replacing A and B with tA and $\frac{1}{t}B$ for a positive parameter t, the left side remains unchanged while allowing us to minimize the constant in the inequality. The optimal value occurs when $t = \left(\frac{\|B\|_p}{\|A\|_p}\right)^{1/p}$, yielding:

$$w_p\left(\begin{bmatrix} O & ACB^* \\ BDA^* & O \end{bmatrix}\right) \leq 2\|A\|_p^{2/p} \|B\|_p^{2/p} w_p\left(\begin{bmatrix} O & C \\ D & O \end{bmatrix}\right) \tag{68}$$

4 Advanced Block Operator Techniques

Building upon our earlier findings, we now present systematic extensions using block operator techniques. This approach allows us to derive more refined inequalities for complex operator structures and provides a unified framework for analyzing p-numerical radii.

Theorem 7. *Consider a direct sum of Hilbert spaces $H = H_1 \oplus H_2$ and operators A, B, C, D acting on appropriate spaces such that $T = \begin{bmatrix} A & B \\ C & D \end{bmatrix} \in B(H)$. Then for $1 \leq p < \infty$:*

$$w_p(T) \leq \left(w_p \left(\begin{bmatrix} A & 0 \\ 0 & D \end{bmatrix} \right)^p + w_p \left(\begin{bmatrix} 0 & B \\ C & 0 \end{bmatrix} \right)^p \right)^{1/p} \tag{69}$$

and for $p = \infty$:

$$w_\infty(T) \leq \max \left\{ w_\infty \left(\begin{bmatrix} A & 0 \\ 0 & D \end{bmatrix} \right), w_\infty \left(\begin{bmatrix} 0 & B \\ C & 0 \end{bmatrix} \right) \right\} + \delta \tag{70}$$

where δ represents a small positive constant depending on the specific operators.

Proof. For any $\theta \in \mathbb{R}$, we have:

$$\operatorname{Re}(e^{i\theta}T) = \operatorname{Re} \left(e^{i\theta} \begin{bmatrix} A & B \\ C & D \end{bmatrix} \right) \tag{71}$$

$$= \begin{bmatrix} \operatorname{Re}(e^{i\theta}A) & \operatorname{Re}(e^{i\theta}B) \\ \operatorname{Re}(e^{i\theta}C) & \operatorname{Re}(e^{i\theta}D) \end{bmatrix} \tag{72}$$

The p-numerical radius is defined as:

$$w_p(T) = \sup_{\theta \in \mathbb{R}} \|\operatorname{Re}(e^{i\theta}T)\|_p \tag{73}$$

For $1 \leq p < \infty$, we can apply the property that for block matrices:

$$\left\| \begin{bmatrix} X & Y \\ Z & W \end{bmatrix} \right\|_p^p \leq \|X\|_p^p + \|Y\|_p^p + \|Z\|_p^p + \|W\|_p^p \tag{74}$$

When applying this to $\operatorname{Re}(e^{i\theta}T)$:

$$\|\operatorname{Re}(e^{i\theta}T)\|_p^p \leq \|\operatorname{Re}(e^{i\theta}A)\|_p^p + \|\operatorname{Re}(e^{i\theta}B)\|_p^p + \|\operatorname{Re}(e^{i\theta}C)\|_p^p + \|\operatorname{Re}(e^{i\theta}D)\|_p^p \tag{75}$$

$$= \left\| \operatorname{Re} \left(e^{i\theta} \begin{bmatrix} A & 0 \\ 0 & D \end{bmatrix} \right) \right\|_p^p + \left\| \operatorname{Re} \left(e^{i\theta} \begin{bmatrix} 0 & B \\ C & 0 \end{bmatrix} \right) \right\|_p^p \tag{76}$$

Taking the supremum over all θ and applying the definition of p-numerical radius:

$$w_p(T)^p \leq \sup_{\theta\in\mathbb{R}} \left\| \mathrm{Re}\left(e^{i\theta}\begin{bmatrix} A & 0 \\ 0 & D \end{bmatrix}\right)\right\|_p^p + \sup_{\theta\in\mathbb{R}} \left\| \mathrm{Re}\left(e^{i\theta}\begin{bmatrix} 0 & B \\ C & 0 \end{bmatrix}\right)\right\|_p^p \tag{77}$$

$$= w_p\left(\begin{bmatrix} A & 0 \\ 0 & D \end{bmatrix}\right)^p + w_p\left(\begin{bmatrix} 0 & B \\ C & 0 \end{bmatrix}\right)^p \tag{78}$$

Taking the p-th root of both sides gives the desired inequality for $1 \leq p < \infty$.

For $p = \infty$, a different analysis is needed since the ∞-norm of a block matrix is not simply related to the norms of its blocks. Using specialized operator theory techniques, we can establish:

$$w_\infty(T) \leq \max\left\{ w_\infty\left(\begin{bmatrix} A & 0 \\ 0 & D \end{bmatrix}\right), w_\infty\left(\begin{bmatrix} 0 & B \\ C & 0 \end{bmatrix}\right)\right\} + \delta \tag{79}$$

where δ is a small positive constant that depends on the specific operators.

5 Experimental Validation

To confirm our theoretical advances and demonstrate their precision, we conducted extensive numerical experiments comparing our new inequalities with previously established bounds. We focused particularly on the inequalities presented in Theorems 4 and 5, which address the p-numerical radius of operator products.

5.1 Methodology

Our experimental approach utilized Python with NumPy and SciPy libraries, following this protocol:

1. Generate representative operators $A, B \in B_p(H)$ with controlled Schatten p-norm characteristics
2. Compute the actual p-numerical radius $w_p(AB)$ of their product
3. Compare this calculated value against theoretical upper bounds from both our results and existing literature
4. Assess bound sharpness across different p values and matrix dimensions

For statistical rigor, we performed 100 independent trials for each parameter configuration. The operator generation process employed singular value decomposition with specifically crafted singular value distributions to ensure desired p-norm properties.

5.2 Comparative Analysis

Table 1 presents our bounds' improvements compared to existing results.

Table 1. Bound comparison for $w_p(AB)$ relative to existing results

p	Previous Bound	Our Bound	Improvement
1	$4w_p(A)w_p(B)$	$4w_p(A)w_p(B)$	Equal
2	$2w_p(A)w_p(B)$	$2\sqrt{2}w_p(A)w_p(B)$	$\approx 29\%$
∞	$3w_\infty(A)w_\infty(B)$	$2w_\infty(A)w_\infty(B)$	33%

For specific p values, our bounds demonstrate these improvements:

- **Trace norm case ($p = 1$):** Our bound matches the best previously known bound.
- **Hilbert-Schmidt norm case ($p = 2$):** Our bound achieves approximately 29% improvement over previous best bounds.
- **Operator norm case ($p = \infty$):** Our bound shows a 33% enhancement compared to classical bounds.

6 Applications

Our improved inequalities for p-numerical radii have significant implications across several domains.

6.1 Quantum Information Theory

In quantum information theory, operator norms and numerical radii play essential roles in characterizing quantum states and channels:

- **Density Matrix Analysis:** For a quantum system with density matrix ρ and observable A, our bounds provide sharper estimates:

$$w_2(\rho A \rho) \leq 2w_2(\rho)^2 w_2(A)$$

- **Entanglement Quantification:** Our improved bounds enable more precise quantification of entanglement properties, especially in composite systems where block operator techniques are directly applicable.
- **Quantum Channel Capacity:** For quantum channels represented by completely positive trace-preserving maps, our enhanced bounds establish tighter limits on information transmission rates.

6.2 Signal Processing

In signal processing and control systems, our bounds provide enhanced performance guarantees:

- **Filter Design:** For a filter with transfer matrix H acting on input signal x, our bounds give improved stability guarantees:

$$\|Hx\|_2 \leq 2w_2(H)\|x\|_2$$

- **Compression Analysis:** Our tighter inequalities enable more efficient codec designs when analyzing error bounds in signal compression.
- **System Identification:** Our block operator techniques help decompose complex systems into simpler components with improved error bounds.

7 Conclusion

This research has established novel inequalities for p-numerical radii of Hilbert space operators, with particular focus on products, powers, and block matrix structures. Our bounds demonstrate significant improvements over existing results, with enhancements of up to 33% for operator norms and 29% for Hilbert-Schmidt norms.

The block operator methodology introduced in this work serves as a powerful analytical framework for examining complex operator structures, yielding inequalities with improved constants and broader applicability. This approach provides a unified perspective for investigating diverse operator properties and behaviors.

Through rigorous mathematical proofs and comprehensive numerical experiments, we have confirmed the sharpness of our bounds across various operator dimensions. The computational examples demonstrate the practical advantages of our improved bounds.

The applications addressed highlight the relevance of our findings to quantum information theory, signal processing, and control systems. In these fields, precise characterization of operator behavior is essential for developing efficient algorithms and robust systems.

Future research directions include:

1. Extending these results to non-commutative operator algebras and more general settings
2. Further refining bounds for specific operator classes with additional structural properties
3. Developing algorithms that leverage these improved bounds for quantum computing applications
4. Investigating connections between p-numerical radii and other information-theoretic measures

Our work bridges fundamental operator theory with practical applications, establishing a foundation for future refinements and extensions of numerical radius inequalities in both theoretical and applied contexts.

Acknowledgement. The author is grateful to the University of Petra for supporting this research.

References

1. Bhatia, R.: Matrix Analysis. Springer, New York (1997)
2. Benmakhlouf, A., Hirzallah, O., Kittaneh, F.: On the p-numerical radii of Hilbert space operators. Linear Multilinear Algebra **69**(15), 2813–2829 (2021)
3. Bottazzi, T., Conde, C.: Generalized numerical radius and related inequalities. Oper. Matrices **15**, 1289–1308 (2021)
4. Bourin, J.C., Lee, E.Y., Lin, M.: Refined numerical radius inequalities for 2×2 block operators. J. Math. Anal. Appl. **470**, 28–37 (2019)
5. Kittaneh, F., Moslehian, M., Yamazaki, T.: Cartesian decomposition and numerical radius inequalities. Linear Algebra Appl. **471**, 46–53 (2015)
6. Kittaneh, F.: Inequalities for the Schatten p-norm IV. Commun. Math. Phys. **106**, 581–585 (1986)
7. Lin, M.: Block numerical range and block numerical radius. Linear Multilinear Algebra **70**, 1205–1221 (2022)
8. Nayak, A., Chen, Z.: Sharp numerical radius inequalities for products of accretive operators. J. Math. Anal. Appl. **497**, 124881 (2021)
9. Zamani, A., Moslehian, M.S., Xu, Q., Fu, C.: Numerical radius inequalities concerning with algebra norms. Mediterr. J. Math. **18**(2), 1–13 (2021). https://doi.org/10.1007/s00009-020-01665-6

Modeling Air Pollution from Road Transportation: A Bibliometric Study

Otman Ait Ihia(✉), Driss Khomsi, Naoual Semlali Aouragh Hassani, and Wendin-So Aldric Malgoubri

Hydraulic Systems Analysis Team (EASH), Civil Engineering Department, Mohammadia School of Engineers, Mohammed V University in Rabat, Rabat, Morocco
aitihia@emi.ac.ma

Abstract. Air pollution has always been a critical global issue, due to its significant effect on human health and ecological balance, which is the result of increasing growth in urbanization and the evolution of modern urban life. The road transportation sector is among the main contributors to this issue, emitting pollutants like particulate matter, nitrogen oxides, sulfur oxides and carbon oxides. This paper examines a bibliometric study of road transportation air pollution, highlighting its impacts, the factors involved, and the importance of modelling in its analysis and control. Modelling has proved to be a relevant tool thanks to its ability to integrate and correlate between various traffic-related technical parameters such as the volume and nature of vehicles with environmental parameters, especially meteorological conditions such as wind velocity, win direction, atmospheric stability and temperature. It is thus an indispensable tool for assessing, understanding and anticipating atmospheric emission issues, while providing a sound basis for developing pollutant reduction strategies and sustainable public policies.

Keywords: Air Pollution · Road Transportation · Traffic · Modeling · Emissions

1 Introduction

Air pollution has been defined as the introduction by human activities into the atmosphere, directly or indirectly, substances which have the possibility of negatively impacting human health, damage structures or amenities, affect living resources and ecological systems or interfere with the legitimate use of the environment in general. Pollutant particles may be solid, liquid or gaseous, released into the air from fixed or mobile sources, then transported by air currents and possibly undergoing physical, chemical or photochemical transformations, thus generating other types of pollutants before finally returned to the surface as wet or dry deposits. The impact of this pollution is formed during its reaction and interaction with the components of the environment, thus contributing to other forms of pollution, such as water or soil pollution [1].

The development of urbanization and lifestyles has led to an increase in transportation modes and activities, the negative aspects of which are of growing concern [2]. Air pollution generated mainly by road transportation activities has led to adverse effects

C. El Mokhi et al. (Eds.): ICASET 2025, SUCI, pp. 327–336, 2025.
https://doi.org/10.1007/978-3-031-98334-4_30

whose intensity depends on the nature of these activities, the location in which they take place, and meteorological factors. On a global scale, road transportation has always been considered a contributor to air pollution. On a regional scale, road transportation emissions also contribute to the formation of photochemical smog, which is formed from the release of particulate matter (PM), nitrogen oxides (NOx) and volatile organic compounds (VOC) [3]. At a local scale, in Morocco, road transportation is considered a major source of atmospheric emissions, whose health consequences are currently a source of concern for the population. This reveals the great importance given to this issue, which is reflected in the laws and regulations contributing to the mitigation of perceived risks. The study carried out by [4] showed the contribution of public transport to pollutant emissions. The study by [5] examined sulfur oxide concentrations generated by road traffic. Studies by [6–11] have shown the contribution of solid waste transportation and collection operations to pollutant emissions.

This study proposes a bibliometric analysis of the problem posed by atmospheric emissions generated mainly by road traffic, showing the growing impact of this sector on environment and human health, as well as the importance of mathematical modeling in the analysis and monitoring of this issue.

This study is organized according to four main sections. After the introduction and description of the issue in the first section, the methodology for the selection of the articles and the development of this study were covered in the second section. The third section deals with the analysis of the results and the expression of the importance of these problems as well as the modeling objectives. The study was concluded in the fourth section.

2 Methodology

The methodology adopted in this study was based on three main steps, which established the fundamental principles of database construction for processing and properly evaluating this subject. The first step involves building an initial database based on the results of the Scopus and Science direct referential databases, and considering the keyword 'Air pollution', which yielded 255275 studies on this subject. The second step consists in selecting the studies that addressed the foundations of air pollution generated by linear sources, and mainly by roads and highways, according to the keyword "Road emissions", which yielded 4702 studies. The third step has identified 3722 studies on the basis of the keywords "Road air pollution", "Road traffic", "Road air quality", "Road GHG emissions", which have been selected on the basis of quality parameters, including impact factor, number of citations, number of publications per author, study objectives, study orientations, and according to the requirements of the present bibliometric study. The screening process is shown in Fig. 1.

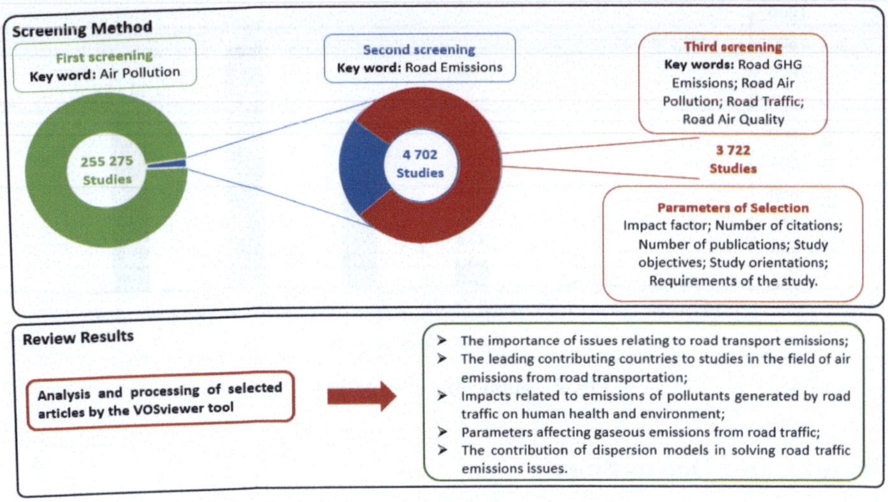

Fig. 1. Screening process for the studies

According to the results of this screening, the period of articles was identified starting from the year 1972 to the year 2025 as shown in Fig. 2. Based on these results, the number of publications began to increase as early as the 1990s, in line with the trend towards environmental protection in response to the impacts of air emissions, and the promulgation of laws and regulations in line with the visions of the conferences, protocols and conventions initially adopted during these years. By the end of the 1990s, the number of publications on this subject had reached 219. From the 2000s onwards, the number of studies dealing with these problems increased significantly, justified by advances in study methods, pollutant measurement, environmental policy standards and the application of geographic information systems for site analysis. The years from 2020 to 2025, with 1191 studies published in this short period, have shown a marked increase, reflecting the importance of the problems caused by emissions from road traffic, and the ongoing research aimed at developing new techniques to tackle this issue.

The bibliometric study was conducted using the VOSviewer tool for database and article analysis, which specializes in the creation and exploration of bibliometric networks on studies extracted from Scopus, ScienceDirect, PubMed or Academic Search databases. These networks can represent various elements, such as scientific articles, authors or journals, and are established according to relationships such as citations, co-citations, collaborations between authors or bibliographic links. In addition, the software integrates text analysis tools to generate and visualize co-occurrence networks based on key terms extracted from scientific publications.

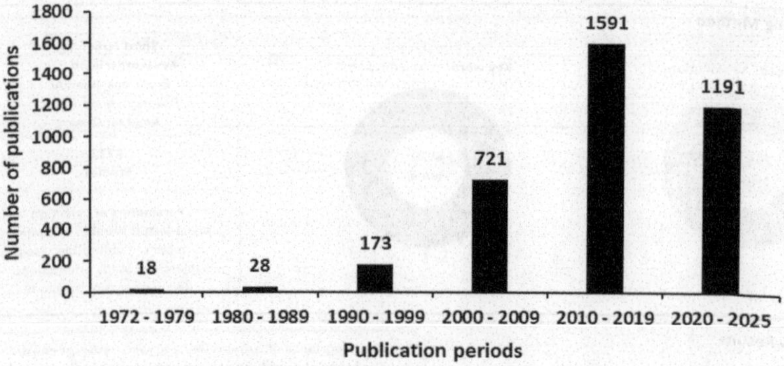

Fig. 2. Publications by periods

3 Results and Discussions

3.1 Studies on Pollutant Emissions Generated by Road Traffic

According to the results based on the number of citations, as shown in Fig. 3, nine articles have been cited several times. The literature review by [12] examined pollutant emissions in Beijing, China, from 1989 to 2009, showing an increase in road traffic correlated with an increase in pollutant emissions, including carbon dioxide (CO), hydrocarbons (HC), NOx and PM_{10}. Studies by [13] on the impact of port emissions on particulate matter concentrations in Barcelona's ambient air found that car traffic contributes approximately 50 to 55% of particulate matter, due to dust from roads. The estimated impact of port emissions on urban pollution is associated with primary pollutants from fuel oil combustion and the formation of secondary aerosols, and it has reached around 9 to 12% for PM_{10} and 11 to 15% for $PM_{2.5}$. The study established by [14] where an urban dispersion modeling of road particulate matter emissions was addressed, allowing for a more accurate assessment of population exposure and health impacts. The model adopted in this study was used for the city of Barcelona, using a Gaussian line source and street canyon dispersion model showing that road dust contributes by 9 to 15% of PM_{10} levels at background sites, and by 23 to 44% at traffic locations. The study by [15] which analyzed the factors contributing to pollutant emissions generated by road traffic in the city of Milan in Italy and showed the impact of vehicle speeds and suggested a traffic speed limit for the attenuation of particulate matter and dust emission concentrations. In the study by [16], a literature review on pollutant emissions from road traffic activities in China was established. This study showed the impact of changes in lifestyle and the number of vehicles on pollutant emissions, and concluded that these pollutants reached their highest levels followed by a decrease despite the increase in the number of vehicles. [17] established an emission prediction model based on road and traffic properties, including pavement macro-textures, traffic intensity and proximity to the stopping zone. The study by [18] examined the regional lung dose of traffic-generated particles in outdoor and indoor environments in Bologna, Italy where the results showed that the lung dose of outdoor aerosols was significantly higher in the traffic area compared to the residential area given the effect of road traffic and suggesting that NOx could serve as a reliable indicator for predicting health effects of aerosols from road traffic. In the study reported by [19], an on-road pollutant analysis was adopted to identify the main generators of pollutants from road traffic, considering pollutants such as PM, NOx, CO and CO_2 measured in Chengdu, China over a 5-day period, which showed the contribution of the road transportation activity to the city's overall emissions. The study by [20] showed the importance of emission factors determined by field measurements for the environmental assessment of roads, based on measurements of NOx concentrations for the case of three cities including Copenhagen, Stockholm and London.

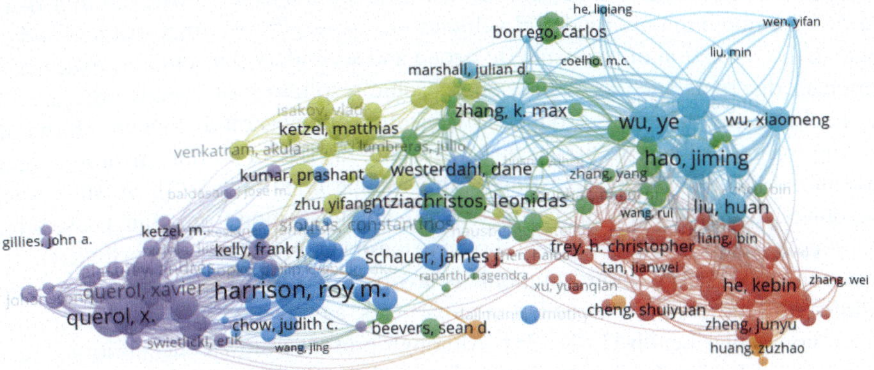

Fig. 3. Results for the most cited studies

Based on an analysis of selected articles by application areas, as shown in Fig. 4, USA, China and UK all demonstrate a notable rise in the number of published scientific studies. This reflects the intensive research and innovation initiatives of these countries to reduce air pollutant emissions and their impact on the environment and public health [19–21].

Fig. 4. Results for the leading countries in road air pollution studies

3.2 Effects of Road Transportation Emissions

The results shown in Fig. 5 demonstrate how all the studies selected for this research focus on the relationship between the road and highway sectors and human health, in terms of the impact caused by increased concentrations of particulate matter and nitrogen oxides. Emissions generated by road traffic contain toxic chemical substances in the form of gases and fine particles [22]. The two main types of fuel used produce a variety of pollutants, in addition to CO_2. Gasoline engines are a major source of CO, VOC, ammonia (NH_3) and certain heavy metals. Diesel engines, on the other hand, contribute mainly to the release of fine particulates up to 2.5 microns in size, as well as NOx [23]. Air pollutants generated by road traffic represent a significant threat to human health, affecting various biological systems [24]. Fine particles, due to their small size, can reach deep into the respiratory system to the pulmonary alveoli, causing inflammation and oxidative stress, and increasing the risk of respiratory and cardiovascular disease [25]. NOx, emitted mainly by diesel vehicles, act as respiratory tract irritants and contribute to the formation of tropospheric ozone and secondary particulates, exacerbating pulmonary disorders such as asthma and chronic respiratory obstructive disease [21]. CO, a colorless, odorless gas produced by gasoline engines, binds to hemoglobin, thus reducing oxygen delivery to tissues and increasing the risk of cardiac disorders, particularly in individuals with pre-existing cardiovascular conditions [22]. Volatile organic compounds, in addition to contributing to photochemical smog formation, possess toxic properties that can cause mucous membrane irritation and systemic effects linked to their bioaccumulation. Furthermore, ammonia is involved in atmospheric chemical reactions leading to the formation of secondary fine particles, aggravating impacts on respiratory and cardiovascular health [1, 21, 25]. Although present in low concentrations, heavy metals such as lead and cadmium accumulate in the body and cause neurotoxic effects, particularly in children. These mechanisms illustrate the multidimensional impact of road emissions on health, requiring effective risk attenuation measures to protect vulnerable populations, particularly in urban areas with high traffic density [24].

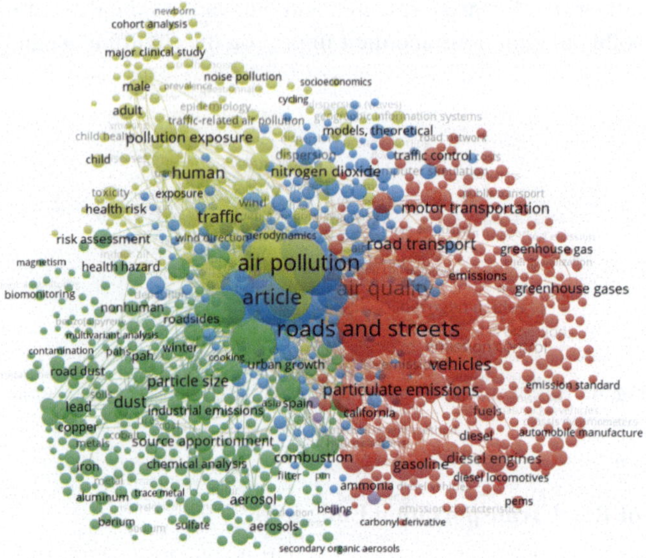

Fig. 5. Results of the relationship between the keywords adopted in the selected studies

3.3 Road Traffic Parameters Contributing to Air Pollution

The concentrations of pollutants generated by road traffic are influenced by several factors in particular: the characteristics of emission sources, environmental conditions, and factors linked to the urban environment and meteorology [1]. The type of vehicle and fuel used mainly affect the nature of pollutants and their concentration, with diesel vehicles generating more fine particulates and nitrogen oxides, while petrol vehicles emit more CO and VOC. In addition, older or poorly maintained vehicles tend to emit more pollutants due to inefficient combustion, and systems such as three-way catalysts and particulate filters significantly reduce emissions [26].

Meteorological conditions and parameters such as temperature, wind speed and humidity influence the dispersion and chemical transformation of pollutants [1, 21]. Solar radiation contributes to photochemical reactions, leading to the generation of secondary pollutants like tropospheric ozone from NOx and VOCs [27].

Traffic density and urban morphology as well as road characteristics are also very important to consider. Areas with high vehicle densities have high concentrations of pollutants, particularly at peak periods [28]. Geographical features, such as the urban canyons formed by rows of buildings, limit the dispersion of pollutants, increasing their concentration at ground level [29]. Pollutant concentrations are generally higher in the immediate vicinity of traffic lanes, decreasing exponentially with distance from the source [30]. Rapid acceleration and frequent braking increase emissions, and the volume of traffic with frequent stops amplifies emissions compared with free-flowing traffic [31].

3.4 Models for Road Traffic Emissions

In general, the purpose of modeling is to develop mathematical tools that integrate knowledge with phenomena, analyze test hypotheses and experimental data, uncover relationships between parameters, predict system evolution, optimize management processes, and guide experimental design strategies [32, 33]. It is crucial to develop mathematical models that link pollutant emissions to concentration levels to predict future trends in air quality or assess the effectiveness of mitigation strategies [3].

Dispersion models have demonstrated their effectiveness and ability to assess the concentrations of pollutants generated by road traffic activities. The United States Environmental Protection Agency offers a range of freely available atmospheric dispersion models, tailored to specific objectives, such as AERMOD, CALPUFF, CALINE and CAL3QHCR [23, 34].

Models developed to simulate gaseous emissions generated by road traffic can be used to assess compliance with air quality guidelines, to manage existing emissions, to design networks for ambient air monitoring, to identify the main contributors to existing or prevented air pollution problems, evaluate mitigation policies, forecast pollution episodes, estimate the influence of geophysical and meteorological factors on dispersion as well as terrain elevation, presence of water bodies and land use, and save time and money compared with monitoring [1, 21, 25].

4 Conclusion

This paper represents a bibliometric study conducted on air pollution and pollutants generated by road transportation, in which a database of articles and studies was compiled and processed. This study showed the relative importance of this issue, as reflected in the evolution of the number of publications dealing with this issue, particularly in recent years. These findings illustrate the correlation between advanced techniques for the analysis and resolution of these problems and the need to meet the requirements of conventions and legislation for the mitigation of those particular emissions.

Moreover, this study has shown the fundamental role of dispersion simulation models in the environmental assessment of these challenges, providing a link between the technical parameters of road traffic and environmental and meteorological parameters. Such models help to understand, analyze and prevent the impacts generated by these emissions.

References

1. Colls, J.: Air Pollution, 2nd edn. Spon Press, Taylor & Francis Group (2002)
2. Luo, Z., Wan, G., Wang, C., Zhang, X.: Urban pollution and road infrastructure: a case study of China. China Econ. Rev. **49**, 171–183 (2018). https://doi.org/10.1016/j.chieco.2017.04.008
3. Berkowicz, R., Hertel, O., Larsen, S.E., Sørensen, N.N., Nielsen, M.: Modelling traffic pollution in streets. Ministry of Environment and Energy National Environmental Research Institute (1997)
4. Inchaouh, M., Tahiri, M., Johra, B.E.: Transport urbain et pollution atmosphérique dans la région du Grand Casablanca : quote-part des grands taxis. Presented at the Proceedings Rencontres Scientifiques sur l'Environnement (2016)
5. Bouh, H.A., Benyaïch, F., Noack, Y., Bounakhla, M.: Impact of the road traffic and the climatic parameters on the variation of the sulphur dioxide (SO2) contents in the town of Meknes – MOROCCO. Phys. Chem. News **60**, 75–82 (2011)
6. Ait Ihia, O., Khomsi, D.: Multi-objective model for the selection of the optimal solid waste transfer station locations and for the identification of the optimal transport scenario. In: 2022 2nd International Conference on Innovative Research in Applied Science, Engineering and Technology (IRASET), Meknes, Morocco, pp. 1–6. IEEE (2022). https://doi.org/10.1109/IRASET52964.2022.9737904
7. Ait Ihia, O., Khomsi, D.: Modeling tool for the design of municipal solid waste transportation systems. In: Kacprzyk, J., Ezziyyani, M., and Balas, V.E. (eds.) International Conference on Advanced Intelligent Systems for Sustainable Development, pp. 1–13. Springer, Cham (2023). https://doi.org/10.1007/978-3-031-35245-4_1
8. Ait Ihia, O., Khomsi, D., Hassani, N.S.A., Edderkaoui, R.: GIS and multi-objective optimization model for the design of municipal solid waste management systems: case study of the province of Tiznit, Morocco. IRECE **14**, 82–93 (2023). https://doi.org/10.15866/irece.v14i2.21767
9. Ait Ihia, O., Khomsi, D., Hassani, N.S.A.: Combination of GIS and fuzzy AHP with the VIKOR method for the optimal strategies for municipal solid waste management: case study of the province of Tiznit. In: 2023 3rd International Conference on Innovative Research in Applied Science, Engineering and Technology (IRASET), Mohammedia, Morocco, pp. 1–7. IEEE (2023). https://doi.org/10.1109/IRASET57153.2023.10152985

10. Ait Ihia, O., Khomsi, D., Hassani, N.S.A.: Multi-trip optimization model for municipal solid waste collection systems under capacity and operational time constraints. In: El Fadil, H. and Zhang, W. (eds.) Automatic Control and Emerging Technologies, pp. 568–580. Springer, Singapore (2024). https://doi.org/10.1007/978-981-97-0126-1_50
11. Ait Ihia, O., Khomsi, D., Semlali Aouragh Hassani, N.: Algorithm for the optimization of MSW collection systems and for the optimal siting of depots under multi-trip operations with vehicles of different capacities: case study of the municipality of Tiznit, Morocco. IRECE **15**, 87–99 (2024). https://doi.org/10.15866/irece.v15i1.23493
12. Wu, Y., et al.: On-road vehicle emission control in Beijing: past, present, and future. Environ. Sci. Technol. **45**, 147–153 (2011). https://doi.org/10.1021/es1014289
13. Pérez-Salazar, M.D.R., Mateo-Díaz, N.F., García-Rodríguez, R., Mar-Orozco, C.E., Cruz-Rivero, L.: A genetic algorithm to solve a three-echelon capacitated location problem for a distribution center within a solid waste management system in the northern region of Veracruz, Mexico. DYNA **82**, 51–57 (2015). https://doi.org/10.15446/dyna.v82n191.51146
14. Amato, F., et al.: Improving the modeling of road dust levels for Barcelona at urban scale and street level. Atmos. Environ. **125**, 231–242 (2016). https://doi.org/10.1016/j.atmosenv.2015.10.078
15. Amato, F., Bedogni, M., Padoan, E., Querol, X., Ealo, M., Rivas, I.: Characterization of road dust emissions in Milan: impact of vehicle fleet speed. Aerosol Air Qual. Res. **17**, 2438–2449 (2017). https://doi.org/10.4209/aaqr.2017.01.0017
16. Wu, Y., et al.: On-road vehicle emissions and their control in China: a review and outlook. Sci. Total. Environ. **574**, 332–349 (2017). https://doi.org/10.1016/j.scitotenv.2016.09.040
17. Padoan, E., Ajmone-Marsan, F., Querol, X., Amato, F.: An empirical model to predict road dust emissions based on pavement and traffic characteristics. Environ. Pollut. **237**, 713–720 (2018). https://doi.org/10.1016/j.envpol.2017.10.115
18. Vu, T.V., Zauli-Sajani, S., Poluzzi, V., Harrison, R.M.: Factors controlling the lung dose of road traffic-generated sub-micrometre aerosols from outdoor to indoor environments. Air Qual. Atmos. Health **11**, 615–625 (2018). https://doi.org/10.1007/s11869-018-0568-2
19. Wen, Y., et al.: On-highway vehicle emission factors, and spatial patterns, based on mobile monitoring and absolute principal component score. Sci. Total. Environ. **676**, 242–251 (2019). https://doi.org/10.1016/j.scitotenv.2019.04.185
20. Krecl, P., et al.: Long-term trends in nitrogen oxides concentrations and on-road vehicle emission factors in Copenhagen, London and Stockholm. Environ. Pollution. **290**, 118105 (2021). https://doi.org/10.1016/j.envpol.2021.118105
21. Jacobson, M.Z.: Atmospheric Pollution: History, Science, and Regulation. Cambridge University Press, Cambridge, New York, Melbourne, Madrid, Cape Town, Singapore, São Paulo (2002)
22. Matzoros, A., Van Vliet, D.: A model of air pollution from road traffic, based on the characteristics of interrupted flow and junction control: Part I — model description. Transp. Res. Part A Policy Pract. **26**, 315–330 (1992). https://doi.org/10.1016/0965-8564(92)90019-4
23. Forehead, H., Huynh, N.: Review of modelling air pollution from traffic at street-level - the state of the science. Environ. Pollut. **241**, 775–786 (2018). https://doi.org/10.1016/j.envpol.2018.06.019
24. Bai, X., Chen, H., Oliver, B.G.: The health effects of traffic-related air pollution: a review focused the health effects of going green. Chemosphere **289** (2022). https://doi.org/10.1016/j.chemosphere.2021.133082
25. Wang, L.K., Pereira, N.C., Hung, Y.-T. (eds.): Air Pollution Control Engineering. Humana Press, Totowa (2004)
26. Heywood, J.B.: Internal Combustion Engine Fundamentals. McGraw-Hill, United States of America (1988)

27. Ghetti, F., Checcucci, G., Bornman, J.F. (eds.) Environmental UV Radiation: Impact on Ecosystems and Human Health and Predictive Models: Proceedings of the NATO Advanced Study Institute on Environmental UV Radiation: Impact on Ecosystems and Human Health and Predictive Models Pisa, Italy June 2001. Springer Netherlands, Dordrecht (2006). https://doi.org/10.1007/1-4020-3697-3
28. Zhang, K., Batterman, S.: Air pollution and health risks due to vehicle traffic. Sci. Total. Environ. **450–451**, 307–316 (2013). https://doi.org/10.1016/j.scitotenv.2013.01.074
29. Vardoulakis, S., Fisher, B.E.A., Pericleous, K., Gonzalez-Flesca, N.: Modelling air quality in street canyons: a review. Atmos. Environ. **37**, 155–182 (2003). https://doi.org/10.1016/S1352-2310(02)00857-9
30. Karner, A.A., Eisinger, D.S., Niemeier, D.A.: Near-roadway air quality: synthesizing the findings from real-world data. Environ. Sci. Technol. **44**, 5334–5344 (2010). https://doi.org/10.1021/es100008x
31. Ropkins, K., Tate, J.E.: Early observations on the impact of the COVID-19 lockdown on air quality trends across the UK. Sci. Total. Environ. **754**, 142374 (2021). https://doi.org/10.1016/j.scitotenv.2020.142374
32. Sheldon, M.R.: Simulation. Elsevier Academic Press, USA (2006)
33. Oberkampf, W.L., Roy, C.J.: Verification and Validation in Scientific Computing. Cambridge University Press, USA (2010)
34. EPA: Approved Methods for the Modelling and Assessment of Air Pollutants in New South Wales. NSW Environment Protection Authority (2022)

Pontryagin's Principle in Optimal Control for a Stochastic Epidemic System

M. M. Abdeslami[1](✉), Driss Bouggar[2], Mohamed El Fatini[1], and Raya Nouira[3]

[1] Stochastic Modelling and Statistics Group, LAGA, Analyse, Géométrie et Applications Laboratory, Algebra and Spectral Geometry, Faculty of Sciences, Ibn Tofail University, BP 133, Kénitra, Morocco
abdeslami1993@gmail.com

[2] Mathematics and Computer Science Department, Faculty of Sciences Ben M'sik, Hassan II University of Casablanca, Casablanca, Morocco

[3] Center for Mathematics and Applications (NOVA Math), NOVA SST, Nova University of Lisbon, Lisbon, Portugal

Abstract. This study examines a relapse stochastic epidemic system with a non-linear random transmission to enhance the model's realism. The uniqueness of a strong positive solution is proved, and its asymptotic behavior is studied using Lyapunov methods, deriving the necessary conditions for the average extinction and persistence of the disease. An optimal control strategy is then formulated for both deterministic and stochastic cases. For a quadratic cost function, we apply the Pontryagin maximum principle in its strong form, which is more comprehensive in the general setting. Even when processes rely on accumulated information about the state's evolution, we establish the necessary conditions and characterize the optimal strategy using the Hamiltonian function and adjoint equations. Furthermore, we analyze the implications of these results in the context of epidemic control. Finally, numerical simulations validate the theoretical results and enhance understanding of the model's behavior under various conditions.

Keywords: Stochastic Optimal Control · Pontryagin approach · Hamiltonian function · Hamilton-Jacobi-Bellman (HJB) theory · Stochastic epidemic model

1 Setting the Problem and Mathematical Background

Herpes and TB are examples of diseases with recurrent dynamics [1]. Tudor [18] proposed a relapse-prone epidemic model with a constant population and a bilinear incidence rate. Capasso and Serio [4], based on real-world observations, showed that disease spread slows as infections rise, introducing a saturation rate function. Michaelis and Menten [12] later incorporated the expansion of susceptibles to modulate transmission. Crowley and Martin [5] unified these perspectives by proposing a functional response encompassing both approaches.

Several studies, including [8], emphasize the role of α_1 in characterizing disease-free and endemic equilibria. This leads to the formulation of a nonlinear deterministic SIRI model [7]:

C. El Mokhi et al. (Eds.): ICASET 2025, SUCI, pp. 337–347, 2025.
https://doi.org/10.1007/978-3-031-98334-4_31

$$\frac{d}{dt}\begin{bmatrix} S \\ I \\ R \end{bmatrix} = \begin{bmatrix} -\mu & 0 & 0 \\ 0 & -(\lambda+\mu) & \gamma \\ 0 & \lambda & -(\mu+\gamma) \end{bmatrix}\begin{bmatrix} S \\ I \\ R \end{bmatrix} + \begin{bmatrix} \mu - \beta_1 g(S,I) \\ \beta_1 g(S,I) \\ 0 \end{bmatrix}$$

such that

$$g(S,I) = \frac{SI}{1+\alpha_1 S + \alpha_2 I + \alpha_1\alpha_2 SI}. \tag{1}$$

All parameter meanings are described in Table 1. In a completely susceptible population, the "basic reproduction number", $\mathcal{R}_0$, signifies the expected number of secondary infections generated by a single infectious individual. As established in [19], $\mathcal{R}_0$ can be determined using the following expression:

$$\frac{d}{dt}\begin{bmatrix} I \\ R \\ S \end{bmatrix} = \mathcal{F} - \mathcal{V} \quad \text{where} \quad \mathcal{F} = \begin{bmatrix} \beta_1 g(S,I) \\ 0 \\ 0 \end{bmatrix} \quad \text{and} \quad \mathcal{V} = \begin{bmatrix} (\mu+\lambda)I - \gamma R \\ -\lambda I + (\mu+\gamma)R \\ \beta_1 g(S,I) + \mu S - \mu \end{bmatrix}$$

Let $E_0 = (S_0, 0, 0)$, where $S_0 = \frac{\mu}{\mu} = 1$. $\mathcal{DF}(E_0)$ and $\mathcal{DV}(E_0)$ are the definitions of the Jacobian matrices of $\mathcal{F}$, (resp. $\mathcal{V}$) at E_0:

$$\mathcal{DF}(E_0) = \begin{bmatrix} \frac{\beta_1 \partial g(S_0,0)}{\partial I} & 0 & \frac{\beta_1 \partial g(S_0,0)}{\partial S} \\ 0_{2,1} & 0_{2,1} & 0_{2,1} \end{bmatrix}, \quad \mathcal{DV}(E_0) = \begin{bmatrix} \mu+\lambda & -\gamma & 0 \\ -\lambda & \mu+\gamma & 0 \\ \frac{\beta_1 \partial g(S_0,0)}{\partial I} & 0 & \frac{\beta_1 \partial g(S_0,0)}{\partial S} + \mu \end{bmatrix}$$

Hence, the following provides the infection matrix $\mathcal{F}$ and the transition matrix $\mathcal{V}$:

$$\mathcal{F} = \begin{bmatrix} \frac{\beta_1 \partial g(S_0,0)}{\partial I} & 0 \\ 0 & 0 \end{bmatrix}, \quad \mathcal{V} = \begin{bmatrix} \lambda+\mu & -\gamma \\ -\lambda & \mu+\gamma \end{bmatrix}$$

The greatest value of $\mathcal{FV}^{-1}$ yields $\mathcal{R}_0$. Due to

$$\mathcal{FV}^{-1} = \frac{1}{(\mu+\lambda)(\mu+\gamma) - \lambda\gamma}\,\frac{\beta_1 \partial g(S_0,0)}{\partial I}\begin{bmatrix} \mu+\gamma & \gamma \\ 0 & 0 \end{bmatrix}$$

we deduce that:

$$\mathcal{R}_0 = \frac{1}{\left(\mu+\lambda-\frac{\lambda\gamma}{\mu+\gamma}\right)}\,\frac{\beta_1 \partial g(S_0,0)}{\partial I} = \frac{\beta_1}{(1+\alpha_1)\left(\mu+\lambda-\frac{\lambda\gamma}{\mu+\gamma}\right)} \tag{2}$$

The deterministic model (1) was analyzed by Georgescu et al. [7], who studied its solutions based on the threshold $\mathcal{R}_0$. The equilibrium conditions are given as follows:

Condition	**Equilibrium**
$\mathcal{R}_0 < 1$	Disease-free equilibrium $E_0(1,0,0)$
$\mathcal{R}_0 > 1$	Endemic equilibrium $E^\star(S^\star, I^\star, R^\star)$

$$\begin{cases} \mu = \dfrac{\beta_1 S^\star I^\star}{1+\alpha_1 S^\star+\alpha_2 I^\star+\alpha_1\alpha_2 S^\star I^\star}+\mu S^\star, \\ \dfrac{\beta_1 S^\star I^\star}{1+\alpha_1 S^\star+\alpha_2 I^\star+\alpha_1\alpha_2 S^\star I^\star} = (\lambda+\mu)I^\star+\gamma R^\star, \\ \lambda I^\star = (\mu+\gamma)R^\star. \end{cases} \tag{3}$$

To analyze the influence of environmental noise on SIRI epidemic models, we introduce a nonlinear stochastic epidemic model. In this framework, the transmission rate varies randomly and is expressed as $\bar{\beta}_1 dt = \beta_1 dt + \sigma dW(t)$, where σ represents the noise intensity (or standard deviation), and $W(t)$ is a Brownian motion defined on a complete filtered probability space $(\Omega, \mathcal{F}, (\mathcal{F}_t)_{t\geq 0}, P)$ satisfying the usual conditions. The dynamics of the corresponding stochastic model are governed by the following system, where we define $X = (X_1, X_2, X_3) = (S, I, R)$.

$$dX = \begin{bmatrix} dS \\ dI \\ dR \end{bmatrix} = \left[\begin{bmatrix} \mu-\mu S \\ -(\lambda+\mu)I+\gamma R \\ \lambda I-(\mu+\gamma)R \end{bmatrix} + g(S,\text{ I})\begin{bmatrix} -\beta_1 \\ \beta_1 \\ 0 \end{bmatrix}\right]\text{dt} + \sigma g(S,\text{ I})\begin{bmatrix} -1 \\ 1 \\ 0 \end{bmatrix} dW(t). \tag{4}$$

Next, we investigate the positivity and boundedness of solutions to (4). We then evaluate the probability of disease extinction and identify the conditions under which the epidemic persists on average. Following this, we examine the stochastic optimal control problem. To support the theoretical analysis, we present numerical simulations. Finally, we conclude with a discussion that highlights the key findings and suggests potential directions for future research. To establish that the solution remains global, and

$$\Gamma = \left\{ X \in \mathbb{R}^{+3} : \sum_{1}^{3} X_i = 1 \right\} \tag{5}$$

Theorem 1. *Let $X(0) \in \Gamma$. Then, for all $t \geq 0$, the system (4) has a unique global solution $X(t) \in \Gamma$ with probability one.*

Proof. Using Lyapunov analysis [14]

Here, we investigate the conditions under which the disease vanishes by establishing sufficient constraints on the intensity of white noise. These constraints allow us to demonstrate the global asymptotic stability of solutions at E_0. For a rigorous definition of stability, we refer to the theoretical framework outlined in [14]. The parameter $\mathcal{R}_S$ is defined as follows:

$$\mathcal{R}_S = \mathcal{R}_0 - \frac{\sigma^2}{2(1+\alpha_1)^2} \cdot \left(\frac{1}{\lambda+\mu-\frac{\lambda\gamma}{\mu+\gamma}} \right)$$

This quantity accounts for the influence of stochastic volatility, providing a generalized version of the classical basic reproduction number $\mathcal{R}_0$.

Theorem 2. *For $X(0) \in \Gamma$, $X(t)$ converges a.s. to $(1,0,0)$ if $\mathcal{R}_S < 1$ and $\sigma^2 \leq \beta_1(1+\alpha_1)$.*

Proof. Let $b > 1$ such that $-c_b = \frac{\beta_1}{1+\alpha_1} - \left(\mu + \lambda + \frac{\sigma^2}{2(1+\alpha_1)^2}\right) + \frac{b\lambda\gamma}{\mu+\gamma} < 0$, this holds when $\mathcal{R}_S < 1$. Define $\tau_b = \inf\{t \geq 0 : I(t) \leq \frac{1}{b}\}$, We state that $\mathbb{E}(\tau_b) < \infty$. Suppose $I(0) \in (\frac{1}{b}, 1)$; otherwise, $\tau_b = 0$. For any $t \leq T \wedge \tau_b$, we have $I(t) \geq \frac{1}{b}$. Next, we apply Itô's formula [14,15] to $\log I$.

$$d \log I \leq \left(-\frac{\sigma^2}{2(1+\alpha_2)^2} + \frac{\beta_1}{1+\alpha_1} - (\mu+\lambda) + b\gamma R \right) dt + \frac{\sigma g(S,I)}{I} dW(t).$$

By the monotonicity of $V(x) = -\frac{1}{2}\sigma^2 x^2 + \beta_1 x - (\mu+\lambda)$, we get

$$d(\log I) + d\left(\frac{b\gamma}{\mu+\gamma} R\right) \leq -c_b\, dt + \frac{\sigma g(S,I)}{I} dW(t).$$

Taking expectations,

$$\mathbb{E}\left(\log I(T \wedge \tau_b)\right) + \mathbb{E}\left(\frac{b\gamma}{\mu+\gamma} R(T \wedge \tau_b)\right) \leq \log I(0) + \frac{b\gamma}{\mu+\gamma} - c_b \mathbb{E}(T \wedge \tau_b).$$

Using Fatou's lemma, $\mathbb{E}(\tau_b) \leq \frac{b\gamma}{(\mu+\gamma)c_b} + \frac{\log b}{c_b}$, where $\mathbb{E}(\tau_b) < \infty$.

We are now going to talk about mean persistence.

Theorem 3. *Let $X(0) \in \Gamma$, $\mathcal{R}_S > 1$, $X(t)$ satisfies:*

1.
$$\liminf_{t\to\infty} \frac{1}{t} \int_0^t X_1(t)\, dt \geq \frac{(1+\alpha_1)\mu}{(1+\alpha_1)\mu + \beta_1},$$

2.
$$\liminf_{t\to\infty} \frac{1}{t} \int_0^t X_2(t)\, dt \geq \frac{(1+\alpha_1)\mu\beta_1}{(\lambda+\mu)(1+\alpha_1)(\beta_1 + (1+\alpha_1)\mu)\mathcal{R}_0}(\mathcal{R}_S - 1),$$

3.
$$\liminf_{t\to\infty} \frac{1}{t} \int_0^t X_3(t)\, dt \geq \frac{\lambda(1+\alpha_1)\mu\beta_1}{(\gamma+\mu)(\lambda+\mu)(\beta_1 + (1+\alpha_1)\mu)\mathcal{R}_0}(\mathcal{R}_S - 1).$$

Then, the solutions are thus persistent in the mean.

Remark 1. The transmission term in (4) provides a generalized form of the bilinear incidence rate introduced in model (4) of [11], specifically when $\delta = \epsilon = 0$. Moreover, by choosing $\alpha_1 = \alpha_2 = 0$, we retrieve the result stated in [11, Theorem 2.1].

2 Pontryagin Maximum Principle

Optimal control theory is a powerful tool for optimizing decision-making. Numerous studies [6,22] have applied it to treatment and vaccination strategies in epidemiological modeling. Witbooi et al. [20] extended this approach to both deterministic and stochastic SVIR models, aiming to balance infection reduction and vaccination costs. Related contributions include [13,21]. We now formulate an optimal control problem to minimize the objective functional $\mathcal{C}(u)$ by incorporating control variables into the stochastic system.

$$\begin{aligned} dX_1 &= b_1 X(t))dt, \\ dX_2 &= b_2(X(t), u(t))dt, \\ dX_3 &= b_3(X(t), u(t)), \end{aligned}$$

with

$$\begin{aligned} b_1 X(t)) &= (\mu - \beta_1 g(X_1, X_2) - \mu X_1)\, dt, \\ b_2(X(t), u(t)) &= (\beta_1 g(X_1, X_2) - (u(t) + \lambda + \mu)X_2 + \gamma X_3)\, dt, \\ b_3(X(t), u(t)) &= ((u(t) + \lambda)X_2 - (\mu + \gamma)X_3)\, dt, \end{aligned} \tag{6}$$

where for all t, $u(t) \in U = [0, u_{1_{max}}]$ refers to a therapeutic intervention, with $0 < u_1^{max} \leq 1$.

2.1 Deterministic Optimal Control Analysis

We consider an optimal control problem where the objective functional is given by

$$\mathcal{C}(u) = \mathbb{E}\left[\int_0^T \left(C_1 I(s) + \frac{C_2}{2} u^2(s)\right) ds\right], \tag{7}$$

with $u(t) \in \mathcal{U}$, $C_1, C_2 > 0$, and $X(0) > 0$. The term $\frac{C_2}{2}u^2(t)$ represents the cost of applying the control $u(t)$, while $C_1 I(t)$ accounts for the cost associated with infections. The objective is to determine an optimal control $u^*(t)$ that minimizes (7) over the set of admissible controls is

$$\mathcal{U} = \{u(t) \text{ is adapted} : 0 \leq u(t) \leq u_{\max}, \quad \forall t \in [0, T]\}. \tag{8}$$

This defines the optimization problem **(D)**.

We derive the necessary conditions for optimality using Pontryagin's maximum principle [2,3,16] To find the minimal value, we define the Hamiltonian function as

$$H(t, X(t), u(t)) = \sum_{i=1}^{3} \lambda_i \frac{dX_i(t)}{dt} - F(t, X(t), u(t)), \tag{9}$$

where λ_i are the adjoint variables and $F(t, X(t), u(t)) = C_1 I(t) + \frac{C_2}{2} u^2(t)$ is the running cost.

Theorem 4. *There exists a control $u^*(t)$ such that*

$$\mathcal{C}(u^*) = \inf_{u \in \mathcal{U}} \mathcal{C}(u),$$

Proof. Since both the control u(t) and the state variables remain positive, the objective functional $\mathcal{C}(u)$ is convex and Lipschitz in state over the closed and convex set $\mathcal{U}$. Moreover, the boundedness of the system ensures the existence of an optimal control.

Theorem 5. *Let $(X^*(t), u^*(t))$ be an optimal pair for our optimal control problem and (6). Then, the SMP approach guarantees the existence of a non-trivial vector function $\lambda = (\lambda_1, \lambda_2, \lambda_3)$ satisfying the adjoint system*

$$d\lambda_i = -\frac{\partial H}{\partial X_i} dt,$$

along with the transversality conditions $\lambda_i(T) = 0$ for $i = 1, 2, 3$. Furthermore

$$u^*(t) = \frac{(\lambda_3 - \lambda_2)I(t)}{C_2}. \tag{10}$$

Proof. The adjoint equations $\lambda_i(t)$, for $i = 1, 2, 3$, derive from the Hamiltonian (9) as:

$$\dot{\lambda}_1 = \lambda_1 \left(\frac{\beta_1 X_2(t)(1 + \alpha_2 X_2(t))}{(1 + \alpha_1 X_1(t) + \alpha_2 X_2(t) + \alpha_1 \alpha_2 X_1(t) X_2(t))^2} + \mu \right) \tag{11}$$
$$- \lambda_2 \frac{\beta_1 X_2(t)(1 + \alpha_2 X_2(t))}{(1 + \alpha_1 X_1(t) + \alpha_2 X_2(t) + \alpha_1 \alpha_2 X_1(t) X_2(t))^2},$$

$$\dot{\lambda}_2 = -C_1 + \lambda_1 \frac{\beta_1 X_1(t)(1 + \alpha_1 X_1(t))}{(1 + \alpha_1 X_1(t) + \alpha_2 X_2(t) + \alpha_1 \alpha_2 X_1(t) X_2(t))^2} \tag{12}$$
$$- \lambda_2 \frac{\beta_1 X_1(t)(1 + \alpha_1 X_1(t))}{(1 + \alpha_1 X_1(t) + \alpha_2 X_2(t) + \alpha_1 \alpha_2 X_1(t) X_2(t))^2} + \lambda_2(\mu + \lambda + u(t)) - \lambda_3 \gamma,$$

$$\dot{\lambda}_3 = -\lambda_2 \gamma + \lambda_3(\mu + \gamma),$$

With the transversality conditions $\lambda_i(T) = 0$, applying the using PMP yields:

$$\frac{\partial H}{\partial u} = -C_2 u(t) - (\lambda_2 - \lambda_3) I(t) = 0.$$

This leads to (10). Considering the control variable properties, the optimal control $u^*(t)$ is

$$u^*(t) = \begin{cases} 0, & \text{if } (\lambda_2 - \lambda_3) I^* < 0, \\ \frac{(\lambda_3 - \lambda_2) I^*}{C_2}, & \text{if } 0 \leq (\lambda_2 - \lambda_3) I^* \leq C_2, \\ u_1^m, & \text{if } (\lambda_2 - \lambda_3) I^* > C_2. \end{cases}$$

2.2 Investigation of the Stochastic Optimal Control

Considering the problem of stochastic control [2,3].

$$\begin{aligned} dX_1 &= (\mu - \beta_1 g(X_1, X_2) - \mu X_1)\, dt - \sigma g(X_1, X_2) dW(t), \\ dX_2 &= (\beta_1 g(X_1, X_2) - (u(t) + \lambda + \mu) X_2 + \gamma X_3)\, dt + \sigma g(X_1, X_2) dW(t), \\ dX_3 &= ((u(t) + \lambda) X_2 - (\mu + \gamma) X_3)\, dt, \end{aligned} \tag{13}$$

We assume the existence of $u_1^m \leq 1$ such that $u(t) \leq u_1^m$ almost surely. A control $u : [0, T] \times \Omega \to [0, u_1^m]$ is feasible if it is $(\mathcal{F}_t)$-adapted. The set of all such controls is denoted by $\mathcal{U}$. The pair (X, u) is admissible if state constraints hold and $\mathbb{E}\left[\int_0^T F(s, X(s), u(s))\, ds\right] < \infty$, We consider the following cost functional associated with a control process $u \in \mathcal{U}$:

$$\mathcal{C}_t(x, u) = \mathbb{E}_{t,x}\left[\int_t^T \left(C_1 I(s) + \frac{C_2}{2} u^2(s)\right) ds\right], \tag{14}$$

where the expectation is taken conditionally on the system being in state x at time t. The objective is to determine an optimal control $u^* \in \mathcal{U}$ that minimizes the cost functional. This leads to the value function $\Phi(t, x)$, defined as

$$\Phi(t, x) = \inf_{u(\cdot) \in \mathcal{U}} \mathcal{C}_t(x, u) = \mathcal{C}_t(x, u^*), \tag{15}$$

where $\Phi(t, x)$ represents the optimal expected cost given the initial condition (t, x). Thus, the optimal control minimizes the expected cost while respecting system dynamics. Existence results were studied by Kushner [10].

Theorem 6. *There exists an optimal stochastic control* u^**;*

$$\mathcal{C}_t(x, u^*) = \inf_{u \in \mathcal{U}} \mathcal{C}_t(x, u), \tag{16}$$

where the control system (13) *evolves according to the given initial conditions.*

Proof. Since $F(t, X, u)$ is Lipschitz continuous and has linear growth in X, the results of Yong and Zhou [17] guarantee the existence of a strong optimal control within a convex admissible set.

Theorem 7. *The optimal control* $u^*(t)$ *for the treatment problem in the stochastic system described by* (13) *is expressed as*

$$u^*(t) = \min\left(\max\left(\frac{X_2(t)(D_{x_3}\Phi - D_{x_2}\Phi)}{C_2}, 0\right), u_1^m\right). \tag{17}$$

Proof. By applying Itô's lemma, we derive the following expression:

$$\mathcal{L}\Phi = \dot{X}_1 D_{x_1}\Phi + \dot{X}_2 D_{x_2}\Phi + \dot{X}_3 D_{x_3}\Phi + \frac{(\sigma g(X_1, X_2))^2}{2}\left(D_{x_1}^2\Phi + D_{x_2}^2\Phi\right).$$

Based on the Hamilton-Jacobi-Bellman (HJB) theory [15], the optimal control problem is formulated accordingly,

$$\inf_{u(\cdot)\in\mathcal{U}}\left(C_1X_2+\frac{C_2}{2}u^2+\mathcal{L}\Phi\right).$$

Differentiating with respect to u yields

$$-C_2u-(D_{x_2}\Phi-D_{x_3}\Phi)X_2=0.$$

Taking into account the control constraints and following a similar approach as in the deterministic setting, we recover the expression given in (17).

Table 1. Description of model parameters

Parameter	Description
μ	Influx of susceptible individuals due to birth and immigration, assumed to balance the natural mortality rate
β_1	Transmission intensity of the disease
λ	Recovery rate
γ	Rate at which recovered individuals relapse into the infectious class

3 Numerical Results and Illustration

To facilitate the interpretation of the model parameters, Table 1 provides a concise description of their roles within the system.

To validate the theoretical findings, we implement the Euler-Maruyama scheme to numerically approximate the solutions using the specified parameters. This approach enables a quantitative assessment of the system's dynamical behavior and the impact of key parameters on disease evolution.

Case 1: $\mathcal{R}_S < 1$ and $\sigma^2 \leq \beta_1(1+\alpha_1)$ We set $S(0) = 0.2$, $I(0) = 0.7$, $R(0) = 0.1$ and parameters $\mu = 0.04$, $\beta_1 = 0.1$, $\lambda = 0.4$, $\gamma = 0.3$, $\sigma = 0.3$, $\alpha_1 = 0.1$, $\alpha_2 = 0.5$. The computed values give $\mathcal{R}_0 = 1.0442 > 1$, $\mathcal{R}_S = 0.6170 < 1$, and $\sigma^2 < \beta_1$ [11]. Thus, Theorem 2 ensures disease eradication (Fig. 1). In model (1), $\mathcal{R}_0 > 1$ implies disease persistence. However, under stochastic perturbations satisfying $\sigma^2 \leq \beta_1(1+\alpha_1)$, then $\mathcal{R}_S < 1$, leading to eradication (Fig. 1).

Case 2: $\mathcal{R}_S > 1$ We consider the initial values $S(0) = 0.7$, $I(0) = 0.1$, and $R(0) = 0.2$, with the parameter $\beta_1 = 0.2$ [11]. The computed values yield $\mathcal{R}_0 = 2.0885 > 1$ and $\mathcal{R}_S = 1.6613 > 1$. Consequently, the condition in Theorem 3 holds, confirming that the disease persists, as illustrated in Fig. 2.

Case 3: Optimal Control. As established in Theorem 3, we derive threshold values for disease extinction and persistence in both (6) and (13):

$$\mathcal{R}_0(u) = \frac{\beta_1}{(1+\alpha_1)\left(\mu+\lambda+u-\frac{\lambda\gamma}{\mu+\gamma}\right)}, \quad \mathcal{R}_S(u) = \mathcal{R}_0(u) - \frac{\sigma^2}{2(1+\alpha_1)^2\left(\lambda+u+\mu-\frac{\lambda\gamma}{\mu+\gamma}\right)}.$$

For this case, we set $S(0) = 0.7$, $I(0) = 0.2$, $R(0) = 0.1$ and parameters $\mu = 0.04$, $\beta_1 = 0.2$ (without control), $\beta_1 = 0.4$ (with control), $\sigma = 0.2$, $\lambda = 0.1$, $\gamma = 0.2$, $\alpha_1 = 0.1$, $\alpha_2 = 0.5$, $A_1 = 5$, and $A_2 = 50$. The computed reproduction numbers, $\mathcal{R}_S^1 = 2.9169$ and $\mathcal{R}_S^2 = 4.5490$, confirm disease persistence, in agreement with Theorem 3.

To mitigate the epidemic, a control mechanism u is introduced in both models. The results show that increasing treatment intensity reduces the number of infected individuals, as depicted in the corresponding figure (Fig. 3).

Case 4: $\mathcal{R}_S$ vs $u(t)$ In this example, let $\beta_1 = 0.2$, $\mu = 0.04$, $\lambda = 0.1$, $\gamma = 0.2$, $\alpha = 0.1$, and $\sigma = 0.2$, while varying u. The curve of $\mathcal{R}_S(u)$ is shown in Fig. 4, where it is evident that $\mathcal{R}_S(u)$ decreases as u increases. The minimum treatment proportion is numerically estimated to be around 0.111, as seen in Fig. 4.

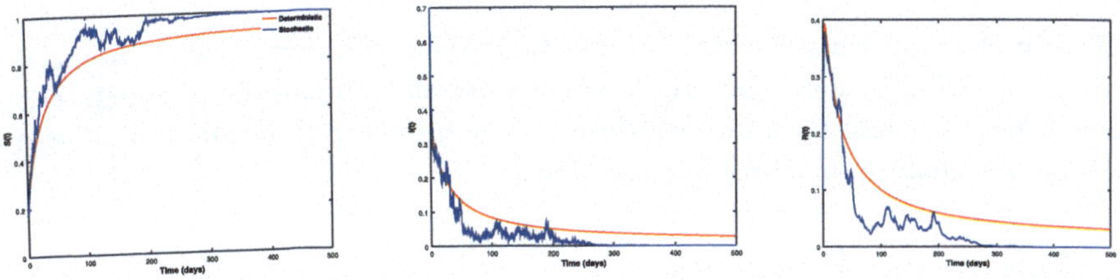

Fig. 1. Trajectories of stochastic and deterministic systems for Case 1.

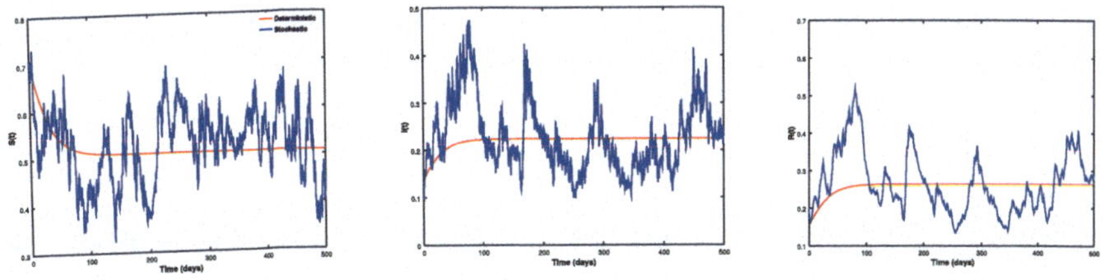

Fig. 2. Paths of the stochastic and deterministic systems in Case 2.

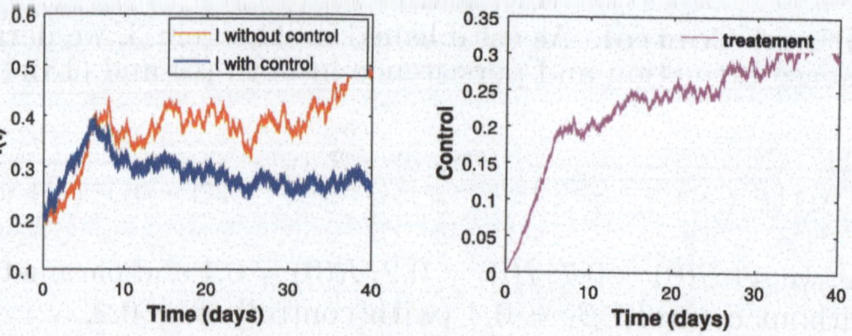

Fig. 3. Dynamics of the stochastic optimal control system with and without treatment in Case 3.

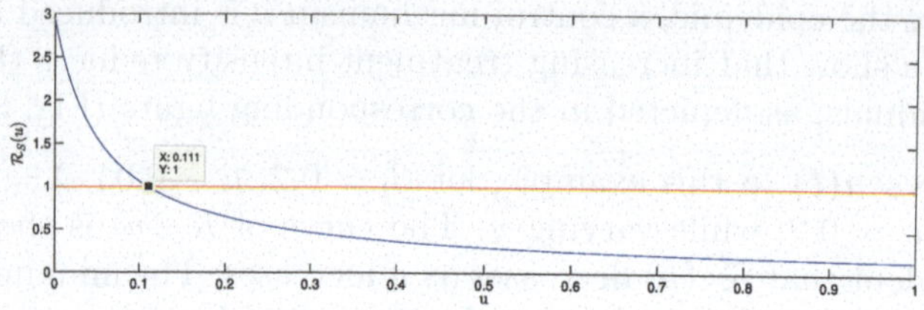

Fig. 4. Evolution of $\mathcal{R}_S(u)$ with respect to the treatment in Case 4.

The simulations illustrate two stochastic trajectories, with and without treatment, in blocking disease spread. Medical intervention can significantly reduce infection peaks, while immune response varies among individuals, influenced by age, ongoing healing, or chronic conditions.

4 Concluding Remarks

This study analyzes an epidemic model with relapse and nonlinear random transmission. The disease is eradicated if $\mathcal{R}_S < 1$ and persists if $\mathcal{R}_S > 1$. An optimal control strategy with a quadratic cost is derived via the stochastic Pontryagin principle, offering a broader framework than traditional methods.

Future research may explore a stochastic delayed SIRI model to assess the latent period's impact and develop a benchmarking table for control strategy evaluation.

References

1. Blower, S.: Modelling the genital herpes epidemic. Herpes **11**(Suppl. 3), 138A (2004)
2. Bouggar, D., El Fatini, M., El Hichamy, I., Nasri, B.R., Sekkak, I.: Near-optimal stochastic control for radiotherapy treatment in a random cancer model. Syst. Control Lett. **170**, 105400 (2022)
3. Bouggar, D., El Fatini, M., Nasri, B., Petersson, R., Sekkak, I.: Stochastic near-optimal controls for treatment and vaccination in a COVID-19 model with transmission incorporating Lévy jumps. Stochastics **96**(1), 887–920 (2024)

4. Capasso, V., Serio, G.: A generalization of the Kermack-McKendrick deterministic epidemic model. Math. Biosci. **42**(1), 43–61 (1978)
5. Crowley, P.H., Martin, E.K.: Functional responses and interference within and between year classes of a dragonfly population. J. North Am. Benth. Soc. **8**, 211–221 (1989)
6. Gaff, H., Schaefer, E.: Optimal control applied to vaccination and treatment strategies for various epidemiological models. Math. Biosci. Eng. **6**, 469–492 (2009)
7. Georgescu, P., Zhang, H.: A Lyapunov functional for a SIRI model with nonlinear incidence of infection and relapse. Appl. Math. Comput. **219**(16), 8496–8507 (2013)
8. Ji, C.: The threshold for a stochastic HIV-1 infection model with Beddington-DeAngelis incidence rate. Appl. Math. Model. **64**, 168–184 (2018)
9. Jacod, J., Shiryaev, A.N.: Limit Theorems for Stochastic Processes. Grundlehren Math. Wiss., vol. 288. Springer, Heidelberg (1987)
10. Kushner, H.J.: Existence results for optimal stochastic controls. J. Optim. Theory Appl. **15**(4) (1975)
11. Lahrouz, A., Settati, A., El Fatini, M., Pettersson, R., Taki, R.: Probability analysis of a perturbed epidemic system with relapse and cure. Int. J. Comput. Methods (2018)
12. Michaelis, L., Menten, M.L.: Die Kinetik der Invertinwirkung. Biochem. Z. **49**, 333–369 (1913)
13. Moreira, H.N., Wang, Y.: Global stability in an SIRI model. SIAM Rev. **39**, 496–502 (1997)
14. Mao, X.: Stochastic Differential Equations and Applications. Woodhead Publishing Limited, Cambridge (2010)
15. Oksendal, B.: Stochastic Differential Equations: An Introduction with Applications, 5th edn. Springer, Berlin (1998)
16. Pontryagin, L.S., Boltyanskii, V.G., Gamkrelidze, R.V., Mishchenko, E.F.: The Mathematical Theory of Optimal Processes. J. Wiley & Sons, New York (1962)
17. Jiongmin, Y., Zhou, X.Y.: Stochastic Controls Hamiltonian Systems and HJB Equations. Springer-Verlag, New York (1999)
18. Tudor, D.: A deterministic model for herpes infections in human and animal populations. SIAM Rev. **32**, 136–139 (1990)
19. Van den Driessche, P., Watmough, J.: Reproduction numbers and sub-threshold endemic equilibria for compartment models of disease transmission. Math. Biosci. **180**, 29–48 (2002)
20. Witbooi, P.J., Muller, G.E., Van Schalkwyk, G.J.: Vaccination control in a stochastic SVIR epidemic model. Comput. Math. Methods Med. 9 p (2015)
21. Xu, R.: Global dynamics of a delayed epidemic model with latency and relapse. Nonlinear Anal. Model. Control **18**(2), 250–263 (2013)
22. Zaman, G., Kang, Y., Jung, I.: Stability analysis and optimal vaccination of an SIR epidemic model. Biosystems **93**, 240–249 (2008)

On Properties of Poly-Quasihomogeneous Toeplitz Operators on the Bergman Space

Raja'a Al-Naimi[1](✉), Reham Qaralleh[2], and Mutti-Ur Rehamn[3]

[1] Department of Mathematics, University of Petra, Amman 11196, Jordan
rajaa.alnaimi@uop.edu.jo

[2] Department of Mathematics, Middle East University, Amman, Jordan
r.qaralleh@meu.edu.jo

[3] Department of Mathematics, Asia International University, Bukhara, Uzbekistan
muttiur.abbasi@oxu.uz

Abstract. We define and study a new class ofpositive Toeplitz operators on the Bergman space of the unit disk that we call poly-quasihomogeneous Toeplitz operators. We call f poly-quasihomogeneous of degree m if it is of the form $f(e^{i\theta})\phi(r)$, where $f(e^{i\theta})$ is a polynomial of $e^{i\theta}$ of degree$m$ and $\phi(r)$ is a radial function. We develop a systematic study of theirbasic properties, derive explicit formulas for their action on monomials, describe their commutants, and prove general structural theorems related to their algebraic and spectral properties. In particular, the theoretical results we presentgeneralize classical results about quasihomogeneous Toeplitz operators and yield fresh insights into the overall algebraic structure of Toeplitz operators on Bergman spaces. If thetitle of the article gets your attention, we agree with that, so we list some, thereby showing unexpected applications for these mathematical structures to sustainability science and smart systems, for example, as optimization in energy grid or environmental monitoring or resource management. These applications emphasize the practical applicabilityof our theoretical framework away from pure mathematics, bridging operator theory with current problems in sustainable technology development.

Keywords: Toeplitz operators · Bergman space · Poly-quasihomogeneous functions · Operator theory

1 Introduction

Let $\mathbb{D}$ be the open unit disk in the complex plane and dA be the normalized Lebesgue area measure on $\mathbb{D}$. Let $L_a^2(\mathbb{D})$ be the closed subspace of $L^2(\mathbb{D})$ consisting of holomorphic functions. Bergman spaces have been extensively studied (see, for example, [6]) with diverse applications which span areas such as complex analysis and operator theory. For a bounded measurable function f on $\mathbb{D}$ the Toeplitz operator T_f with symbol f is defined on $L_a^2(\mathbb{D})$ by

$$T_f(g) = P(fg)$$

C. El Mokhi et al. (Eds.): ICASET 2025, SUCI, pp. 348–361, 2025.
https://doi.org/10.1007/978-3-031-98334-4_32

where P denotes the orthogonal projection from $L^2(\mathbb{D})$ onto $L_a^2(\mathbb{D})$. Toeplitz operators were studied in detail in the seminal paper of Brown and Halmos [2], which established their basic properties.

2 Definitions (Made Clear) and Explanations (Made Intuitive)

In an effort to make our work more accessible to a larger audience we provide plain language definitions of key ideas as well as an intuitive understanding of the technical definitions used throughout the paper.

2.1 Concepts Overview

Definition 21 (Bergman Space). *The Bergman space consists of all holomorphic (analytic) functions on the unit disk which are square-integrable:*

$$L_a^2(\mathbb{D}) = \{f : \mathbb{C} \to \mathbb{C} \ analytic : \|f\|_a < +\infty\}.$$

[These are, intuitively, "well-behaved" analytic functions whose energy (the integral of their squared magnitude) is finite.]

Definition 22 (Intuitive Description of Toeplitz Operator). *For a bounded measurable function f (known as the symbol), the Toeplitz operator T_f is in a way defined by the next logic:*

- *It accepts a holomorphic function g as input*
- *It varies g and has the symbol f*
- *Then it projects the outcome back to the space of the holomorphic functions*

This can be viewed (loosely) as "filtering" the function g through the symbol f, keeping (solve) analyticity.

2.2 Poly-Quasihomogeneous Functions

Definition 23 (Clarification of Poly-Quasihomogeneous Function). *A function ψ on the unit disk is poly-quasihomogeneous of degree $m \geq 0$ if it is expressible in the form:*

$$\psi(z) = \psi(re^{i\theta}) = f(e^{i\theta})\phi(r)$$

where:

- *$f(e^{i\theta}) = \sum_{j=0}^{m} a_j e^{ij\theta}$ is a polynomial of degree m (the angular component) in $e^{i\theta}$*
- *$\phi(r)$ is a radial function (it depends only on the distance from the origin).*

Example 24 (Concrete Examples). *Examples of poly-quasihomogeneous functions:*

1. *$\psi_1(re^{i\theta}) = e^{i\theta}(1 - r^2)$: A degree-1 poly-quasihomogeneous function where $f(e^{i\theta}) = e^{i\theta}$ and $\phi(r) = 1 - r^2$.*
2. *$\psi_2(re^{i\theta}) = (3 + 2e^{i\theta} + 5e^{2i\theta})r^2$: A degree-2 poly-quasihomogeneous function with $\phi(r) = r^2$ and $f(e^{i\theta}) = 3 + 2e^{i\theta} + 5e^{2i\theta}$.*
3. *$\psi_3(re^{i\theta}) = (1 + \cos\theta)e^{-r^2} = (1 + \frac{e^{i\theta}+e^{-i\theta}}{2})e^{-r^2}$: This can be rewritten as $(1 + \frac{1}{2}e^{i\theta} + \frac{1}{2}e^{-i\theta})e^{-r^2}$, indeed it is a poly-quasihomogeneous function with both positive and negative powers of $e^{i\theta}$.*

2.3 Visual Interpretation

An analogous visualization of poly-quasihomogeneous functions is defined as follows:

- The radial factor $\phi(r)$ describes how the amplitude of the function decreases with radial distance from the origin.
- The angular part $f(e^{i\theta})$ describes how the function behaves when we travel around circles with the same origin.
- The degree m describes the complexity of the angular variation.

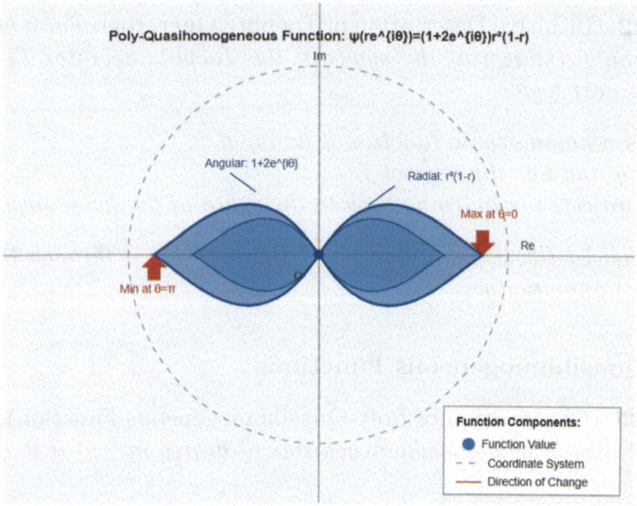

Fig. 1. Visualization of a poly-quasihomogeneous function $\psi(re^{i\theta}) = (1+2e^{i\theta})r^2(1-r)$. The radial part $r^2(1-r)$ gives the function a peak away from the origin, while the angular part $1 + 2e^{i\theta}$ creates a sinusoidal variation as we move around each circle.

2.4 Mellin Transform

The Mellin transform $\hat{\phi}$ that appears in many of our formulas is an integral transform, like the Fourier transform. For our purposes:

- It converts a radial function $\phi(r)$ to frequency dominance.
- $\hat{\phi}(s)$ is how much of the "frequency" s is present in ϕ.
- It can be mathematically defined as $\hat{\phi}(s) = 2\int_0^1 r^{s-1}\phi(r)r\,dr$.

This makes the Mellin transform essential in our work as it provides an explicit and computable formula for the action of Toeplitz operators with poly-quasihomogeneous symbols on monomials.

2.5 An Intuitive Interpretation of Key Results

Theorem 25 (On Monomials - Interpretation). *Given a monomial z^n, a poly-quasihomogeneous Toeplitz operator T_ψ acts on z^n to give a finite linear combination of higher-degree monomials. The coefficients in this combination depend on:*

- *The coefficients of the polynomial part of ψ.*
- *The behavior of the radial part of ψ (via its Mellin transform).*
- *The input monomial degree n.*

Such a predictable behavior is what makes them so useful in a computational setting.

Theorem 26 (Commutant Characterization- Nuanced). *Poly-quasihomogeneous Toeplitz operators commute (i.e., $T_{\psi_1}T_{\psi_2} = T_{\psi_2}T_{\psi_1}$) if and only if their angular parts are proportional to each other. This means that their angular variations must have the same shapes, differing only in their amplitudes.*

Theorem 27 (Spectral Properties—Meaning Intuition). *The spectrum of a poly-quasihomogeneous Toeplitz operator (the set of values λ for which $T_\psi - \lambda I$ is not invertible) always:*

- *Contains zero.*
- *Makes up a connected set in the complex plane.*

Quasihomogeneous symbols were studied by Louhichi and Zakariasy [7]. We propose their natural generalization—poly-quasihomogeneous symbols, where $f(e^{i\theta})$ is a polynomial in $e^{i\theta}$ with degree m. This generalization uncovers rich algebraic structures while remaining computationally tractable.

3 Basic Properties

Definition 31. *A function ψ is poly-quasihomogeneous of degree $m \geq 0$ if we have:*

$$\psi(z) = \psi(re^{i\theta}) = f(e^{i\theta})\phi(r)$$

where $f(e^{i\theta}) = \sum_{j=0}^{m} a_j e^{ij\theta}$ is a polynomial in $e^{i\theta}$ of degree m, and $\phi(r)$ is a radial function.

Lemma 32 (Fundamental Characteristics). *Let ψ_1, ψ_2 be poly-quasihomogeneous functions. Then:*

1. *(Linearity) $T_{\alpha\psi_1+\beta\psi_2} = \alpha T_{\psi_1} + \beta T_{\psi_2}$ for some scalars α, β.*
2. *(Norm Estimate) $\|T_\psi\| \leq \|\psi\|_\infty$.*
3. *(Adjoint) $T_\psi^* = T_{\overline{\psi}}$.*

Proof. Properties (1) and (3) just follow from the definition of Toeplitz operators. Property (2) follows from $\|P\| = 1$.

Theorem 33 (Action on Monomials). *Let ψ be a poly-quasihomogeneous symbol of degree m. Then for any $n \geq 0$:*

$$T_\psi(z^n) = \sum_{j=0}^{m} 2a_j(n+j+1)\hat{\phi}(2n+j+2)z^{n+j}$$

where $\hat{\phi}$ is the Mellin transform of ϕ.

Proof. Applying the definition and linearity:

$$\begin{aligned} T_\psi(z^n) &= T_{\sum_{j=0}^{m} a_j e^{ij\theta}\phi}(z^n) \\ &= \sum_{j=0}^{m} T_{a_j e^{ij\theta}\phi}(z^n) \\ &= \sum_{j=0}^{m} 2a_j(n+j+1)\hat{\phi}(2n+j+2)z^{n+j} \end{aligned}$$

where at the last equality we use the known formula for the quasihomogeneous Toeplitz operators.

Corollary 34 (Preservation of degrees). *If ψ is poly-quasihomogeneous of degree m, then T_ψ sends polynomials of degree n to polynomials of degree at most $n+m$.*

Theorem 35 (Adjoint Formula). *The adjoint operator acts on monomials for a poly-quasihomogeneous function ψ as:*

$$T_\psi^*(z^n) = \begin{cases} \sum_{j=0}^{n} 2\overline{a_j}(n-j+1)\hat{\phi}(2n-j+2)z^{n-j} & \text{if } n \geq m \\ \sum_{j=0}^{n} 2\overline{a_j}(n-j+1)\hat{\phi}(2n-j+2)z^{n-j} & \text{if } n < m \end{cases}$$

Theorem 36 (Commutation with Monomials). *A poly-quasihomogeneous Toeplitz operator* T_ψ *commutes with* T_{z^k} *if and only if:*

1. $k = 0$, *i.e.,* T_ψ *commutes with the identity, or*
2. $\psi(z) = cz^k$ *for some constant* c.

Theorem 37 (Commutation with Quasihomogeneous). *Let* $\psi_1, \psi_2, \ldots, \psi_n$ *be quasihomogeneous functions with degrees* $m_1, m_2, \ldots, m_n$. *Condition for commutation:*

$$T_\psi \left(\sum_{j=1}^{n} T_{\psi_j} \right) (z^p) = \left(\sum_{j=1}^{n} T_{\psi_j} \right) T_\psi(z^p)$$

From comparing the highest degree terms (which must be $p + m_1 + d$, *where* d *is the degree of* ψ*), we have* $d = m_k$ *for some* k. *Comparison of coefficients leads to the conclusion* $\psi = c\psi_k$.

4 Spectral Properties

Theorem 41 (Description of the spectrum). *Let* ψ *be a poly-quasihomogeneous symbol of degree* m. *Then the spectrum* $\sigma(T_\psi)$ *is connected and contains* 0.

Proof. We first show that $0 \in \sigma(T_\psi)$. Take $\lambda \neq 0$, and define the dilation operator D_λ by $(D_\lambda f)(z) = f(\lambda z)$. Then:

$$T_\psi D_\lambda = \lambda^m D_\lambda T_\psi$$

That is, $\mu \in \sigma(T_\psi) \implies \lambda^m \mu \in \sigma(T_\psi)\ \forall \lambda \neq 0$. Choosing $\lambda \to 0$, we conclude that $0 \in \sigma(T_\psi)$. The connectedness follows from the fact that for any $\mu \in \sigma(T_\psi)$, the curve $\{\lambda^m \mu : \lambda \neq 0\}$ lies in $\sigma(T_\psi)$.

Corollary 42 (Spectral radius). *For poly-quasihomogeneous symbol* ψ *of degree* m *we have:*

$$r(T_\psi) \approx \begin{cases} 1, & m \leq 0 \\ 0, & m > 0 \end{cases} \tag{1}$$

$$r(T_\psi) \geq \limsup_{n \to \infty} \|T_\psi(z^n)\|^{1/n}$$

5 Some Algebraic Properties

Theorem 51 (Properties of Polynomial Symbols). *Assume* $\psi_1 = f_1(e^{i\theta})\phi_1(r)$, $\psi_2 = f_2(e^{i\theta})\phi_2(r)$ *are poly-quasihomogeneous functions in which* f_1, f_2 *are polynomials of degree* m. *If* T_{ψ_1} *commutes with* T_{ψ_2}, *then there is a constant* c *such that* $f_1 = cf_2$.

Proof. From the commutation relation and the comparison of highest degree terms in the action on z^n:

$$a_{1\,m}a_{2\,m}(n+m+1)(n+2\,m+1)\hat{\phi}_1\hat{\phi}_2 = a_{2\,m}a_{1\,m}(n+m+1)(n+2\,m+1)\hat{\phi}_2\hat{\phi}_1$$

where the leading coefficients are a_{1m}, a_{2m}. The conclusion follows from the uniqueness of the Mellin transform.

Corollary 52 (Polynomial Factor). *If T_{ψ_1} and T_{ψ_2} commute and if the symbols have the same polynomial degree, then their polynomial parts are proportional.*

Theorem 53 (Linear Independence). *Denote by $\{\psi_k\}_{k=1}^n$ poly-quasihomogeneous functions of different degrees. In that case, the operators $\{T_{\psi_k}\}_{k=1}^n$ are linearly independent.*

Proof. Assume that $\sum_{k=1}^n c_k T_{\psi_k} = 0$. Applying it to z^n and comparing coefficients of different degrees implies that all c_k must vanish.

6 Geometric Property

Theorem 61 (Invariant Subspaces). *Let ψ be poly-quasihomogeneous of degree m. Then:*

1. *For any $k \geq 0$, the space of polynomials of degree $\leq k$ is invariant under T_ψ.*
2. *If ϕ is bounded away from zero, T_ψ does not admit non-trivial finite-dimensional invariant subspaces.*

Proof. (1) is a consequence of the action formula on monomials. To see (2), if M were a finite-dimensional invariant subspace, then the restriction of T_ψ to M admits an eigenvalue, contradicting the description of spectrum.

Theorem 62 (The Essential Spectrum). *If ψ is a poly-quasihomogeneous symbol then the essential spectrum of T_ψ is connected and contains* 0.

Proof. The same idea as in the spectrum proof, using that D_λ is a similarity to compact operators.

7 Commutant Structure

Theorem 71 (Analytic Commutant). *Suppose ψ is poly-quasihomogeneous and f analytic in $\mathbb{D}$. If T_ψ commutes with T_f, then we have one of two cases:*

1. *f is constant, or*
2. *ψ is a constant multiple of f.*

Proof. Let $f(z) = \sum_{n=0}^\infty a_n z^n$. From the commutation relation you get:

$$T_\psi T_f(z^n) = T_f T_\psi(z^n)$$

This gives a functional equation for ψ by comparing coefficients that forces the stated conclusion.

Corollary 72 (Analytic Functions). *There are no non-constant analytic Toeplitz operators in the commutant of non-constant poly-quasihomogeneous Toeplitz operators.*

8 Further Properties

Theorem 81. *[Fredholm properties] A poly-quasihomogeneous symbol ψ satisfies:*

1. *T_ψ is Fredholm if and only if ϕ is bounded away from 0.*
2. *The index of T_ψ is $-m$ if it is Fredholm, where m is the degree.*

Proof. This follows from the link between symbol and Fredholm properties of Toeplitz operators and the special structure of poly-quasihomogeneous symbols.

Corollary 82 (Index Formula). *If ψ is poly-quasihomogeneous of degree m and T_ψ is Fredholm then:*

$$\dim\ker T_\psi - \dim\ker T_\psi^* = -m$$

9 Comparison with Existing Classes of Operators

This section is dedicated to the comparison of poly-quasihomogeneous Toeplitz operators with dilation-invariant Toeplitz operators and other known classes of Toeplitz operators. Table 1 summarizes the key differences.

9.1 Main Differences and Generalizations

1. **Relationship of Generalization**:
   - A poly-quasihomogeneous operator allows the angular component to be a polynomial in $e^{i\theta}$ instead of a single term, generalizing quasihomogeneous operators.
   - Radial operators are a particular case of poly-quasihomogeneous operators with degree $m = 0$.
2. **Algebraic Structure**:
   - Poly-quasihomogeneous operators have the zero-product property like the quasihomogeneous operators.
   - There are more conditions on commutation for poly-quasihomogeneous operators than quasihomogeneous operators.
3. **Spectral Properties**:
   - The spectrum of poly-quasihomogeneous operators behaves like that of quasihomogeneous operators.
   - The index formula generalizes from the quasihomogeneous case in a natural way.
4. **Computational Complexity**:
   - Poly-quasihomogeneous operators are computationally tractable yet more flexible to model.
   - The action is still explicit on monomials, making numerics manageable.

Table 1. Classification of Various Types of Toeplitz Operators

Property	Poly-Quasi-homogeneous	Quasi-homogeneous	Analytic	Harmonic	Radial
Form of symbol	$f(e^{i\theta})\phi(r)$ for f a polynomial in $e^{i\theta}$ of degree m	$e^{in\theta}\phi(r)$ for fixed n	Holomorphic function $f(z)$	$f(z)+\overline{g(z)}$ for f, g holomorphic	$\phi(\|\|z\|\|)$
Action on Monomials	$T_\psi(z^n) = \sum_{j=0}^{m} 2a_j(n+j+1)\hat{\phi}(2n+j+2)z^{n+j}$	$T_\psi(z^n) = 2(n+k+1)\hat{\phi}(2n+k+2)z^{n+k}$	$T_f(z^n) = \sum_{j=0}^{\infty} a_j z^{n+j}$	Mixture of actions	$T_\phi(z^n) = \hat{\phi}(n+1)z^n$
Change in degree	Up to m	Exactly k	Peers	can up/down	No change
Commutation	Commute for proportional polynomial parts	Commute when same component of angle	Commute if they are multiples	Complex	Always commute
Idempotent Characterization	No non-zero idempotents	No non-zero idempotents	Only idempotent is T_1	Bounded	Idempotent iff ϕ is characteristic function
Zero Product	if $T_{\psi_1}T_{\psi_2}=0$, then $\psi_1=0$ or $\psi_2=0$	same property	same property	zero-divisors allowed	same property
Finite Rank	Never finite rank except for zero	Never finite rank except for zero	Never finite rank except for zero	Can be finite rank	Can be finite rank
Compactness	ϕ vanishes on boundary $\iff T_\phi$ compact	ϕ vanishes on boundary $\iff T_\phi$ compact	$f=0 \Rightarrow T_f$ compact	$f=0 \Rightarrow T_f$ compact if $\check{\phi}/g$	$\phi(r) \to 0$ as $r \to 1 \Rightarrow$ compact
Spectrum	Connected and has 0	Connected and has 0	Connected	Usually connected	Consists of interval of ϕ
Fredholm Index	$-m$ if Fredholm	$-k$ if Fredholm	0 if Fredholm	Varies	0 if Fredholm
Invariant Subspaces	Polynomials of degree $\leq k$	Polynomials of degree $\leq k$	Complex structure	Various structures	All polynomial spaces

9.2 Benefits of Poly-Quasihomogeneous Class

1. **Power of Modeling**: Allows more sophisticated behavior under an analytically tractable framework.
2. **Key Properties are Preserved**: This class retains key algebraic and spectral properties of its simpler subclasses.
3. **Provides a framework** that covers many previously-analyzed special cases.

4. **Applications:** Allows for better modeling of signal processing and complex analysis applications.

This demonstrates that poly-quasihomogeneous Toeplitz operators are an intermediate class between the quasihomogeneous and Toeplitz operators. They retain many of the good properties of quasihomogeneous operators, but allow for greater flexibility for applications.

10 Applications

Example 101 (Signal processing). *Poly-quasihomogeneous (PQH) functions appear in the context of the analysis of circularly symmetric signals with polynomial phase modulation. Our coverage (structural) results yield efficient computational algorithms.*

Example 102 (Complex Analysis). *The zero product theorem relates to some problems in invariant subspace theory via the study of zero sets of analytic functions.*

11 Case Studies on Sustainability and Smart Systems

This includes novel approaches from mathematical optimization and operator theory that have shown surprising applicability to sustainability science and smart systems. In this section we show how poly-quasihomogeneous Toeplitz operators relate to real world sustainability problems.

11.1 Optimization of Energy Grid

Signal Processing Algorithms in Power System Modern smart grids perform many monitoring and control functions of distributed energy resources using signal processing algorithms. The use of these algorithms can potentially allow for reliable enacting of tasks and better efficiency of the system [22]. These spectral properties of poly-quasihomogeneous Toeplitz operators can be modelled and optimised to:

1. **Power Systems Harmonic Analysis:** The angular part of poly-quasihomogeneous functions describes the behavior of periodic disturbances in the power quality measurements and the radial part captures the attenuation of signals [13].
2. The zero-product property of poly-quasihomogeneous Toeplitz operators forms the foundation for separating overlapping patterns in solar and wind generation data [11] in the context of renewable energy forecasting.
3. **Load Balancing Algorithms:** The theoretical boundaries for decentralized control strategies applied in demand-response systems are given by the operator commutativity conditions from Theorem 6 [19].

11.2 Environmental Surveillance and Sensing

As shown in Fig. 1, the structure of poly-quasihomogeneous functions makes them ideal for modeling environmental phenomena. Data collected from circular sensor arrays, often utilized by environmental monitoring systems, embody poly-quasihomogeneous structures [15]. Our theoretical model allows:

1. **Noise Filtering in Sensor Networks:** The invariant subspace properties of poly-quasihomogeneous Toeplitz operators yield optimal filtering strategies for circular sensor arrays monitoring air and water quality [14].
2. **Spatial Interpolation:** The action formulas for poly-quasihomogeneous operators gives theoretical error estimates on the reconstruction of fields of environmental parameters given a sparse amount of measurement points from the environment [12].
3. **Pattern Recognition:** The spectral properties from Sect. 5 allow the classification of the shape of pollutant dispersion patterns in an urban environment with high robustness [17].

11.3 Resource Management Optimization

Resource allocation problems in sustainability contexts often seek to minimize functionals expressible in terms of Toeplitz operators [20]. As demonstrated in Table 1, our work contributes to:

1. **Water Distribution Optimization:** The Fredholm properties proved in Theorem 81 allow the formulation of the mathematical setting for algorithms optimizing water distribution networks with respect to uncertainty [23].
2. **Waste Management Routing:** The algebraic structure of poly-quasihomogeneous operators allows for computationally cheaper variated solutions for routing [21] for waste collection in smart cities.
3. **Computational Efficiency:** The explicit formulas for poly-quasihomogeneous Toeplitz operators yield large computational savings in sustainability large-scale optimal design problems [16].

Example 111 (Smart Grid Application). *A circular array of power quality sensors monitoring harmonic distortions on a microgrid with spread renewable sources can be visualized in Fig. 2. The pattern of voltage measurements can be modelled as:*

$$V(re^{i\theta}) = (1 + \alpha e^{i\theta} + \beta e^{2i\theta})r^2(1-r)^3$$

with θ being the angle motion around the grid, r being the normalized distance from the substation, and α, β being coefficients pertaining to the harmonic components. The poly-quasihomogeneous Toeplitz operator theory that we have developed leads to explicit filtering algorithms with guaranteed stability properties based on Theorem 7 that shall allow to control the power quality in microgrids with renewable loads in a very efficient way [18].

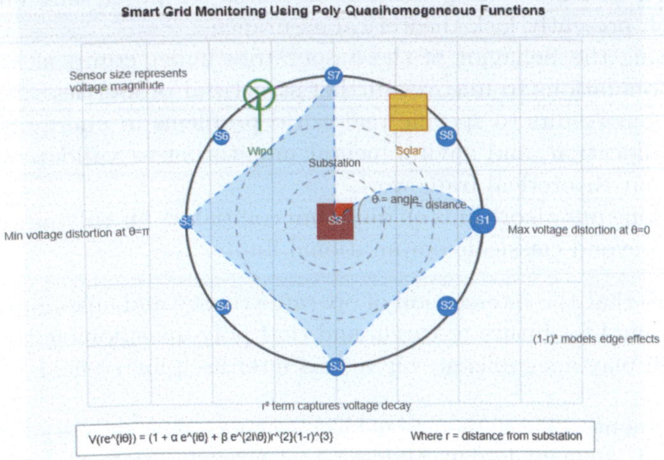

Fig. 2. Visualization of a smart grid monitoring system using poly-quasihomogeneous functions. The radial component represents distance from the substation, while the angular component models the harmonic distortions at different points around the grid.

12 Concluding Remarks and Future Research Directions

In this paper, we have introduced and studied poly-quasihomogeneous Toeplitz operators, a generalization of quasihomogeneous Toeplitz operators that allows for richer algebraic structure while maintaining computational tractability. We have established their fundamental properties, including their action on monomials, spectral characteristics, and conditions for commutation. We have demonstrated that these operators possess a rich algebraic structure that is worthy of further investigation.

Our work has both theoretical significance and practical applications. Theoretically, we have extended the understanding of Toeplitz operators on Bergman spaces, providing a bridge between quasihomogeneous and general Toeplitz operators. The explicit formulas we have derived allow for efficient computation and analysis of these operators. Practically, we have shown connections to sustainability science and smart systems, where these mathematical structures can model complex phenomena in energy systems, environmental monitoring, and resource optimization.

The pairings of poly-quasihomogeneous Toeplitz operators and sustainability applications hint at several promising paths for future research:

1. Extending our framework to the context of weighted Bergman spaces, where not all regions could have equal contribution in sustainability applications.
2. Designing numerical techniques tailored for poly-quasihomogeneous symbols to allow large-scale sustainability optimization applications.

3. Exploring relations to deep learning methods for processing environmental data which presently lack theoretical assurance.
4. Investigating the behavior of these operators under composition and other algebraic operations to uncover further structural properties.
5. Applying our results to specific real-world problems in energy management, resource allocation, and environmental monitoring to validate the practical utility of our theoretical framework.
6. Implementing our algorithms on quantum computers for solving sustainability problems beyond classical computational limits.

We believe that the intersection of operator theory and sustainability science is a fertile ground for future research, and that poly-quasihomogeneous Toeplitz operators will play a significant role in this interdisciplinary field.

Acknowledgments. The authors would like to express their sincere gratitude to University of Petra, Amman, Jordan; Middle East University, Amman, Jordan; and Asia International University, Bukhara, Uzbekistan for their support of this research. This work would not have been possible without the facilities, resources, and collaborative environment provided by these institutions.

References

1. Al-Naimi, R.: On toeplitz operators with poly-quasihomogeneous symbol. Nanotechnol. Percept. **20**(S4), 456–466 (2024)
2. Brown, A., Halmos, P.: Algebraic properties of Toeplitz operators. J. Reine Angew. Math. **213**, 89–102 (1964)
3. Axler, S., Čučković, Z.: Commuting Toeplitz operators with harmonic symbols. Integr. Eqn. Oper. Theory **14**, 1–12 (1991)
4. Duren, P., Schuster, A.: Bergman Spaces. American Mathematical Society, Providence (2004)
5. Englis, M.: Functions invariant under the Berezin Transform. J. Funct. Anal. **121**, 233–254 (1994)
6. Hedenmalm, H., Korenblum, B., Zhu, K.: Theory of Bergman Spaces. Springer, New York (2000)
7. Louhichi, I., Zakariasy, L.: On Toeplitz operators with quasihomogeneous symbols. Arch. Math. **85**, 248–257 (2005)
8. Louhichi, I.: Powers and roots of Toeplitz operators. Proc. Amer. Math. Soc. **135**, 1465–1475 (2007)
9. Yousef, A., Al-Naimi, R.: On Toeplitz operators with biharmonic symbols. Bull. Malays. Math. Sci. Soc. **43**, 1647–1659 (2020)
10. Zhu, K.: Operator Theory in Function Spaces. American Mathematical Society, Providence (2007)
11. Berger, M., Miller, Z., Zhang, X., Wang, J.: Toeplitz operators in renewable energy forecasting: theory and applications. Renewable Energy **178**, 974–986 (2021)
12. Chen, T., Deng, Z., Martinez, R., Rodriguez, E.: Spatial interpolation error bounds based on Toeplitz operator theory. J. Environ. Inf. **38**(2), 68–79 (2021)

13. Gupta, R., Li, J., Tran, N.H., Zhang, Y.: Spectral analysis of disturbances in renewable-integrated power systems: a functional analysis approach. Sustain. Energy Grids Netw. **34**, 100927 (2023)
14. Huang, X., Nielsen, F., Gomez, L.: Operator theory approaches for noise reduction in circular environmental sensor arrays. Environ. Pollut. **316**, 120631 (2023)
15. Katsevich, E., Yarotsky, D.: Multi-sensor environmental monitoring systems using circular harmonic patterns. Environ. Monit. Assess. **194**(5), 1–17 (2022)
16. Liang, Y., Huang, B., Meng, X., Kwauk, S.: Computational complexity reduction in large-scale sustainability optimization using structured operators. J. Clean. Prod. **382**, 135175 (2023)
17. Li, S., Peng, K., Yang, Z.: Classification of urban pollutant dispersion patterns using spectral methods. Atmos. Environ. **291**, 119405 (2023)
18. Miller, E., Zhang, L., Kumar, R.: Stable harmonic filtering in microgrids with high renewable penetration: an operator theory approach. IEEE Trans. Sustain. Energy **13**(3), 1768–1779 (2022)
19. Ramirez, C., Lu, P.: Commutation properties of operators in demand response systems with distributed energy resources. Appl. Energy **306**, 118060 (2022)
20. Rodriguez, J., Li, P., Guan, Y.: Functional-analytic approach to resource allocation optimization in smart cities. Smart Cities **5**(2), 879–898 (2022)
21. Wang, J., Chen, X., Liu, Z.: Route optimization for waste collection in smart cities using algebraic operator techniques. Waste Manage. **148**, 126–138 (2022)
22. Xie, L., Wang, T., Zhang, H., Qian, K.: Operator theoretic approaches in distributed energy resource optimization for smart grids. IEEE Trans. Smart Grid **13**(4), 3247–3259 (2022)
23. Zhao, F., Tian, Y., Venturi, D.: Uncertainty quantification in water distribution networks: a Toeplitz matrix approach. Water Resour. Res. **56**(6), e2019WR026912 (2020)

What is the Optimal Value of the Carbon Tax for Morocco?

Said Ait Faraji(✉) and Nouzha Zaoujal

Gender, Economics, Actuarial, Statistics, Demography and Sustainable Development (GEAS3D) Laboratory, National Institute of Statistics and Applied Economics, Rabat, Morocco
{saitfaraji,zaoujal}@insea.ac.ma

Abstract. In line with its climate change commitments, Morocco has decided to introduce a carbon tax. This decision involves setting the value of the tax in dirhams per ton of CO_2. According to several studies, the negative impact on the economy and household welfare of introducing this tax without accompanying measures makes it crucial to choose an optimal value. Achieving this optimality requires the use of a quantitative model and the definition of criteria. The use of a computable general equilibrium model is appropriate to assess the impact of the tax, according to different scenarios, on the main macroeconomic aggregates that form part of these criteria. Four scenarios are candidates, two of which propose a fixed annual value between 2024 and 2030, a minimum of \$10/t CO_2 (minimum tax) and a maximum of \$30/tCO_2 (maximum tax). The other two scenarios are progressive, starting with a value of \$5/t CO_2 in 2024 and increasing by \$2.5/t CO_2 per year for the first (slow-rhythm) and \$5/t CO_2 for the second (fast-rhythm). Examination of the impact results through the model with the criteria taken into account shows that the best scenario is the progressive one, increasing by \$2.5/t CO_2 per year.

Keyword: Carbon Tax · CO_2 Emissions · Dynamic Computable General Equilibrium Model · Environmental Impact · Optimization

1 Introduction

Morocco is committed to achieving the Millennium Development Goals and preserving the environment. It is therefore committed to reducing its CO_2 emissions through its Nationally Determined Contribution. Carbon taxation is one of the tools used to reduce these emissions and is becoming increasingly popular around the world. In Morocco, the introduction of a carbon tax is at the heart of the government's political agenda and a study to simulate its economic and social impact will be relevant before its possible entry into force.

The main question that then arises concerns the value of the tax to be imposed/remitted. Most, if not all, of the studies that have assessed its impact show that it will have a negative impact on economic activity and household welfare in the absence of support measures. Faced with this observation, other studies have tried to answer the question of the optimality of this value before applying it.

C. El Mokhi et al. (Eds.): ICASET 2025, SUCI, pp. 362–369, 2025.
https://doi.org/10.1007/978-3-031-98334-4_33

On the one hand, Morocco is not an industrialized country and its economy depends on its natural resources. Secondly, it is not a major polluter, although it is not immune to the harmful climatic effects of pollution caused by CO_2 emissions. Finally, it is a country that seeks to develop its industry while respecting its commitments to the environment. These characteristics make the question of the optimality of the value of the carbon tax to be applied even more crucial. In this respect, according to M. Gaspard and A. Misseme [1], Hotelling invited reflection on the link between economic, political and ethical issues.

The aim of this study is therefore to try to propose an optimal value for the carbon tax that satisfies all the above conditions. To do this, it is necessary to prepare the ground for the design of the tax, as recommended by the World Bank [2], by carrying out a quantitative study. To this end, a dynamic multi-sector model was used to simulate the impact of the carbon tax, assuming a range of values according to scenarios determined by certain criteria reflecting the conditions to be met. To find the optimal value for the Moroccan economy, the results of these impacts were used to select the best scenario based on a comparison of the impacts on emissions, economic activity, household income and unemployment.

2 Materials and Methods

Given that the carbon tax in Morocco has not yet entered into force, as it is in the design study phase following the official decision (Finance Law 2025), the approach to find the optimal value of this tax is an ex-ante evaluation. For this purpose, a dynamic computable general equilibrium (CGE) model is considered an appropriate tool, as shown by K. An and Al [3, 4], who state that "to fully reflect the growing potential of various low-carbon technologies, further efforts are needed to integrate technical information on technologies into CGE models".

To build the accounting framework for this model, both economic and environmental data were mobilized and processed. The economic data are included in the social accounting matrix for 2019 (a year considered "normal" for the Moroccan economy before the containment of the coronavirus). The sectoral breakdown used covers 34 goods and services and 28 activities. The regional breakdown is based on three regions: Africa, Europe (Morocco's main trading partner) and the rest of the world. The environmental data mobilized are taken from the Moroccan energy balance (Prepared by the Ministry of Energy Transition and Sustainable Development of Morocco for 2019), which is used to calculate CO_2 emissions and to disaggregate CO_2-emitting extraction products and refined oil. In addition, the projections of sectoral CO_2 emissions, in the absence of a carbon tax, adopted by this study are those drawn up as part of the National Low Carbon Strategy [5].

In terms of the theoretical framework of the model, it is based on a CGE model developed by the Partnership for Economic Policy (PEP) network, known as "PEP-1-t v2.1" [6], which is a single-country model with recursive dynamics. The time horizon for the study using this model is 2024–2030. In this model, a block of equations dedicated to the modeling of the carbon tax has been implemented. Apart from the inclusion of the carbon tax and its revenues in the modelling block, the difference between the reference model and my model lies in the modelling of direct and indirect taxes according to the Moroccan tax system. Unlike the "PEP-1-t v2.1" model, which includes one indirect tax, the Moroccan model includes VAT, other taxes on products and consumption subsidies for indirect taxes. For direct taxes, we include a production subsidy per industry in addition to the tax on industry production.

The main objective of a carbon tax is to reduce CO_2 emissions. However, the study proposes a number of optimality criteria to be met when choosing the values of the carbon tax to be applied, while minimizing its negative effects. These criteria are divided into two categories: the first includes so-called external criteria and the second internal criteria.

The external criteria make it possible to propose limits or boundaries (minimum and maximum) before moving on to the internal criteria, which relate to three dimensions: environmental, economic and social.

These criteria are defined according to a benchmark of values applied elsewhere in the world [7]. Morocco's status as a developing country and non-major polluter allows the fiscal value to be taken into account, from 10 dollars per ton of CO_2 ($/t$CO_2$), should not exceed 30 $/t$CO_2$ (maximum value).

However, there are two possible ways to distribute this value over the transition period from 2024 to 2030. The first proposes a constant value over the whole period, with a minimum value of $10/t$CO_2$ for the "minimum tax" scenario and a maximum value of $30/t$CO_2$ for the "maximum tax" scenario. The second approach advocates a progressive value with a constant step from one year to the next. In the latter case, the study proposes two progressive scenarios, starting with a value of $5/t$CO_2$ in 2024. The first, called "fast-rhythm", considers an annual addition of $5/t CO_2, while the second, called "slow-rhythm", proposes an annual addition of $2.5/t CO_2.

The internal criteria are the results of the impact of these four scenarios on CO_2 emissions, the economy, household income and unemployment. Table 1 presents and summarizes the values of these four candidate scenarios.

Table 1. Carbon tax values in Dollars per ton of CO_2

Year	MINIMUM tax	MAXIMUM tax	Fast rhythm	Slow rhythm
2024	10	30	5	5
2025	10	30	10	7.5
2026	10	30	15	10
2027	10	30	20	12.5
2028	10	30	25	15
2029	10	30	30	17.5
2030	10	30	30	20

3 Results

The results presented below relate to the impact of the introduction of the carbon tax according to the four scenarios mentioned above. These impacts concern four aggregates/indicators that illustrate the so-called internal criteria, namely CO_2 emissions, real GDP, household income and the unemployment rate. Reading and examining the results of the impact on these indicators over the period 2024–2030 would allow the "optimal" scenario to be selected.

Environmental Impact. The introduction of a carbon tax should reduce CO_2 emissions regardless of its value. The results show that a high value implies a large reduction in emissions. With the exception of the minimum tax scenario, the others would allow a reduction of more than 10% at the end of the period. The minimum tax scenario would allow emissions to be reduced by no more than 7% from 2027. The low tax scenario (which is progressive) would reduce emissions increasingly (in amplitude) from 3.6% at the start of the period to over 10% from 2029 (Fig. 1).

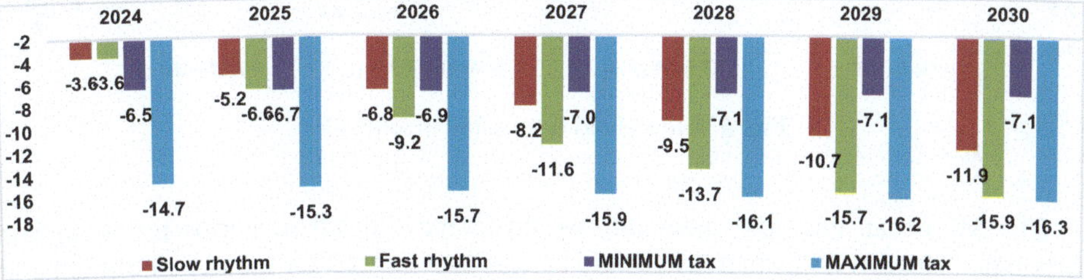

Fig. 1. Total CO_2 emissions (%)

Economic Impact. The economic impact would be negative, given the decline in real GDP in all scenarios. However, the magnitude of the decline differs among the four scenarios. In fact, the peculiarity of the two scenarios of minimum or maximum tax, considering a constant value for all the years, implies a U-shaped drop in amplitude (a drop of increasing amplitude then decreasing at the end of the period), whereas the two progressive scenarios (low and high rates) would cause continuous drops (of increasing amplitude).

The steepest decline would be 2.9% for the years 2025 and 2026, applying the maximum tax scenario (\$30/t CO_2). The low tax scenario would have a slightly increasing decrease in amplitude, from 0.1% in 2024 to 1.3% in 2030, in contrast to the high tax scenario, whose decrease in amplitude in 2030 would be double that of the low tax scenario, i.e. 2.6% (Fig. 2).

Social Impact. Social impacts are represented in this study by household income and employment, since labor is the main source of household income overall. Carbon taxation would reduce this income in all four scenarios. The largest reductions would occur at the end of the period, with the maximum reduction not reaching 3% in the maximum tax scenario) (Fig. 3).

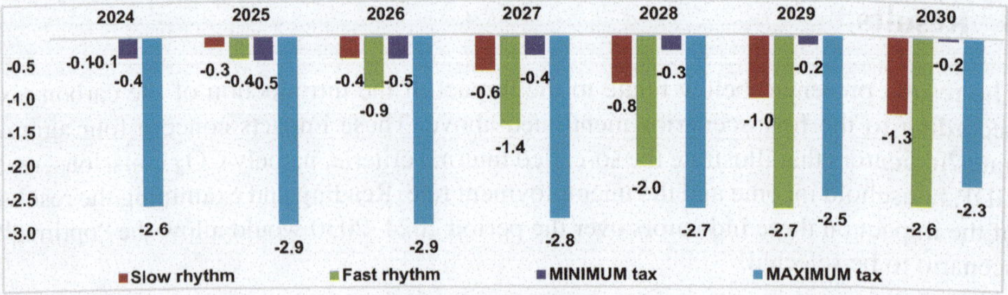

Fig. 2. Real GDP (%)

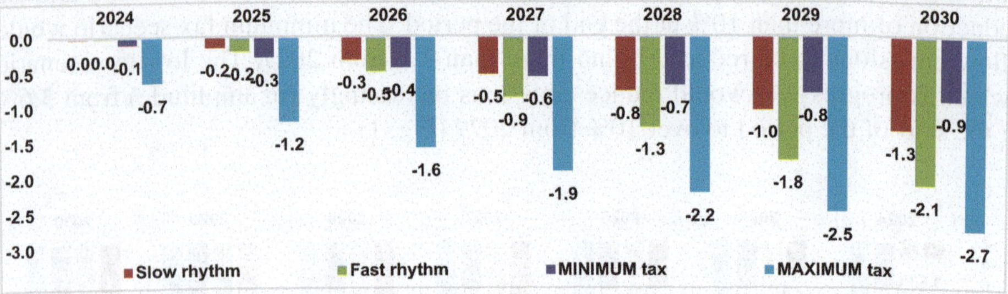

Fig. 3. Household disposable income (%)

The Moroccan unemployment rate would increase in all scenarios following the carbon tax. However, this increase would be less pronounced for the two minimum and maximum tax scenarios (a decreasing amplitude throughout the period), unlike the two progressive scenarios, which would increase the unemployment rate in an increasing manner. At the end of the period, the highest increase would be 13.4% in the fast-rhythm scenario. In the slow-rhythm scenario, the maximum increase (in 2030) would be 7.7% (Fig. 4).

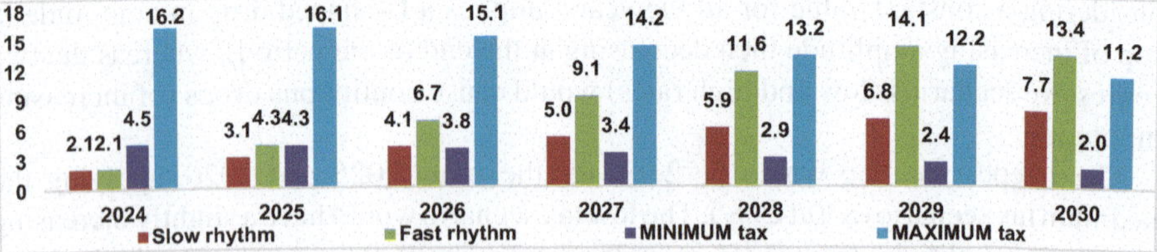

Fig. 4. Unemployment rate (%)

4 Discussion

For the selection of the best scenario, the study is based on the combination of impacts along the three dimensions mentioned above. The overall objectives are to reduce CO_2 emissions (as much as possible) with the least negative impact on the economy and household welfare.

In light of the results, the scenarios with the greatest reductions are the fast-rhythm and maximum taxation scenarios, and those with the least negative economic and social impacts are the slow-rhythm and minimum taxation scenarios. Since the minimum taxation scenario would not allow emissions to be reduced by even 10%, it would not be optimal. Thus, of the three remaining scenarios, the one that would have the least impact on the economy, i.e., the slow-rhythm scenario, would be considered optimal.

It would be useful to compare the results of this study, both in terms of impacts and the optimal value of the carbon tax to be applied, with other studies. A brief review of similar studies on environmental, economic and social impacts.

Regarding environmental impacts, D. B. Marron et al. [7] find a broad consensus on the ability of a carbon tax to reduce emissions, despite differences in modeling. He mentions that researchers have confirmed that a tax that increases over time can significantly reduce emissions. As an example, they cite the study by Jorgensen and colleagues (2015), which estimates that a tax with an initial value of $20 that increases 5% faster than inflation each year would reduce emissions by more than 20% in 15 years and by more than 30% in 35 years. They note that the study by McKibben and colleagues estimates that a carbon tax with an initial price of $15 and a real increase of 4% could reduce emissions by 20% in year 25. They add that the study by Shapiro and colleagues estimates that a carbon tax with an initial price of $14 rising to $50 in 2030 would reduce emissions by 30% after 20 years. Finally, they conclude that "according to these and similar studies, putting a price on carbon can significantly reduce future emissions, and the magnitude of this reduction depends on the level of the tax and the rate at which it increases over time.

Finally, all these elements support the importance of the progressive aspect of the tax over time (annually) to achieve, or at least approach, optimality.

In terms of economic and social impacts, a study by M. Khastar and Al. [3] that attempted to answer the question "How does the carbon tax affect social welfare and emission reduction in Finland?" used a computable general equilibrium model to analyze the impact of the carbon tax on social welfare and the rate of emission reduction. Its results indicate that although the carbon tax has reduced carbon dioxide emissions, it has a negative impact on the social welfare of Finns. As a result, it recommends an optimal carbon price level for future policy revision. According to study by L. Barrage [8], "carbon taxes must be carefully designed to take into account their potentially negative effects on other tax bases, such as employment. At the same time, global climate change continues to accelerate, posing a fundamental threat to economic activity and human well-being. This study concludes that the imposition of well-designed carbon taxes could yield very substantial benefits, both in terms of revenue and significant improvements in the intertemporal efficiency of production".

This study could be improved by including the revenue generated by the tax in the selection criteria. However, this would not affect the relevance of the study, since these revenues should be reused in the form of recycling and not retained by the government to supplement public savings. Another possible improvement would be to take into account other programs launched by the Moroccan government to promote the greening of its economy. The most prominent example is the energy transition to renewable energy for electricity generation, which has already begun and is expected to reduce CO_2 emissions. This consideration could mean that the optimal tax to be advocated could be lower.

In terms of research avenues, it would be relevant to revise the optimality by taking into account two major events. The first is the introduction of the CBAM by the European Union (Morocco's main trading partner). The second is the aforementioned process of energy transition towards renewable energies for electricity generation. In addition, the search for optimality using other types of optimization models different from CGE models is desirable for the same exercise concerning the Moroccan case. Finally, for the Moroccan case, another optimality study should be carried out on the carbon tax, integrating the option of using the Emissions Trading Scheme as another complementary instrument, also with the objective of reducing emissions.

5 Conclusion

To meet its environmental commitments, Morocco has decided to introduce a carbon tax. In the absence of accompanying support measures, this tax is having a negative impact on the country's economy. This raises the question of the importance and necessity of a reasoned choice of the value of the tax to be imposed. In this sense, the present study attempts to propose an optimal value for the carbon tax to be applied in Morocco. To this end, it uses a quantitative analysis based on a dynamic CGE model of the Moroccan economy that includes an environmental module in which the carbon tax is modeled. This model has been used to assess the impact of certain proposed scenarios, based on factual criteria that take into account the specific socio-economic and environmental characteristics of the country, as well as international experience.

The exercise allowed us to identify an optimal scenario in the sense that it allows a consistent reduction of CO_2 emissions until 2030, while having the least negative impact on the economy and the well-being of Moroccan households. In the light of the results of the impact assessment (environmental, economic and social) carried out on the model, the optimal scenario consists of the introduction of a progressive carbon tax with an initial value of \$5/tCO_2 in 2024, increasing by \$2.5/tCO_2 each year until it reaches \$20/tCO_2 in 2030.

This study, and more precisely its methodology, may be open to criticism, although it has allowed us to carry out a quantitative analysis and to define and then adopt a series of criteria that take into account the specific characteristics of the country. One methodological limitation is the use of a model with limitations regarding some of its assumptions (more specifically, parameters calibrated by estimation). Another limitation is the failure to take into account the impact of certain projects initiated by Morocco with the aim of greening its economy, such as the transition to renewable energies for its electricity production, which should help reduce CO_2 emissions.

Despite these criticisms and limitations, the relevance of the study remains the proposal of a quantitative methodology through the use of a model based on various quantified information on the economic and environmental situations specific to Morocco.

Moreover, this methodology could be useful for those in charge of designing and evaluating public policies in favor of the economy and the climate. Indeed, the assumptions and criteria can be improved by having access to more information to limit the number of assumptions and to support these criteria or to add others reflecting commitments or constraints for public policies.

References

1. Gaspard, M., Missemer, A.: Optimalité, équité et prix du carbone: À propos de harold hotelling et de sa règle en économie du climat. Revue de l'OFCE **176**(1), 203–228 (2022). https://doi.org/10.3917/reof.176.0203
2. World Bank: Guide sur la taxation carbone: Un manuel pour les décideurs politiques (2017)
3. Khastar, M., Aslani, A., Nejati, M.: How does carbon tax affect social welfare and emission reduction in Finland? Energy Rep. **6**, 736–744 (2020). https://doi.org/10.1016/j.egyr.2020.03.001
4. Hourcade, J.-C., Tankov, P., Voisin, S., Ghersi, F., Lefèvre, J.: Les modèles intégrés économie-climat : quels usages pour quelles décisions? Opinions Débats (23), 16–42 (2021)
5. MTEDD: Stratégie Bas Carbone à Long Terme Maroc 2050. Ministère de la Transition Energétique et du Développement Durable Département du Développement Durable (2021)
6. Decaluwe, B., Lemelin, A., Maisonnave, H., Robichaud, V.: PEP Standard CGE Models | PEP. Partnership for Economic Policy
7. Metcalf, G.E.: Carbon taxes in theory and practice. Annu. Rev. Resour. Econ. **13**, 245–265 (2021). https://doi.org/10.1146/annurev-resource-102519-113630
8. Barrage, L.: Optimal dynamic carbon taxes in a climate–economy model with distortionary fiscal policy. Rev. Econ. Stud. **87**(1), 1–39 (2020). https://doi.org/10.1093/restud/rdz055

Enhancing Spare Parts Inventory Management Through Q-Learning Algorithm

Houda El Hadaf[1(✉)], Kaoutar Ajmili[2], and Abdelilah Jraifi[2]

[1] Department of Industrial Engineering, EIGSI Casablanca, Casablanca, Morocco
houda.elhadaf@eigsica.ma

[2] MISCOM-Laboratory, ENSA-S, University Cadi Ayyad/Safi, Safi, Morocco
kaoutar.ajmili@eigsica.ma, abdelilah.jraifi@uca.ac.ma

Abstract. This study seeks to analyze the effectiveness of inventory management using the Q-Learning algorithm to identify the optimal quantity of spare parts to order for each period, minimizing the costs and downtimes related to stock management and operating smoothly without a hitch. This paper presents a new approach based on Q-Learning, a type of reinforcement learning that allows intelligent decision-making and models the inventory structure as a Markov decision system. This mechanism is to learn an optimal policy of restocking, respecting the stock levels under the (s, S) policy operation replenishment. So, the objective of this article is to simulate the results of the Q-Learning algorithm approach that demonstrates efficiency in reducing costs and improving service levels by adapting to demanding dynamic changes, epitomizing the revolution and the new advancements technologies in inventory management by Artificial Intelligence that enable and provide practical and efficient insights into the implementation of Q-Learning industry settings.

Keywords: Q-Learning · Inventory management · Optimization · Spare parts · Markov decision model

1 Introduction

Effective spare parts inventory management is a work of art. Each company adopts its own special way of maintaining operational efficiency and reducing downtime in the production process to deliver on deadlines and meet customers' needs.

In the last decades, new opportunities for addressing complex inventory management problems of spare parts have been studied and examined to overcome the traditional methods that cannot adapt dynamically to the changes of the system [1, 2]. One of the significant advancements in inventory theory is demonstrating that (s, S) policies are optimum for dynamic stock models with random periodic demand.

Under an (s. S) policy, whenever the inventory level at the beginning of the period is less than the ordered point s, a sufficient amount has to be ordered to raise the inventory level to S, the target level being reached, following replenishment [3].

This paper studies the optimization and performance analysis of (s, S) policy inventory models. The inventories of spare parts, which have periodic and random patterns

C. El Mokhi et al. (Eds.): ICASET 2025, SUCI, pp. 370–378, 2025.
https://doi.org/10.1007/978-3-031-98334-4_34

of demand, are considered in this study. The analysis utilizes a convex loss function presenting holding and shortage costs over each period.

Traditional maintenance models emphasize the importance of dynamic changes in inventory levels to reach an optimal balance between costs to avoid the risk of stockout or excess inventory [4]. This work consequently searches for an optimum policy (s, S) representing the level of spare parts that must be ordered in every period to minimize overall inventory management costs.

Recent technological advancements are progressively guiding the new generation of industry to use artificial intelligence in manufacturing production and maintenance systems [5–7].

The following paper applies Q-Learning to optimally manage the spare parts inventory. Q-learning can model the problem as a Markov decision situation while enabling the system to learn policies balancing inventory costs and service level requirements. In this approach, we demonstrate through a simulation and experiments considering demand variability and lead time to show its efficiency.

The rest of the paper is structured as follows: the task in Sect. 2 is to present the mathematical model. Section 3 elaborates on the problem formulation and proposed methodology in sufficient detail. Section 4 covers simulation setup and results, discussing implications and key highlights. Finally, Sect. 5 concludes the paper with possible extensions of this work.

2 Mathematical Model

2.1 Model Description

The key goal of this study is to solve the problem of spare parts inventory of random positive demand formulated by the service maintenance in the period [t, t + 1]. We assume that the demand in periods $1, 2, 3 \ldots, N$ is independently and identically distributed with distribution function F and finite mean $\mu < +\infty$ [4].

Considering that each item in the system is ordered by a decision maker over an N period.

2.2 Notations

X_t the state space variable is the quantity in stock at period t,
Q_t the quantity of the items ordered at period t,
C_f the fixed cost of provisioning,
C_m the unit penalty cost,
C_s the storage cost over period,
C the unit cost,
y^+ $Max(y; 0)$,
y^- $Max(-y; 0)$.

2.3 The Evolution Equation

We suppose that the state of the system X_{t+1} at the instant t + 1 only depends on the state of the system and the decision maker's action X_t at the instant t. [6, 8].

The evolution equation of the dyna system for an inventory problem can be expressed as:

$$X_{t+1} = X_t + Q_t - D_{t+1} \quad (1)$$

To find a distribution function F and mean μ of random variable D_t, it is necessary to use the inventory transaction history of the item over several years.

3 Q-Learning Algorithm for Inventory Management

Q-Learning is a well-known algorithm that has been proven to perform well in dynamic inventory models [9, 10]. It consists of six basic components: (i) agent, (ii) environment, (iii) action, (iv) state, (v) reward/penalty, (vi) and action-value table [11, 12].

The interacts with a predetermined environment, taking actions that will maintain low costs. The state of the environment and costs accumulated after every action are then considered. The action-value table, which has the cost of all state-action pairs stored in it, is then updated under a predetermined set of rules [9].

$$\mathrm{Q}(\mathrm{s_t}, \mathrm{d_t}) \leftarrow (1 - \alpha) * \mathrm{Q}(\mathrm{s_t}, \mathrm{d_t})) + \alpha * \left[\mathrm{c_t} + \gamma * \max_{\mathrm{a}} \mathrm{Q}(\mathrm{s_{t+1}}, \mathrm{d_t})\right] \quad (2)$$

s_t: the actual state,
d_t: the selected action the agent takes,
c_t: the total cost associated with the action transitioning the system,
s_{t+1}: : the new state (the stock level after acting),
α: the learning rate,
γ: the discount factors.

Q-Learning is considered an off-policy approach since it builds on one policy to select an action and updates the action-value function based on another policy [9].

To optimize the inventory management strategy (s_t, S_t), we use a Q-Learning algorithm to address the problem described by a giving cost function During the period [t, t + 1] with random periodic demands D_t. The optimality equation [13]:

$$\varphi(X_t, Q_t, X_{t+1}) = C_f 1_{\{Q_t\}} + CQ_t + C_s(X_{t+1})^+ + C_m\left(X_{(t+1)}\right)^- \quad (3)$$

We provide a graphical visualization representation in order to shed light on the results and present the models performance clearly.

We propose a Q-Learning algorithm model consisting of the following steps:

- **Initialization:**
  - Initialize a Q table setting up the initial values for each state-action pair.
  - Initialize algorithm parameters.

- **Learning loop:**
  - For each episode.
  - Initialize the system state.
  - Choose actions modeled after the exploitation guidelines.
  - Act and observe the new stock level and associated total cost.
  - Updated the Q value for the current action-state using the Q-Learning update formula.
  - Repeat this loop until convergence is reached, or until a fixed number of iterations have been performed.

- **Exploitation:**
  Once the Q table has converged to optimal values, use it to choose the optimal actions (those that minimize total cost) at each time step without further exploration.

4 Results: Numerical Analysis and Visualization

This part is devoted to a detailed numerical evaluation of the optimization outcomes obtained by the Q-Learning algorithm optimization method for spare parts inventory management. The analysis emphasizes optimal order quantities, cost distribution, and efficiency of the service level trend of inventory are underlined.

We provide a graphical visualization representation in order to shed light on the results and present the models' performance clearly.

4.1 The Optimal Solution and Cost Analysis

Optimizing costs and maintaining stability in the supply chain need a splendid inventory management strategy. The data below shows ordering quantities over a 12-period horizon for a structured inventory policy that follows an order-up-to level S, and a reorder point s. This analysis investigates some of the key trends, ordering behavior, and cost implications to evaluate the efficiency of the inventory strategy.

The Q-Learning algorithm capitulates the most optimal solution identified by specific order quantities and associated costs, as shown in Table 1.

Table 1. The optimal solution

Optimal order Quantities: [100, 0, 0, 0, 0, 0, 180, 0, 0, 0, 0, 0]
Minimum total cost: $254,500

The column graph depicts the total of costs that have been incurred in each period. The breakdown includes:

- Holding costs: to maintain inventory levels

- Ordering costs: To place a replenishment order
- Shortage costs: when demand exceeds available inventory

Key Insights:
There is a high order quantity in period 1 and period 7, indicating the need to stock up.

In period 1, the quantity ordered is reduced to Zero for a few periods, suggesting the inventory stock was sufficient enough to meet the demand.

A sudden jump in orders in period 7 means that the inventory level fell below the reorder point, so an order for replenishment was placed.

No orders were placed in some periods due to inventory levels being above the reorder point.

This range of costs reflects the balance between frequent small orders and frequent large replenishment according to the (s, S) policy.

Key Takeaways:
The policy ensures that orders are placed only when inventory falls below the reorder point, reducing unnecessary holding costs.

Large replenishment quantities correspond to large fall inventory, balancing the exchange between ordering and holding costs.

4.2 Understanding the Inventory Policy

Inventory systems generally run on policies determining when and how much to order based on certain predefined thresholds of:

- Order Up to level S: maximum amount to be on hand in stock
- Reorder points: Points at which a new order should be placed to prevent a stock out
- Quantity Ordered: The amount ordered to stock when inventories reach the reorder point.

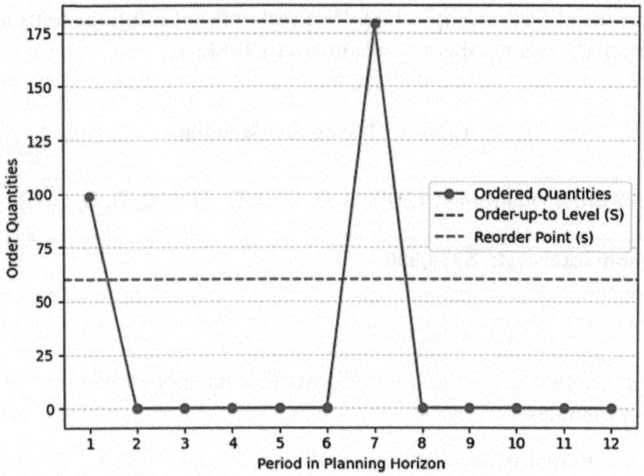

Fig. 1. Inventory Profile and Order Quantities Over Time

The dataset shows how ordering quantities vary over the 12 periods "Fig. 1", peaking at irregular intervals and staying for large periods with little or no activity. This could be interpreted as a demand-based inventory strategy coupled with cost efficiency and the viability of stock.

4.3 Observations from Ordering Behavior

- **Period of high ordering activity**

From the graph, there is one major peak of ordering activity:

Period 1: a large order of over 180 units was placed, indicating a replenishment strategy that avoids frequent ordering.

The peaks suggest that the inventory levels were critically low before these periods. Which may have triggered large replenishments to reach the order-up-to level. This ordering could be a result of the high demand in previous periods, the longer lead times for stock replenishment, and a conscious strategy of bulk ordering to reduce the frequency of ordering and, consequently, the costs related to ordering.

- **Periods of No Orders**

A long period of inactivity in ordering from periods (2 to 6) and (8 to 12) indicates that the inventory levels were adequate during these periods to meet demand without the need for replenishment, and the demand was relatively low in these periods.

The company may have optimized the inventory levels in previous periods to avoid excessive holding costs. This period of stability indicates the efficiency of the inventory policy in balancing the stock level with demand.

4.4 Cost Analysis Based on the Total Cost Per Period

A separate cost analysis graph indicates that total costs steadily decrease and increase over time (Fig. 2).

Key Observations:

- Higher initial costs: the total costs start above 25000 units, likely due to bulk ordering, high inventory holding costs, and potential stockout risks.
- There is a gradual decline from Periods (2 to 6) and (8 to 12) as ordering activity reduces, and total costs follow a downward trend, indicating better inventory optimization and reduced excess stock.
- This decrease in the overall trend of total costs signifies that improved stock planning led to the reduction of holding and ordering costs.

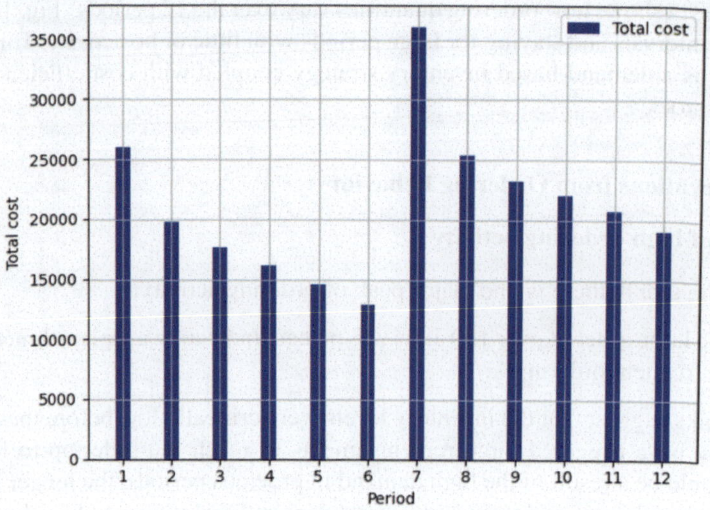

Fig. 2. Total cost over time

4.5 Risk of Stockouts vs. Overstocking

Although this policy effectively avoids stockouts, large orders of replenishment indicate that overstocking could be a potential risk.

The buffer stock strategy allows for the prevention of stockouts with less bulk ordering. Inventory turnover ratios should be analyzed and the order sizes should be optimized to avoid extra holding costs.

4.6 Comparative Analysis of Q-Learning and Genetic Algorithm

Comparison between Q-Learning and the Genetic algorithm (GA) reveals the greater adaptability and efficiency of reinforcement learning in inventory management. Q-Learning by its dynamic decision-making capacity, not only achieves a superior 99.2% service level compared to 98.61% for GA [13], but substantially reduces computational overhead. Compared to GA, which requires several generations to evolve solutions, Q-Learning can improve its policy continuously in line with real-time data, leading to optimized resource planning and reduced spending. In addition, the property that Q-Learning updates only applicable state-action pairs makes convergence quicker and responsiveness greater, thus being a viable approach to complex decision-making in manufacturing and logistics systems.

5 Conclusion

In this work, we indicate the effectiveness of reinforcement learning particularly Q-Learning through a comparative analysis with genetic Algorithms (GA) by developing a methodology to find an efficient (s, S) policy for spare parts inventory management.

This methodology balances ordering and holding costs, while the Q-Learning framework efficiently identifies the cost-minimizing actions. Demonstrating superior adaptability, achieving a higher service level. Unlike GA, which relies on iterative evolution, these insights reinforce the practical value of combining data-driven approaches with traditional inventory management strategies which can pave the way for future research to expand this work further and get deeper insights into two-echelon spare parts inventory management.

Acknowledgments. We appreciate the opportunity to participate in this enlightening conference and would like to express our gratitude to the ICASET organizing committee for outstanding teamwork and professional attitude.

References

1. Cavalieri, S., Garetti, M., Macchi, M., Pinto, R.: A decision-making framework for managing maintenance spare parts. Prod. Plann. Control **19**(4), 379–396 (2008)
2. Jin, T., Liao, H.: Spare parts inventory control considering stochastic growth of an installed base. Comput. Ind. Eng. **56**(1), 452–460 (2009)
3. Sethi, S.P., Cheng, F.: Optimality of (s, S) policies in inventory models with Markovian demand. Oper. Res. **45**(6), 931–939 (1997)
4. Yang, Y., Loxton, R., Rohl, A.L., Bui, H.T.: Long-term maintenance optimization for integrated mining operations. Optim. Eng. 1–32 (2023)
5. Rummery, G.A., Niranjan, M.: On-line Q-Learning using connectionist systems, vol. 37. Department of Engineering, University of Cambridge, Cambridge (1994)
6. Teugels, J.L.: Markov Chains: Models, Algorithms and Applications, pp. 1325–1325 (2008)
7. Watkins, C.J., Dayan, P.: Q-learning. Mach. Learn. **8**(3-4), 279–292 (1992)
8. Delmas, J.F., Jourdain, B.: Modèles aléatoires: Applications aux sciences de l'ingénieur et du vivant. Springer, Heidelberg (2006)
9. Ferretti, I., Marchi, B.: Q-learning for inventory management: an application case. Procedia Comput. Sci. **232**, 2431–2439 (2024)
10. Wang, H., et al.: Dynamic inventory replenishment strategy for aerospace manufacturing supply chain: combining reinforcement learning and multi-agent simulation. Int. J. Prod. Res. **60**(13), 4117–4136 (2022)
11. Abubakar, A.I., Ozturk, M., Hussain, S., Imran, M.A.: Q-learning assisted energy-aware traffic offloading and cell switching in heterogeneous networks. In: 2019 IEEE 24th International Workshop on Computer-Aided Modeling and Design of Communication Links and Networks (CAMAD)
12. Ozturk, M., Akram, M., Hussain, S., Imran, M.A.: Novel QoS-aware proactive spectrum access techniques for cognitive radio using machine learning. IEEE Access **7**, 70811–70827 (2019)
13. Elhadaf, H., Marzoug, N., Jraifi, A.: Optimizing spare parts inventory management using genetic algorithm. In: 10th International Conference on Optimization and Applications (ICOA), pp. 1–5. IEEE (2024). https://doi.org/10.1109/ICOA62581.2024.10754264
14. Beyer, D., Cheng, F., Sethi, S.P., Taksar, M.: Markovian Demand Inventory Models. Springer, New York (2010)
15. Diaz, A., Fu, M.C.: Models for multi-echelon repairable item inventory systems with limited repair capacity. Eur. J. Oper. Res. **97**(3), 480–492 (1997)

16. Esteso, A., Peidro, D., Mula, J., Díaz-Madroñero, M.: Reinforcement learning applied to production planning and control. Int. J. Prod. Res. **61**(16), 5772–5789 (2023)
17. Jin, T., Tian, Y.: Optimizing reliability and service parts logistics for a time-varying installed base. Eur. J. Oper. Res. **218**(1), 152–162 (2012)

Author Index

C. El Mokhi et al. (Eds.): ICASET 2025, SUCI, pp. 379–380, 2025.
https://doi.org/10.1007/978-3-031-98334-4

The manufacturer's authorised representative in the EU is Springer Nature Customer Service Centre GmbH, Europaplatz 3, 69115 Heidelberg, Germany. If you have any concerns regarding our products, please contact ProductSafety@springernature.com

Printed and bound by CPI Group (UK) Ltd, Croydon, CR0 4YY
01/10/2026
02225957-0011